Student Resource Manual

for

Introductory Chemistry

An Active Learning Approach

Second Edition

Student Resource Manual
for
Introductory Chemistry
An Active Learning Approach
Second Edition

Mark S. Cracolice
University of Montana

Edward I. Peters

Australia • Canada • Mexico • Singapore • Spain • United Kingdom • United States

Printed in the United States of America
1 2 3 4 5 6 7 07 06 05 04 03

Printer: Globus Printing

ISBN: 0-534-40696-3

For more information about our products, contact us at:
Thomson Learning Academic Resource Center
1-800-423-0563

For permission to use material from this text, contact us by:
Phone: 1-800-730-2214
Fax: 1-800-731-2215
Web: http://www.thomsonrights.com

Cover image: Tony Stone/Paul Morrell

Brooks/Cole—Thomson Learning
10 Davis Drive
Belmont, CA 94002-3098
USA

Asia
Thomson Learning
5 Shenton Way #01-01
UIC Building
Singapore 068808

Australia/New Zealand
Thomson Learning
102 Dodds Street
Southbank, Victoria 3006
Australia

Canada
Nelson
1120 Birchmount Road
Toronto, Ontario M1K 5G4
Canada

Europe/Middle East/South Africa
Thomson Learning
High Holborn House
50/51 Bedford Row
London WC1R 4LR
United Kingdom

Latin America
Thomson Learning
Seneca, 53
Colonia Polanco
11560 Mexico D.F.
Mexico

Spain/Portugal
Paraninfo
Calle/Magallanes, 25
28015 Madrid, Spain

Contents: Part I

Solutions and Resources for Textbook Questions

Contents: Part II

Active Learning Workbook Answers

Preface

To the Student

The primary purpose of this *Student Resource Manual to accompany Introductory Chemistry: An Active Learning Approach* is to provide you with a convenient guide that explains how to arrive at the solutions for the Questions, Exercises, and Problems found at the end of each chapter in the textbook and to give you answers to the Questions, Exercises, and Problems in the *Active Learning Workbook.* Complete solutions for textbook problems are provided in the text, but this guide gives you even more information.

Part I of this manual consists of solutions for the textbook problems with additional information to help you understand and diagnose errors on those problems that you initially solve incorrectly. Most answers in this section of the manual have two parts: a complete solution, which is sometimes exactly the same as in the textbook and sometimes more, and a reference to a place in the text where you can go for additional study, as needed. For example, compare the textbook solution to Question 38 in Chapter 20 to the *Student Resource Manual* solution:

Textbook Solution: $\frac{[HC_7H_5O_2]}{[C_7H_5O_2^-]} = \frac{[H^+]}{K_a} = \frac{10^{-4.80}}{6.5 \times 10^{-5}} = 0.24$

Student Resource Manual Solution:
GIVEN: pH = 4.80; $K_a = 6.5 \times 10^{-5}$ *WANTED:* $[HC_7H_5O_2]/[C_7H_5O_2^-]$

EQUATIONS: $HC_7H_5O_2(aq) \rightleftharpoons H^+(aq) + C_7H_5O_2^-(aq)$ $K_a = \frac{[H^+][C_7H_5O_2^-]}{[HC_7H_5O_2]}$

$$\frac{K_a}{[H^+]} = \frac{[H^+][C_7H_5O_2^-]}{[H^+][HC_7H_5O_2]} \qquad \frac{K_a}{[H^+]} = \frac{[C_7H_5O_2^-]}{[HC_7H_5O_2]} \qquad \frac{[H^+]}{K_a} = \frac{[HC_7H_5O_2]}{[C_7H_5O_2^-]}$$

$$\frac{[HC_7H_5O_2]}{[C_7H_5O_2^-]} = \frac{[H^+]}{K_a} = \frac{10^{-4.80}}{6.5 \times 10^{-5}} = 0.24$$

See the Ionization Equilibria subsection on Pages 573-576.

Note how this manual gives you additional information, including a complete setup, a step-by-step analysis of the algebra, and a reference to the textbook for further study or review. The additional information will allow you to diagnose where you are going wrong if you do not arrive at the correct solution in your initial attempt at a problem. *We strongly urge you to use this manual as a resource only after you have first attempted to solve each problem on your own.* A great deal of learning occurs when you first answer a question wrong and then mentally fix the error. Almost no learning occurs if you read a solution before you attempt a problem on your own.

Part II of this manual is composed of answers for the Questions, Exercises, and Problems in the *Active Learning Workbook for Introductory Chemistry: An Active Learning Approach.*

Each chapter of the *Student Resource Manual* is self-contained on a separate set of pages. This design was chosen so that you can tear out the perforated pages and place them in your homework notebook. You won't need to carry around the whole book, but rather, you can carry just the solutions for the questions on which you are presently working.

We hope that this manual provides you with a useful tool to assist you in your study of chemistry!

Chapter 2

Matter and Energy

1. Macroscopic: a, e. Microscopic: c. Particulate: b, d.

   The different levels at which matter is studied are explained on Page 16 (Goal 1).

2. Chemists use models, real or imaginary or both, to represent the invisible particles that make up matter.

   Models are discussed on Pages 16–18 (Goal 2).

3. Ice consists of water molecules that vibrate or shake in fixed positions relative to one another. The ice has a fixed shape. When the ice melts, the molecules remain together but move freely among each other at the bottom of the container that holds them, assuming the shape of the container. As a gas, the molecules separate, moving freely throughout a closed container or escaping into the atmosphere from an open container.

   See Figure 2.5, Page 20.

4. The particles in a solid occupy fixed positions relative to each other and cannot be poured, but different pieces of solids can move relative to each other. The slogan emphasizes that this brand of table salt has solid pieces small enough to move freely relative to one another, but not so small that they stick together in humid weather.

   The pouring action of solid particles at the macroscopic level is somewhat analogous to chemists' model of the pouring action of liquids at the particulate level.

5. Liquid volume is usually slightly greater than solid; gas volume is very much larger than liquid.

   The relative volumes of gases, liquids, and solids are discussed on Page 19.

6. Physical: a, c, d. Chemical: b, e.

   Physical properties are discussed on Page 21. Chemical properties are discussed on Page 22. They are summarized in Table 2.1 on Page 22.

7. Chemical: a, d, e. Physical: b, c.

   Physical changes are discussed on Page 21. Chemical changes are discussed on Page 22. They are summarized in Table 2.1 on Page 22.

8. Physical—the particles simply change state.

   The models of two-atom molecules are the same before and after. On the left, they appear to be in fixed positions, which is the solid state, and on the right, the particles are independent of one another, which is the gaseous state.

9. The change is chemical because the reactant diatomic molecules are destroyed and become atoms.

   The models of two-atom molecules are change to one-atom particles. A change in the particulate-level arrangement of atoms is a chemical change.

10. If both products are the same pure substance, they must be identical; thus, purchasing the new product is the better choice.

    The definition of *pure substance* is discussed on Page 24.

11. The photograph alone does not provide enough information to answer the question. You probably know, however, that dishwashing liquid is a mixture.

    The definition of *mixture* is discussed on Page 24.

12. Tap water is a mixture of water and dissolved minerals and gases. Distilled water is pure water.

    Natural water, or natural water that is treated for household use, is not pure.

13. All are mixtures.

    Mixture is defined on Page 24. All boxes show combinations of different types of particles.

14. Pure substances: a, c. Mixtures: b, d.

    All of the particles are identical in a pure substance: it is *one* kind of matter.

15. a is a mixture because different substances are visible. b could be a pure substance in two different states, but it is probably a mixture. c could be either a pure substance or a mixture because it may be one kind of matter or two or more types of matter with similar appearances.

    Figure 2.9 on Page 25 explains why a pure substance cannot be distinguished from a mixture solely based on its macroscopic appearance.

16. Pure substances: a, d. Mixtures: b, c.

    Table salt can be purchased as pure sodium chloride (a pure substance), but it is often sold as a mixture of sodium chloride and potassium iodide (iodized salt), and if you know this, you should have classified it as a mixture. Steam is water in the gaseous state and therefore pure water (a pure substance) as long as it is not mixed with air, as is often the case with steam formed during cooking.

17. Yes, the terms homogeneous and heterogeneous refer to the macroscopic appearance of a sample. A container filled with ice and liquid water is heterogeneous in appearance but is also pure, as long as in both phases the water is pure.

    The terms *homogeneous* and *heterogeneous* are discussed in Page 24.

18. b, d, e.

    Unstirred paint has visibly different phases, so if you were thinking of that, you would be correct in classifying it as heterogeneous.

19. The cylinder appears the same throughout, so it is a homogeneous substance.

Both pure substances and mixtures may be homogeneous.

20. Pick out table tennis balls or ball bearings based on size and appearance; use magnetic properties of steel to pick up ball bearings with a magnet; use the lower density of balls and higher density of steel to float the balls in water. These methods are based on physical properties.

Density is mass per unit volume; it is formally introduced in Section 3.8. If you said that ball bearings would sink in water and table-tennis balls would float, you were comparing the relative densities of ball bearings, water, and table-tennis balls.

21. Place the mixture in water. Salt will dissolve and sand will not. Filter out the sand. Evaporate the water to obtain pure salt. Dissolving and recrystallizing the salt is a physical change because the salt did not undergo a chemical change (you can taste the salt dissolved in the water, so it is still salt.)

Dissolving is sometimes mistaken to be a chemical change. When a solid such as sugar or salt is dissolved in water, the original solid can be recovered by distillation or simple evaporation of the water. The substance is therefore not destroyed, and reacting substances are destroyed in a chemical change. It undergoes a phase change from the solid state to being dissolved in water.

22. Elements: b, e, f. Compounds: a, c, d.

On Page 28 we state, "At present, you may use the number of words in the name of a chemical to predict whether it is an element or a compound."

23. Elements: a, e. Compounds: b, c, d.

The chemical formula of a compound has symbols for two or more different elements. The chemical formula of an element may have more than one atom of that element, but there is only one elemental symbol in the formula.

24. Elements: a, d, e. Compounds: b, c.

All atoms are the same in elements.

25. Elements: a, b, e. Compounds: c, d.

In c and d, there are two different kinds of atoms. These are therefore particulate-level models of compounds.

26. (a) Elements: 2, 3. Compounds: 1, 4, 5. (b) In general, if there are two or more words in the name, the substance is a compound. However, many compounds are known by one-word common names, and one-word names for many carbon-containing compounds are assembled from prefixes, suffixes, and special names for recurring groups.

The name dimethylhydrazine is a compound name made up of three parts: di-methyl-hydrazine. You will find that multisyllabic single-word names are always compounds.

27. Only a compound with one taste can be decomposed into a gas and a solid having a different taste.

If the solid was an element, it would be a pure substance, one kind of matter. If some was changed into the gas phase, the remaining solid would be the same pure substance with the same taste.

28.	G, L, S	P, M	Hom, Het	E, C
Limestone (calcium carbonate)	S	P	Hom	C
Lead	S	P	Hom	E
Freshly squeezed orange juice	L & S	M	Het	
Oxygen	G	P	Hom	E
Butter in the refrigerator	S	M	Hom	

Figure 2.18 on Page 32 summarizes the classification system for matter.

29. Attraction: a, c. Repulsion: b.

The two sentences in italics near the top of Page 33 summarize the electrical character of matter.

30. Reactants: KOH, $HCHO_2$. Products: $KCHO_2$, H_2O.

Reactants and products in chemical equations are summarized in the Chemical Equations subsection of Section 2.8, Page 33.

31. Product element: H_2. Reactant compound: H_2O.

H_2O and NaOH are compounds because they are made up of atoms of more than one element. Na and H_2 are elements, composed of one type of atom. Reactants are on the left of the arrow; products on the right.

32. Exothermic: c, d, e. Endothermic: a, b.

The terms *exothermic* and *endothermic* are explained in the Energy in Chemical Change subsection on Pages 33-34.

33. Kinetic energy increases.

Potential energy and kinetic energy are discussed on Page 34.

34. The masses would be the same. Products of a chemical reaction have the same total mass as reactants.

The Law of Conservation of Mass is explained on Page 35.

35. Potential energy of the water above the dam changes into kinetic energy as it falls into the turbine. The turbine drives a generator in which kinetic energy is change into the electrical energy that produces the heat and light energy given off by the bulb.

Potential energy and kinetic energy are discussed on Page 34. See Figure 2.21 on Page 36 for an illustration of similar energy conversions.

37. True: e, f, i, j, l. False: a, b, c, d, g, h, k.

a) The fact that paper burns is a chemical property.
b) Particles of matter are moving no matter the state.
c) A homogeneous substance has a uniform appearance throughout.
d) Compounds are pure substances.
g) Two positively charged objects repel each other. Two negatively charged objects repel each other.
h) Mass is conserved in any non-nuclear chemical change.
k) There is no absolute power relationship between potential and kinetic energy.

38. Almost everything you use is either a result of or involves human-made chemical change in some way.

Chemical change is described on Page 22.

39. Yes, nitrogen and oxygen in air are a mixture of two elements.

Element and *compound* are discussed on Page 28. Compounds made from nitrogen and oxygen include nitrogen monoxide, NO, nitrogen dioxide, NO_2, and a number of other N_xO_y compounds.

40. (a) The powder is neither an element nor a compound, both of which have a fixed composition. (b) The contents of the box are heterogeneous because samples from different locations have different compositions. (c) The contents must be a mixture of varying composition.

Element and *compound* are discussed on Page 28. *Homogeneous* and *heterogeneous* are discussed on Page 24. *Pure substance* and *mixture* are found on Page 24.

41. The sources of usable energy now available are limited. If we change them into forms that cannot be used, we are threatened with an energy shortage in the future.

The Law of Conservation of Energy is discussed on Page 35.

42. (a) Neither: The distinction between homogeneous and heterogeneous is not a particulate property. (b) The sample is a mixture because it consists of two different particle types. (c) The particles are compounds because they consist of more than one type of atom. (d) The particles are molecules because they are made up of more than one atom. (e) The sample is a gas because the particles are independent of one another.

Homogeneous and *heterogeneous* are discussed on Page 24. *Pure substance* and *mixture* are discussed on Page 24. *Element* and *compound* are discussed on Page 28. *Atom* is defined on Page 28 at the end of the first paragraph in Section 2.6. *Molecule* is defined in the first paragraph on Page 30: "Other elements exist in nature as stable, distinct, and independent molecules, which are made up of two or more atoms." The states of matter are the subject of Section 2.2, Pages 18-20.

43. (a) Reactants: AB, CD. Products: AD, CB. (b) A chemical change is shown. (c) The masses o the product particles are the reactant particles are equal. (d) The energy in the container is the same before and after the reaction.

Reactants and *products* are discussed in the Chemical Equations subsection on Page 33. *Chemical and physical changes* are compared on Pages 21-22. *The Law of Conservation of Mass* and *The Law of Conservation of Energy* are explained on Page 35.

44. Pure substance. A mixture would have changed boiling temperature during distillation because of a change in composition of the mixture.

Consider Figure 2.8 on Page 24. Notice how water (red curve), a pure substance, remains at a constant temperature of 100°C as it boils. All pure substances have a distinct, constant boiling temperature. Contrast that with the solution (blue) curve. The boiling temperature increases during the boiling process; it is not constant. The boiling temperature of a solution (homogeneous mixture) increases because the composition of the mixture is changing as one component of the mixture boils off.

45. (a) This is a physical change. (b) The number of particles is not conserved. (c) New particles appear from nowhere. (d) This could be the product of a chemical change.

Atoms are not created or destroyed in a chemical change. They are reordered into new molecular arrangements. The reactants consist of 10 green atoms, 5 black atoms, and 10 white atoms. The products of a chemical change must be made up of these atoms in a new arrangement.

46. Your illustration will resemble the phases in Figure 2.5. According to kinetic molecular theory, particles move faster at higher temperatures, so particles are more likely to separate from one another.

See Figure 2.5 on Page 20. Kinetic molecular theory is discussed on Page 19.

Chapter 3

Measurement and Chemical Calculations

1. (a) $34,100,000 = 3.4100000 \times 10^7 = 3.41 \times 10^7$
(b) $0.00556 = 5.56 \times 10^{-3}$
(c) $303,000 = 3.03000 \times 10^5 = 3.03 \times 10^5$

See the *Procedure: How to Change a Decimal Number to Standard Exponential Notation* on Page 48.

2. (a) $2.86 \times 10^{-9} = 0.00000000286$
(b) $8.27 \times 10^6 = 8,270,000$
(c) $9.88 \times 10^{-13} = 0.000000000000988$

On Page 49, we state, "To change a number written in exponential notation to ordinary decimal form, simply perform the indicated multiplication."

3. (a) $(7.44 \times 10^7)(1.44 \times 10^4) = 1.07 \times 10^{12}$
(b) $(9.68 \times 10^{-4})(8.59 \times 10^{-7}) = 8.32 \times 10^{-10}$
(c) $(4.34 \times 10^{-9})(9.72 \times 10^3) = 4.22 \times 10^{-5}$
(d) $(8.84 \times 10^5)(6.76 \times 10^4)(7.83 \times 10^{-7}) = 4.68 \times 10^4$

Perform the indicated operation with your calculator. If you are having difficulty, the best source of help is the instruction book that came with your calculator. If you do not have it, Table 3.1 on Page 50 gives calculator procedures for typical calculators.

4. (a) $\frac{7.38 \times 10^7}{2.90 \times 10^3} = 2.54 \times 10^4$
(b) $\frac{1.15 \times 10^5}{9.51 \times 10^{-4}} = 1.21 \times 10^8$
(c) $\frac{5.55 \times 10^{-6}}{6.98 \times 10^4} = 7.95 \times 10^{-11}$
(d) $\frac{8.63 \times 10^{-7}}{8.47 \times 10^{-5}} = 0.0102$ or 1.02×10^{-2}

Perform the indicated operation with your calculator. If you are having difficulty, the best source of help is the instruction book that came with your calculator.

5. (a) $\frac{(4.44 \times 10^5)(4.75 \times 10^{-9})}{1.38 \times 10^8} = 1.53 \times 10^{-11}$
(b) $\frac{287(7.07 \times 10^{-8})}{(3.76 \times 10^6)(7.30 \times 10^4)} = 7.39 \times 10^{-17}$

See the section on Chain Calculations on Pages A-3 to A-5. This type of calculation occurs frequently in introductory chemistry, so if you are having trouble with these calculations, be sure to *learn it now!*

6. (a) $9.04 \times 10^{-3} + 4.17 \times 10^{-2} = 0.0507 = 5.07 \times 10^{-2}$
   (b) $9.15 \times 10^{13} - 8.8 \times 10^{12} = 8.27 \times 10^{13}$

   Perform the indicated operation with your calculator. If you are having difficulty, the best source of help is the instruction book that came with your calculator.

7. *GIVEN:* 3 mi *WANTED:* ft

   PER/PATH: mi $\xrightarrow{5280 \text{ ft/mi}}$ ft

   $$3 \text{ mi} \times \frac{5280 \text{ ft}}{\text{mi}} = 1.5840 \times 10^{4} \text{ ft}$$

   See Section 3.3 on Dimensional Analysis, Pages 51 to 58. Note, in particular, the *Procedure: How to Solve a Problem by Dimensional Analysis* box on Page 58.

8. *GIVEN:* 1.5×10^{11} m *WANTED:* min

   PER/PATH: m $\xrightarrow{3.00 \times 10^{8} \text{ m/s}}$ s $\xrightarrow{60 \text{ s/1 min}}$ min

   $$1.5 \times 10^{11} \text{ m} \times \frac{1 \text{ s}}{3.00 \times 10^{8} \text{ m}} \times \frac{1 \text{ min}}{60 \text{ sec}} = 8.3 \text{ min}$$

   It is assumed that you know the 60 s ≡ 1 min conversion. See Section 3.3 on Dimensional Analysis, Pages 51 to 58. Note, in particular, the *Procedure: How to Solve a Problem by Dimensional Analysis* box on Page 58.

9. *GIVEN:* 63 hr *WANTED:* cm

   PER/PATH: hr $\xrightarrow{91 \text{ cm/24 hr}}$ cm

   $$63 \text{ hr} \times \frac{91 \text{ cm}}{24 \text{ hr}} = 2.4 \times 10^{2} \text{ cm}$$

   See Section 3.3 on Dimensional Analysis, Pages 51 to 58. Note, in particular, the *Procedure: How to Solve a Problem by Dimensional Analysis* box on Page 58.

10. *GIVEN:* $69.95 C *WANTED:* $ A

    PER/PATH: $ C $\xrightarrow{\$1.18 \text{ C/\$1 A}}$ $ A

    $$\$69.95 \text{ C} \times \frac{\$1.00 \text{ A}}{\$1.18 \text{ C}} = \$59.28 \text{ American}$$

    See Section 3.3 on Dimensional Analysis, Pages 51 to 58. Note, in particular, the *Procedure: How to Solve a Problem by Dimensional Analysis* box on Page 58.

11. *GIVEN:* 31 days (in the month of January) *WANTED:* s

    PER/PATH: days $\xrightarrow{24 \text{ hr/day}}$ hr $\xrightarrow{60 \text{ min/hr}}$ min $\xrightarrow{60 \text{ s/min}}$ s

    $$31 \text{ days} \times \frac{24 \text{ hr}}{\text{day}} \times \frac{60 \text{ min}}{\text{hr}} \times \frac{60 \text{ s}}{\text{min}} = 2{,}678{,}400 \text{ s}$$

    It is assumed that you know the time conversions 24 hr ≡ 1 day, 60 min ≡ 1 hr, and 60 s ≡ 1 min, as well as the fact that there are 31 days in the month of January. See Section 3.3 on Dimensional Analysis, Pages 51 to 58. Note, in particular, the *Procedure: How to Solve a Problem by Dimensional Analysis* box on Page 58.

12. The mass of the rock is the same wherever it is. The weight of the rock is the same whether it is in the lake or on the beach—or anywhere in the gravitational field on the earth's surface. In the lake the upward buoyancy force cancels some of the downward weight force. That makes the rock seem lighter.

    Mass and weight are discussed in a subsection on Pages 58-59. The second paragraph in this subsection discusses Goal 4 in detail.

13. The gram

    The SI unit of mass is the kilogram, but the basic metric unit is the gram. See the paragraph that begins on the bottom of Page 58 and continues to Page 59.

14. *Kilounit* is a general term that refers to 1000 units of any measurement unit. Kilogram is a specific example—1000 grams.

    The metric prefix *kilo-* and its application is explained in the second paragraph (Larger and smaller metric units...) on Page 59.

15. 100 cm = 1 m

    Table 3.2 on Page 59 lists metric prefixes.

16. A nm is a nanometer. It is a very short distance, one billionth of a meter, or about 4/100,000,000 of an inch.

    Table 3.2 on Page 59 lists metric prefixes.

17. $276\text{ g} \times \frac{1\text{ kg}}{1000\text{ g}} = 0.276\text{ kg}$

    $31.9\text{ g} \times \frac{100\text{ cg}}{\text{g}} = 3.19 \times 10^{3}\text{ cg}$

    $191\text{ mg} \times \frac{1\text{ g}}{1000\text{ mg}} \times \frac{1\text{ kg}}{1000\text{ g}} = 1.91 \times 10^{-4}\text{ kg}$

    Table 3.2 on Page 59 lists metric prefixes.

18. $25.9\text{ km} \times \frac{1000\text{ m}}{\text{km}} = 2.59 \times 10^{4}\text{ m}$

    $4.27\text{ mm} \times \frac{1\text{ m}}{1000\text{ mm}} = 0.00427\text{ m}$

    $9.46\text{ cm} \times \frac{1\text{ m}}{100\text{ cm}} \times \frac{1000\text{ mm}}{\text{m}} = 94.6\text{ mm}$

    Table 3.2 on Page 59 lists metric prefixes.

19. $231 \text{ mL} \times \frac{1 \text{ L}}{1000 \text{ mL}} = 0.231 \text{ L}$

$5.06 \text{ L} \times \frac{1000 \text{ mL}}{\text{L}} = 5.06 \times 10^3 \text{ mL} = 5.06 \times 10^3 \text{ cm}^3$

$60.1 \text{ mL} = 60.1 \text{ cm}^3$

In the Volume subsection on Page 60, we state that 1 mL = 1 cm^3. This conversion is used frequently in introductory chemistry. Table 3.2 on Page 59 lists metric prefixes.

20. $0.194 \text{ Gg} \times \frac{1 \times 10^9 \text{ g}}{\text{Gg}} = 1.94 \times 10^8 \text{ g}$

$5.66 \text{ nm} \times \frac{1 \text{ m}}{1 \times 10^9 \text{ nm}} = 5.66 \times 10^{-9} \text{ m}$

$0.00481 \text{ Mm} \times \frac{1 \times 10^6 \text{ m}}{\text{Mm}} \times \frac{100 \text{ cm}}{\text{m}} = 4.81 \times 10^5 \text{ cm}$

Table 3.2 on Page 59 lists metric prefixes.

21. 5, 2, 3, uncertain—3 or 4, 3, 5, 3, 5

The Counting Significant Figures subsection on Pages 63-65 discusses how to determine the number of significant figures in a measured quantity. Pay particular attention to the *Summary: The Number of Significant Figures in a Measurement* box.

22. 52.2 mL; 18.0 g; 78.5 mg; 2.36×10^7 μm; 0.00420 kg

Rounding off is discussed in a subsection on Page 66. See the *Procedure: How to Round Off a Calculated Number* box on the same page.

23. 2.86 g + 3.9 g + 0.896 g + 246 g = 254 g

See Pages 66-67 for the Addition and Subtraction subsection within the Significant Figures section. It is summarized in the *Procedure: How to Round Off a Sum or Difference* box on Page 66.

24. 101.209 g – 94.33 g = 6.88 g

See Pages 66-67 for the Addition and Subtraction subsection within the Significant Figures section. It is summarized in the *Procedure: How to Round Off a Sum or Difference* box on Page 66.

25. 2 L × 31.4 g/L = 62.8 g; 7.37 L × 31.4 g/L = 231 g

The significant figure rule for multiplication and division is discussed in a subsection on Pages 67-69. It is summarized in the *Procedure: How to Round Off a Product or Quotient* box on Page 68.

26. $D \equiv \frac{m}{V} = \frac{(84.64\text{ g} - 32.344\text{ g})}{50.6\text{ mL}} = \frac{52.30\text{ g}}{50.6\text{ mL}} = 1.03\text{ g/mL}$

See the Addition/Subtraction and Multiplication/Division Combined subsection on Pages 70-71. You may also want to review the Addition/Subtraction Procedure box on Page 66 and the Multiplication/Division Procedure box on Page 68, if necessary.

27. *GIVEN:* 19.3 L *WANTED:* gal

PER/PATH: $\text{L} \xrightarrow{1\text{ gal}/3.785\text{ L}} \text{gal}$

$19.3\text{ L} \times \frac{1\text{ gal}}{3.785\text{ L}} = 5.10\text{ gal}$

GIVEN: 0.461 qt *WANTED:* L

PER/PATH: $\text{qt} \xrightarrow{1.06\text{ qt/L}} \text{L}$

$0.461\text{ qt} \times \frac{1\text{ L}}{1.06\text{ qt}} = 0.435\text{ L}$

See Table 3.3, Metric-USCS Conversion Factors (Page 71), for the needed *PER* expressions.

28. *GIVEN:* 26.4 g *WANTED:* oz

PER/PATH: $\text{g} \xrightarrow{1\text{ oz}/28.3\text{ g}} \text{oz}$

$26.4\text{ g} \times \frac{1\text{ oz}}{28.3\text{ g}} = 0.933\text{ oz}$

See Table 3.3, Metric-USCS Conversion Factors (Page 71), for the needed *PER* expression.

29. *GIVEN:* 44.4 carats *WANTED:* g and oz

PER/PATH: $\text{carats} \xrightarrow{1\text{ carat}/200\text{ mg}} \text{mg} \xrightarrow{1000\text{ mg/g}} \text{g} \xrightarrow{1\text{ oz}/28.3\text{ g}} \text{oz}$

$44.4\text{ carats} \times \frac{200\text{ mg}}{\text{carat}} \times \frac{1\text{ g}}{1000\text{ mg}} = 8.88\text{ g}$

$44.4\text{ carats} \times \frac{200\text{ mg}}{\text{carat}} \times \frac{1\text{ g}}{1000\text{ mg}} \times \frac{1\text{ oz}}{28.3\text{ g}} = 0.314\text{ oz}$

The 1000 mg/g conversion is found in Table 3.2 (Page 59), and it should be memorized The 1 oz = 28.3 g conversion is given in Table 3.3 (Page 71).

30. *GIVEN:* 922 lb *WANTED:* kg

PER/PATH: $\text{lb} \xrightarrow{2.20\text{ lb}/1\text{ kg}} \text{kg}$

$922\text{ lb} \times \frac{1\text{ kg}}{2.20\text{ lb}} = 419\text{ kg}$

See Table 3.3, Metric-USCS Conversion Factors (Page 71), for the needed *PER* expression.

31. *GIVEN:* 7.5 lb *WANTED:* mass in metric units (assume kg)

PER/PATH: lb $\xrightarrow{2.20 \text{ lb/kg}}$ kg

$$7.5 \text{ lb} \times \frac{1 \text{ kg}}{2.20 \text{ lb}} = 3.4 \text{ kg}$$

We made the choice of using kilograms for mass in metric units because it yields the most convenient numbers. If you answered in grams, 3.4×10^3 g, your answer is also correct. See Table 3.3, Metric-USCS Conversion Factors (Page 71), for the needed *PER* expression.

32. *GIVEN:* 135 cm *WANTED:* in.

PER/PATH: cm $\xrightarrow{1 \text{ in./}2.54 \text{ cm}}$ in.

$$135 \text{ cm} \times \frac{1 \text{ in.}}{2.54 \text{ cm}} = 53.1 \text{ in.}$$

See Table 3.3, Metric-USCS Conversion Factors (Page 71), for the needed *PER* expression.

33. *GIVEN:* 2351 mi *WANTED:* km

PER/PATH: mi $\xrightarrow{5280 \text{ ft/mi}}$ ft $\xrightarrow{1 \text{ ft/}30.5 \text{ cm}}$ cm $\xrightarrow{100 \text{ cm/m}}$ m $\xrightarrow{1000 \text{ m/km}}$ km

$$2351 \text{ mi} \times \frac{5280 \text{ ft}}{\text{mi}} \times \frac{30.5 \text{ cm}}{\text{ft}} \times \frac{1 \text{ m}}{100 \text{ cm}} \times \frac{1 \text{ km}}{1000 \text{ m}} = 3.79 \times 10^3 \text{ km}$$

The textbook solution assumes that you know the 1 mi = 1.61 km conversion. The approach shown here assumes that you know the USCS relationship 1 mi = 5280 ft. The 1 ft = 30.5 cm conversion is from Table 3.3 (Page 71), and the metric conversions are from Table 3.2 (Page 59).

34. *GIVEN:* 0.02 in. *WANTED:* mm

PER/PATH: in. $\xrightarrow{2.54 \text{ cm/in.}}$ cm $\xrightarrow{10 \text{ mm/cm}}$ mm

$$0.02 \text{ in.} \times \frac{2.54 \text{ cm}}{\text{in.}} \times \frac{10 \text{ mm}}{\text{cm}} = 0.5 \text{ mm}$$

The definition 1 in. ≡ 2.54 cm is from Table 3.3 (Page 71). Since 1000 mm = 1 m and 100 cm = 1 m, 1000 mm = 100 cm. Dividing both by 100, 10 mm = 1 cm. If you solved this in two steps, cm to m and m to mm, your approach is equally acceptable.

35. *GIVEN:* 50.0 L *WANTED:* gal

PER/PATH: L $\xrightarrow{3.785 \text{ L/gal}}$ gal

$$50.0 \text{ L} \times \frac{1 \text{ gal}}{3.785 \text{ L}} = 13.2 \text{ gal}$$

See Table 3.3, Metric-USCS Conversion Factors (Page 71), for the needed *PER* expression.

36. Celsius	Fahrenheit	Kelvin
4	40	277
317	603	590
−13	9	260
−44	−47	229
440	824	713
−192	−314	81

$T_{°F} - 32 = 1.8\ T_{°C}$ (Equation 3.7) and $T_K = T_{°C} + 273$ (Equation 3.8) are used for conversions. See Section 3.7 on Pages 72-76.

37. *GIVEN:* 56°C *WANTED:* $T_{°F}$
EQUATION: $T_{°F} - 32 = 1.8\ T_{°C}$
$T_{°F} = 1.8\ T_{°C} + 32 = 1.8(56) + 32 = 133°F$

$T_{°F} - 32 = 1.8\ T_{°C}$ (Equation 3.7) is used for conversions. See Section 3.7 on Pages 72-76.

38. *GIVEN:* 68°F *WANTED:* $T_{°C}$
EQUATION: $T_{°F} - 32 = 1.8\ T_{°C}$

$$T_{°C} = \frac{T_{°F} - 32}{1.8} = \frac{68 - 32}{1.8} = 20°C$$

$T_{°F} - 32 = 1.8\ T_{°C}$ (Equation 3.7) is used for conversions. See Section 3.7 on Pages 72-76.

39. *GIVEN:* 60°C *WANTED:* $T_{°F}$
EQUATION: $T_{°F} - 32 = 1.8\ T_{°C}$
$T_{°F} = 1.8\ T_{°C} + 32 = 1.8(60) + 32 = 140°F$

$T_{°F} - 32 = 1.8\ T_{°C}$ (Equation 3.7) is used for conversions. See Section 3.7 on Pages 72-76.

40. mi $\propto$ gal; mi = k (gal); k = mi/gal; Yes: gas mileage

See Section 3.8 on proportionality and density. In particular, proportionality is discussed on Pages 76-77.

41. k = 189.6 mi/8.0 gal = 24 mi/gal

See Section 3.8 on proportionality and density. In particular, proportionality is discussed on Pages 76-77.

42. P $\propto$ n; P = k (n); atmospheres per mole

See Section 3.8 on proportionality and density. In particular, proportionality is discussed on Pages 76-77.

43. *GIVEN:* 1.51×10^3 g; 865 cm^3 *WANTED:* Density (assume g/cm^3)
EQUATION: $D \equiv m/V$

$$D \equiv \frac{m}{V} = \frac{1.51 \times 10^3 \text{ g}}{865 \text{ cm}^3} = 1.75 \text{ g/cm}^3$$

We assumed that density should be determined in g/cm^3 because mass and volume units were given as g and cm^3, respectively. The defining equation for density is given on Page 77 as Equation 3.9.

44. *GIVEN:* 4.60 cm × 10.3 cm × 13.2 cm; 4.92 kg *WANTED:* Density in g/cm^3
EQUATION: $D \equiv m/V$

$$D \equiv \frac{m}{V} = \frac{4.92 \text{ kg}}{4.60 \text{ cm} \times 10.3 \text{ cm} \times 13.2 \text{ cm}} \times \frac{1000 \text{ g}}{\text{kg}} = 7.87 \text{ g/cm}^3$$

The defining equation for density is given on Page 77 as Equation 3.9.

45. *GIVEN:* 35.3 mL; 0.790 g/mL *WANTED:* mass (assume g)
PER/PATH: $\text{mL} \xrightarrow{0.790 \text{ g/mL}} \text{g}$

$$35.3 \text{ mL} \times \frac{0.790 \text{ g}}{\text{mL}} = 27.9 \text{ g}$$

The use of density as a conversion factor between mass and volume or volume and mass is discussed in Section 3.8, Pages 76-79.

46. *GIVEN:* 227 g; 0.92 g/mL *WANTED:* volume (assume mL)
PER/PATH: $\text{g} \xrightarrow{0.92 \text{ g/mL}} \text{mL}$

$$227 \text{ g} \times \frac{1 \text{ mL}}{0.92 \text{ g}} = 2.5 \times 10^2 \text{ mL}$$

We assumed that volume should be expressed in mL because mL was the volume unit in the given density. The use of density as a conversion factor between mass and volume or volume and mass is discussed in Section 3.8, Pages 76-79.

48. True: a, b, d, f, h, j, l, m, q, r. False: c, e, g, i, k, n, o, p, s, t.

c) The exponential notation form of a number smaller than 1 has a negative exponent.
e) There are 1000 units in a kilounit.
g) There is 1 mL in a cubic centimeter.
i) Celsius degrees (100 between the freezing and boiling points of water) are larger than Fahrenheit degrees (180 between the freezing and boiling points of water).
k) The expression 76.2 g (uncertain in the tenths place) does not mean the same as 76.200 g (uncertain in the thousandths place).
n) The number of significant figures in a product is the same as the number in the quantities multiplied which has the least number of significant figures; this is never more than the number of significant figures in any of the quantities multiplied.
o) *PER* means that one quantity is directly proportional to another.
p) No matter whether or not a quantity in the answer to a problem is familiar, it is necessary to check to make sure the answer is reasonable.
s) Using units in a problem that is solved by algebra has many advantages.
t) A Fahrenheit temperature can be changed to a Celsius temperature by algebra.

49. Centimeters are closest to familiar inches.

Sample calculations for a 5 foot, 8 inch person:

$5 \text{ feet} \times \frac{12 \text{ in.}}{\text{ft}} = 60 \text{ in.}$; $60 \text{ in.} + 8 \text{ in.} = 68 \text{ in.}$

$68 \text{ in.} \times \frac{2.54 \text{ cm}}{\text{in.}} = 173 \text{ cm}$ $173 \text{ cm} \times \frac{1 \text{ m}}{100 \text{ cm}} = 1.73 \text{ m}$

$173 \text{ cm} \times \frac{10 \text{ dm}}{100 \text{ cm}} = 17.3 \text{ m}$ $173 \text{ cm} \times \frac{1000 \text{ mm}}{100 \text{ cm}} = 1.73 \times 10^3 \text{ mm}$

50. $8.5 \text{ in.} \times \frac{2.54 \text{ cm}}{1 \text{ in.}} = 22 \text{ cm}$; $11 \text{ in.} \times \frac{2.54 \text{ cm}}{1 \text{ in.}} = 28 \text{ cm}$

The definition 1 in. ≡ 2.54 cm is from Table 3.3 (Page 71).

51. *GIVEN:* 126 cans; 21 cans/lb; 2.7 g/cm^3 *WANTED:* cm^3 and g

PER/PATH: $\text{cans} \xrightarrow{21 \text{ cans/lb}} \text{lb} \xrightarrow{1 \text{ lb}/454 \text{ g}} \text{g} \xrightarrow{2.7 \text{ g/cm}^3} \text{cm}^3$

$126 \text{ cans} \times \frac{1 \text{ lb}}{21 \text{ cans}} \times \frac{454 \text{ g}}{1 \text{ lb}} \times \frac{1 \text{ cm}^3}{2.7 \text{ g}} = 1.0 \times 10^3 \text{ cm}^3$

$126 \text{ cans} \times \frac{1 \text{ lb}}{21 \text{ cans}} \times \frac{454 \text{ g}}{1 \text{ lb}} = 2.7 \times 10^3 \text{ g}$

The conversion 1 lb = 454 g is from Table 3.3 (Page 71). The use of density as a conversion factor between mass and volume or volume and mass is discussed in Section 3.8, Pages 76-79.

52. $7 \text{ oz} \times \frac{1 \text{ lb}}{16 \text{ oz}} = 0.4 \text{ lb}$; $6.4 \text{ lb} \times \frac{454 \text{ g}}{\text{lb}} = 2.9 \times 10^3 \text{ g} = 2.9 \text{ kg}$

We assume that you know the USCS definition 1 lb ≡ 16 oz. The conversion 1 lb = 454 g is from Table 3.3 (Page 71).

$2.9 \times 10^3 \text{ g} \times \frac{1 \text{ kg}}{1000 \text{ g}} = 2.9 \text{ kg}$

53. *GIVEN:* 12.0 fl oz; 64.4 lb/ft^3; 7.48 gal/ft^3; 4 qt/gal; 32 fl oz/qt *WANTED:* g

PER/PATH: $\text{fl oz} \xrightarrow{32 \text{ fl oz/qt}} \text{qt} \xrightarrow{4 \text{ qt/gal}} \text{gal} \xrightarrow{7.48 \text{ gal/ft}^3} \text{ft}^3 \xrightarrow{64.4 \text{ lb/ft}^3} \text{lb} \xrightarrow{1 \text{ lb}/454 \text{ g}} \text{g}$

$12.0 \text{ fl oz} \times \frac{1 \text{ qt}}{32 \text{ fl oz}} \times \frac{1 \text{ gal}}{4 \text{ qt}} \times \frac{1 \text{ ft}^3}{7.48 \text{ gal}} \times \frac{64.4 \text{ lb}}{1 \text{ ft}^3} \times \frac{454 \text{ g}}{\text{lb}} = 366 \text{ g}$

The conversion 1 lb = 454 g is from Table 3.3 (Page 71). The use of density as a conversion factor between mass and volume or volume and mass is discussed in Section 3.8, Pages 76-79.

54. $$33°\text{F} \times \frac{100 \text{ Celsius degrees}}{180 \text{ Fahrenheit degrees}} = 18°\text{C}$$

Section 3.7 on Temperature, Pages 72-76, explains the relationship between Celsius and Fahrenheit degrees. Note in particular Figure 3.6 on Page 74. The range between the boiling and freezing points of water is 180 Fahrenheit degrees and 100 Celsius degrees, which establishes the conversion factor used in the solution to this problem. Also review the note in the margin on Page 74.

55. *GIVEN:* 0.25 cup; 0.86 g/cm^3; 1 cup/0.25 qt *WANTED:* g

PER/PATH: $\text{cup} \xrightarrow{1\text{ cup}/0.25\text{ qt}} \text{qt} \xrightarrow{1.06\text{ qt/L}} \text{L} \xrightarrow{1000\text{ mL/L}} \text{mL} \xrightarrow{1\text{ cm}^3/1\text{ mL}} \text{cm}^3 \xrightarrow{0.86\text{ g/cm}^3} \text{g}$

$$0.25 \text{ cup} \times \frac{0.25 \text{ qt}}{\text{cup}} \times \frac{1 \text{ L}}{1.06 \text{ qt}} \times \frac{1000 \text{ mL}}{\text{L}} \times \frac{1 \text{ cm}^3}{1 \text{ mL}} \times \frac{0.86 \text{ g}}{\text{cm}^3} = 51 \text{ g}$$

The conversion 1.06 qt = 1 L is from Table 3.3 (Page 71). The 1000 mL = 1 L conversion is one that should be memorized from Table 3.2 (Page 59). You should also memorize the relationship 1 mL = 1 cm^3 (Page 60). The use of density as a conversion factor between mass and volume or volume and mass is discussed in Section 3.8, Pages 76-79.

Chapter 4

Introduction to Gases

1. Kinetic refers to the fact that gas particles are in motion. Kinetic energy is energy of motion.

   The definition of the term *kinetic* and an explanation of kinetic molecular theory are found at the beginning of Section 2.2, Pages 18-19.

2. Pressure results from the large numbers of particle collisions with the container walls.

   See Item 1 in the list of the main features of the ideal gas model on Pages 93-94.

3. Because gas molecules are widely spaced, air is compressible, which makes for a soft and comfortable ride. The low density of air contributes little to the mass of an automobile. The constant motion of the molecules makes the gas fill the tire, exerting pressure uniformly in all directions. There is no loss of pressure because of the elastic molecular collisions in a gas.

   Widely spaced: Item 3 in the list of the main features of the ideal gas model on Page 94.
   Low density: Item 3 in the list of characteristics of gases on Page 93.
   Constant motion: Item 1 in the list of the main features of the ideal gas model on Pages 93-94.
   Elastic molecular collisions: Item 2 in the list of the main features of the ideal gas model on Page 94.

4. This is evidence that gas particles are moving.

   See Item 1 in the list of the main features of the ideal gas model on Pages 93-94.

5. Because gas molecules are more widely spaced than liquid molecules, the gas is less dense and therefore rises through the liquid.

   See Item 3 in the list of the main features of the ideal gas model on Page 94.

6. Gas particles are widely separated compared with the same number of particles close together in the liquid state.

   See Item 3 in the list of the main features of the ideal gas model on Page 94.

7. Particles move in straight lines until they hit something—eventually the walls of the container, thereby filling it.

   See Item 1 in the list of the main features of the ideal gas model on Pages 93-94.

8. Quantity, volume, pressure, and temperature.

   See the beginning of Section 4.3, *Gas Measurements*, Pages 94-95.

9. Barometer operation is explained in Figure 4.6.

   Figure 4.6 is on Page 96.

10.

atm	1.84	0.946	0.959	0.984
psi	27.0	13.9	14.1	14.5
in. Hg	55.1	28.3	28.7	29.4
cm Hg	1.40×10^2	71.9	72.9	74.8
mm Hg	1.40×10^3	719	729	748
torr	1.40×10^3	719	729	748
Pa	1.86×10^5	9.59×10^4	9.72×10^4	9.97×10^4
kPa	186	95.9	97.2	99.7
bar	1.86	0.958	0.971	0.997

Sample calculations:

$$1.84 \text{ atm} \times \frac{14.69 \text{ psi}}{\text{atm}} = 27.0 \text{ psi}$$

$$1.84 \text{ atm} \times \frac{29.92 \text{ in Hg}}{\text{atm}} = 55.1 \text{ in. Hg}$$

$$55.1 \text{ in. Hg} \times \frac{2.54 \text{ cm}}{1 \text{ in}} = 1.40 \times 10^2 \text{ cm Hg}$$

$$1.40 \times 10^2 \text{ cm Hg} \times \frac{10 \text{ mm}}{\text{cm}} = 1.40 \times 10^3 \text{ mm Hg}$$

1 mm Hg = 1 torr so 1.40×10^3 mm Hg = 1.40×10^3 torr

$$1.84 \text{ atm} \times \frac{1.013 \times 10^5 \text{ Pa}}{\text{atm}} = 1.86 \times 10^5 \text{ Pa}$$

$$1.86 \times 10^5 \text{ Pa} \times \frac{1 \text{ kPa}}{1000 \text{ Pa}} = 186 \text{ kPa}$$

$$186 \text{ kPa} \times \frac{1 \text{ bar}}{100 \text{ kPa}} = 1.86 \text{ bar}$$

Conversions among pressure units are given in Equations 4.2, 4.3, and 4.4 on Page 96. The conversion from in. Hg to cm Hg is based on the definition of an inch, 1 in. ≡ 2.54 cm, which is found in Table 3.3, Page 71. The cm Hg to mm Hg conversion is derived from the fact that 100 cm = 1 m = 1000 mm. So 100 cm = 1000 mm, and dividing both sides by 100, 1 cm = 10 mm. The 1 bar = 100 kPa relationship is given in the sentence that defines the bar on Page 96.

11. 752 torr – 284 torr = 468 torr

See Figure 4.7 on Page 97, which describes open-end manometers. In this case, the pressure of the trapped gas is *less* than atmospheric pressure, so the pressure difference is *subtracted* from the atmospheric pressure.

12. Absolute zero is the lowest temperature. At absolute zero, particles have no transnational or rotational motion. They only vibrate.

See the discussion of absolute zero in the Temperature subsection on Page 98.

13. Yes, if the water temperature is –8°C. Water freezes at 0°C. 273 – 8 = 265 K.

Figure 3.6 on Page 74 shows that the freezing point of water is 0°C. Equation 3.8 on Page 74 is the first time the $T_{°C}$-T_K relationship is introduced, and it is repeated as Equation 4.5 on Page 98.

14. $273 - 78 = 195$ K

The $T_{°C}$-T_K relationship is given in Equation 4.5 on Page 98.

15. $240 - 273 = -33°C$

The $T_{°C}$-T_K relationship is given in Equation 4.5 on Page 98.

16.

	Volume	Temperature	Pressure
Initial Value (1)	1.20 L	15 + 273 = 288 K	Constant
Final Value (2)	V_2	40 + 273 = 313 K	Constant

$$V_2 = V_1 \times \frac{T_2}{T_1} = 1.20\text{ L} \times \frac{313\text{ K}}{288\text{ K}} = 1.30\text{ L}$$

Section 4.4, Pages 98-103, explains the Volume-Temperature (Charles's) Law.

17.

	Volume	Temperature	Pressure
Initial Value (1)	14.2 m^3	42 + 273 = 315 K	Constant
Final Value (2)	13.1 m^3	T_2	Constant

$$T_2 = T_1 \times \frac{V_2}{V_1} = 315\text{ K} \times \frac{13.1\text{ m}^3}{14.2\text{ m}^3} = 291\text{ K};\ 291 - 273 = 18°C$$

Section 4.4, Pages 98-103, explains the Volume-Temperature (Charles's) Law.

18. Reducing the volume increases the pressure and breaks the balloon.

Section 4.5, Pages 103-106, explains the Volume-Pressure (Boyle's) Law.

19.

	Volume	Temperature	Pressure
Initial Value (1)	5.83 L	Constant	2.18 atm
Final Value (2)	V_2	Constant	5.03 atm

$$V_2 = V_1 \times \frac{P_1}{P_2} = 5.83\text{ L} \times \frac{2.18\text{ atm}}{5.03\text{ atm}} = 2.53\text{ L}$$

Section 4.5, Pages 103-106, explains the Volume-Pressure (Boyle's) Law.

20.

	Volume	Temperature	Pressure
Initial Value (1)	3.19 L	Constant	644 torr
Final Value (2)	4.00 L	Constant	P_2

$$P_2 = P_1 \times \frac{V_1}{V_2} = 644\text{ torr} \times \frac{3.19\text{ L}}{4.00\text{ L}} = 514\text{ torr}$$

Section 4.5, Pages 103-106, explains the Volume-Pressure (Boyle's) Law.

21.

	Volume	Temperature	Pressure
Initial Value (1)	4.30 L	56 + 273 = 329 K	868 torr
Final Value (2)	6.36 L	12 + 273 = 285 K	P_2

$$P_2 = P_1 \times \frac{V_1}{V_2} \times \frac{T_2}{T_1} = 868 \text{ torr} \times \frac{4.30 \text{ L}}{6.36 \text{ L}} \times \frac{285 \text{ K}}{329 \text{ K}} = 508 \text{ torr}$$

Section 4.6, Pages 106-109, explains the Combined Gas Law.

22.

	Volume	Temperature	Pressure
Initial Value (1)	1 L	135 + 273 = 408 K	844 torr
Final Value (2)	0.790 L	T_2	748 torr

$$T_2 = T_1 \times \frac{P_2}{P_1} \times \frac{V_2}{V_1} = 408 \text{ K} \times \frac{748 \text{ torr}}{844 \text{ torr}} \times \frac{0.790 \text{ L}}{1 \text{ L}} = 286 \text{ K}; \; 286 - 273 = 13°\text{C}$$

Section 4.6, Pages 106-109, explains the Combined Gas Law.

23. Standard temperature and pressure for gases, 0°C and 1 atm.

See Page 108 for the definition of standard pressure and temperature, STP.

24.

	Volume	Temperature	Pressure
Initial Value (1)	8.42 L	35 + 273 = 308 K	725 torr
Final Value (2)	V_2	273 K	760 torr

$$V_2 = V_1 \times \frac{P_1}{P_2} \times \frac{T_2}{T_1} = 8.42 \text{ L} \times \frac{725 \text{ torr}}{760 \text{ torr}} \times \frac{273 \text{ K}}{308 \text{ K}} = 7.12 \text{ L}$$

Section 4.6, Pages 106-109, explains the Combined Gas Law.

25.

	Volume	Temperature	Pressure
Initial Value (1)	6.29 L	273 K	1 atm
Final Value (2)	V_2	–35 + 273 = 238 K	1.86 atm

$$V_2 = V_1 \times \frac{P_1}{P_2} \times \frac{T_2}{T_1} = 6.29 \text{ L} \times \frac{1 \text{ atm}}{1.86 \text{ atm}} \times \frac{238 \text{ K}}{273 \text{ K}} = 2.95 \text{ L}$$

Section 4.6, Pages 106-109, explains the Combined Gas Law.

26.

	Volume	Temperature	Pressure
Initial Value (1)	2.33 L	273 K	1 atm
Final Value (2)	6.19 L	17 + 273 = 290 K P_2	

$$P_2 = P_1 \times \frac{V_1}{V_2} \times \frac{T_2}{T_1} = 760 \text{ torr} \times \frac{2.33 \text{ L}}{6.19 \text{ L}} \times \frac{290 \text{ K}}{273 \text{ K}} = 304 \text{ torr}$$

Section 4.6, Pages 106-109, explains the Combined Gas Law.

27.

	Volume	Temperature	Pressure
Initial Value (1)	4.47 L	273 K	760 torr
Final Value (2)	6.05 L	T_2	552 torr

$$T_2 = T_1 \times \frac{P_2}{P_1} \times \frac{V_2}{V_1} = 273 \text{ K} \times \frac{552 \text{ torr}}{760 \text{ torr}} \times \frac{6.05 \text{ L}}{4.47 \text{ L}} = 268 \text{ K}; \; 268 - 273 = -5°\text{C}$$

Section 4.6, Pages 106-109, explains the Combined Gas Law.

29. True: a, d, f, h. False: b, c, e, g.

b) The large distances between gas particles ensure us that attractions between these molecules are negligible.
c) Gauge pressure is always equal to or greater than absolute pressure. There is no pressure, neither absolute nor gauge, in a vacuum.
e) For a fixed amount of gas at constant pressure, if temperature increases, volume increases.
g) At a given temperature, the number of degrees Celsius is smaller than the number of kelvins, $T_{°C} = T_K - 273$.

30. When gas particles are pushed close to each other, the intermolecular attractions become significant. The molecules are no longer independent. The ideal gas model is violated, so the gas does not behave ideally.

See Section 2.2, States of Matter, on Pages 18-20. Then review Item 5 on the list of main features of the ideal gas model on Page 94.

31.

	Volume	Temperature	Pressure
Initial Value (1)	1.91 L	Constant	959 torr
Final Value (2)	1.91 + 2.45 = 4.36 L	Constant	P_2

$$P_2 = P_1 \times \frac{V_1}{V_2} = 959 \text{ torr} \times \frac{1.91 \text{ L}}{(1.91 + 2.45) \text{ L}} = 4.20 \times 10^2 \text{ torr}$$

Section 4.5, Pages 103-106, explains the Volume-Pressure (Boyle's) Law.

32.

	Volume	Temperature	Pressure
Initial Value (1)	Constant	19 + 273 = 292 K	4.26 atm
Final Value (2)	Constant	42 + 273 = 315 K	P_2

$$P_2 = P_1 \times \frac{T_2}{T_1} = 4.26 \text{ atm} \times \frac{315 \text{ K}}{292 \text{ K}} = 4.60 \text{ atm}$$

Although we did not introduce a "Pressure-Temperature Law," one can be derived from the Combined Gas Law. $\frac{P_1V_1}{T_1} = \frac{P_2V_2}{T_2}$; at constant volume (a steel cylinder has constant volume), $V_1 = V_2$ and $\frac{P_1}{T_1} = \frac{P_2}{T_2}$. Cross multiplying, $P_1T_2 = P_2T_1$. Therefore, $P_2 = P_1 \times \frac{T_2}{T_1}$.

33.

	Volume	Temperature	Pressure
Initial Value (1)	Constant	–18 + 273 = 255 K	355 + 14.7 = 370 psi
Final Value (2)	Constant	23 + 273 = 296 K	P_2

$$P_2 = P_1 \times \frac{T_2}{T_1} = (355 + 14.7) \text{ psi} \times \frac{296 \text{ K}}{255 \text{ K}} = 429 \text{ psi absolute; } 429 - 14.7 = 414 \text{ psi gauge}$$

Gauge pressure is discussed on Page 97. Absolute pressure = Atmospheric pressure + Gauge Pressure. See the discussion for Question 32 regarding the pressure-temperature relationship.

34.

	Volume	Temperature	Pressure
Initial Value (1)	$0.140\ m^3$	$33 + 273 = 306$ K	125 psi gauge
Final Value (2)	V_2	$13 + 273 = 286$ K	751 torr

$$751 \text{ torr} \times \frac{14.69 \text{ psi}}{760 \text{ torr}} = 14.5 \text{ psi} = \text{atmospheric pressure}$$

Initial absolute pressure = 125 + 14.5 = 140 psi

$$V_2 = V_1 \times \frac{P_1}{P_2} \times \frac{T_2}{T_1} = 0.140\ m^3 \times \frac{(125\ +\ 14.5) \text{ psi}}{751 \text{ torr}} \times \frac{760 \text{ torr}}{14.69 \text{ psi}} \times \frac{286 \text{ K}}{306 \text{ K}} = 1.26\ m^3$$

Equations 4.2 and 4.4 on Page 96 give the relationship between pressure in torr and psi: 760 torr = 1 atm = 14.69 psi. Gauge pressure is discussed on Page 97. Absolute pressure = Atmospheric pressure + Gauge Pressure.

35. Volume at start: $350\ cm^3$. Volume at end: $350 - 309 = 41\ cm^3$. Volume is inversely proportional to pressure. Compression ratio $= \frac{350\ cm^3}{41\ cm^3} = 8.5$

Section 4.5, Pages 103-106, explains the Volume-Pressure (Boyle's) Law.

Chapter 5

Atomic Theory: The Nuclear Model of the Atom

1. See the summary in Section 5.1.

   Section 5.1, Dalton's Atomic Theory, describes the theory. The *Summary: Dalton's Atomic Theory* box on Page 115 summarizes the five main features of the theory, and Figure 5.1 on the same page illustrates four of the five features.

2. In a chemical reaction, the atoms in the reactant compounds are rearranged to form the product compounds. Since atoms are not destroyed or created, the total mass must be the same before and after the reaction.

   The Law of Conservation of Mass is discussed in a subsection on Page 35.

3. They are the same because atoms cannot be created or destroyed.

   The second item in the list of main features of Dalton's atomic theory states that atoms cannot be created or destroyed.

4. The Law of Multiple Proportions for this case states that the same mass of chlorine, 10.0 g, will combine with masses of mercury in a ratio of simple whole numbers. $56.6 \div 28.3 = 2$, so the law is confirmed by these data.

   The Law of Multiple Proportions is discussed on Page 116. It is also explained in Figure 5.2 on the same page.

5. Dalton's postulate that atoms are indivisible was first found to be incorrect because of the discovery of subatomic particles.

   See Section 5.2 on Subatomic Particles, Pages 116-117.

6. Most of the volume of a gold atom is empty space.

   See Section 5.3 on The Nuclear Atom, Pages 117-119. Note, in particular, Figure 5.4 and the analogy given in the first paragraph on Page 118.

7. See the discussion in Section 5.3, including the summary box and the captions to Figures 5.3 and 5.4.

   The captions to Figures 5.3 and 5.4 are the primary sources of information about the Rutherford scattering experiment.

8. The Rutherford experiment showed that almost all of the mass of an atom is concentrated in an extremely dense nucleus and that most of the atom's volume is empty space occupied by electrons having very small mass. When the massive proton and neutron were found, it was assumed they made up that dense nucleus.

   The captions to Figures 5.3 and 5.4 are the primary sources of information about the Rutherford scattering experiment.

9. The number of protons is the same as the number of electrons. Neutrons are usually equal to or more than the number of protons or electrons.

In Section 5.4 on Isotopes (Pages 119-121), we state that atoms are electrically neutral. The number of negatively charged electrons must therefore be equal to the number of positively charged protons in each atom. You may have observed from working Example 5.1 that the number of neutrons is greater than or equal to the number of protons for all isotopes in that problem, but that information is not stated directly in the textbook.

10. ^{12}C: Z = 6, 6 protons, 6 neutrons, A = 12, nuclear charge = 6+;
^{13}C: Z = 6, 6 protons, 7 neutrons, A = 13, nuclear charge = 6+.

See Section 5.4 on Isotopes, Pages 119-121.

11.

Name of Element	Nuclear Symbol	Atomic Number	Mass Number	Protons	Neutrons	Electrons
copper	$^{65}_{29}Cu$	29	65	29	36	29
chromium	$^{52}_{24}Cr$	24	52	24	28	24
cobalt	$^{60}_{27}Co$	27	60	27	33	27
silver	$^{107}_{47}Ag$	47	107	47	60	47
lead	$^{207}_{82}Pb$	82	207	82	125	82

See Section 5.4 on Isotopes, Pages 119-121.

12. Exactly 1/12 the mass of a carbon-12 atom.

The definition of an amu is given near the bottom of Page 121.

13. Boron occurs in nature as a mixture of atoms that have different masses.

See Section 5.5 on Atomic Mass, Pages 121-123.

14. Atomic mass is a weighted average that takes into account the percentage distribution of isotopes of an element, not an arithmetic average of the atomic masses of the isotopes.

The definition of atomic mass is given on Page 122. See Section 5.5 on Atomic Mass, Pages 121-123.

15. 0.1978×10.0129 amu + $(1.0000 - 0.1978) \times 11.00931$ amu = 1.981 amu + 8.832 amu = 10.813 amu. The extra significant figure appears because adding two numerals in the *units* column produces a numeral in the *tens* column.

See Section 5.5 on Atomic Mass, Pages 121-123. The addition/subtraction significant figure rule is explained on Page 66-67. The multiplication/division significant figure rule is explained on Pages 67-69.

16. 0.6909×62.9298 amu + 0.3091×64.9278 amu = 63.55 amu, Cu, copper

See Section 5.5 on Atomic Mass, Pages 121-123.

17. 0.3707 × 184.9530 amu + 0.6293 × 186.9560 amu = 186.2 amu, Re, rhenium

See Section 5.5 on Atomic Mass, Pages 121-123.

18. 0.6788 × 57.9353 amu + 0.2623 × 59.9332 amu + 0.0119 × 60.9310 amu + 0.0366 × 61.9283 amu + 0.0108 × 63.9280 amu = 58.73 amu, Ni, nickel

See Section 5.5 on Atomic Mass, Pages 121-123.

19. Be, Mg, Ca, Sr, Ba, Ra; 11 to 18

See Section 5.6, The Periodic Table, on Pages 123-126. *Groups* are explained on Page 125, as are *periods.*

20. a and d, same period; b and c, same family

See Section 5.6, The Periodic Table, on Pages 123-126. *Groups,* which are also called *chemical families,* are explained on Page 125, as are *periods.*

21. Z = 24, 52.00 amu; Z = 50, 118.7 amu; Z = 77, 192.2 amu

Figure 5.7 on Page 125 shows how to find the atomic mass of an element from information given in the periodic table.

22. K, 39.10 amu; S, 32.07 amu

Figure 5.7 on Page 125 shows how to find the atomic mass of an element from information given in the periodic table.

23.

Name of Element	Atomic Number	Symbol of Element
Sodium	11	Na
Lead	82	Pb
Aluminum	13	Al
Iron	26	Fe
Fluorine	9	F
Boron	5	B
Argon	18	Ar
Silver	47	Ag
Carbon	6	C
Copper	29	Cu
Beryllium	4	Be
Krypton	36	Kr
Chlorine	17	Cl
Hydrogen	1	H
Manganese	25	Mn
Chromium	24	Cr
Cobalt	27	Co
Mercury	80	Hg

Section 5.7, Elemental Symbols and the Periodic Table, on Pages 126-129, discusses the methods you can use to memorize 35 name-symbol pairs.

25. True: b (Dalton apparently did not make any specific comment about the diameter of an atom, but he did propose that all atoms of an element are identical in every respect. This would include diameters.), d, f, j, k, l, q, r, s. False: a, c, e, g, h, i, m, n, o, p.

a) Dalton proposed that atoms of different elements combine in a ratio of small whole numbers.
c) The mass of a neutron is about the same as the mass of a proton.
e) Nearly all of the mass of an atom is concentrated in the nucleus.
g) The mass of the proton is about 1 amu; the mass of the electron is nearly zero. A proton has a positive charge; an electron has a negative charge.
h) Isotopes of an element have different masses.
i) The atomic number of an element is the number of protons in the nucleus of an atom of that element.
m) Isotopes of different elements that exhibit the same mass number exhibit dissimilar chemical behavior. They are different elements.
n) The mass number of a carbon-12 atom is 12. The mass of a carbon-12 atom is 12 amu.
o) Periods are arranged horizontally in the periodic table.
p) The atomic mass of the second element in the right column of the periodic table (Ne) is 20.18 amu.

26. What was left had to have a positive charge to account for the neutrality of the complete atom.

The electron and its charge on discussed in the first paragraph of Section 5.2, Subatomic Particles, on Page 116. Item 4 in the *Summary: The Nuclear Model of the Atom* box on Page 117 describes the electrical neutrality of the atom.

27. Sixteen grams of oxygen combines with 46 grams of sodium in sodium oxide, and 32 grams of oxygen combines with 46 grams of sodium in sodium peroxide. The ratio 16/32 reduces to 1/2, a ratio of small, whole numbers.

The Law of Multiple Proportions is described in the first paragraph on Page 116. It is also explained in Figure 5.2 on the same page.

28. $0.7215 \times 84.9118 \text{ amu} + (1 - 0.7215) \times x \text{ amu} = 85.4678 \text{ amu}$;
$x = 86.91 \text{ amu}$

See Section 5.5, Atomic Mass, on Pages 121–123. The solution also calls upon your algebra skills. The sum of the fractional composition of components of a mixture must equal 1. $72.15 \div 100 = 0.7215$, the fraction of the total composed of the known isotope. The fraction of the unknown isotope must be $1 - 0.7215$.

29. Different e/m ratios for positively charged particles from different elements indicate that, unlike the electron, all positively charged particles are not alike. Either the charge, the mass, or both must vary from element to element. This suggests the presence of at least two particles in varying number ratios. One or both must have a positive charge; others could be electrically neutral.

See Section 5.2, Subatomic Particles, Pages 116–117.

30. 12.09899 amu. The difference in masses of the nuclear parts and the sum of the masses of protons and neutrons is what is responsible for nuclear energy in an energy-mass conversion. This is discussed in Chapter 21.

The difference between the sum of the masses of the subatomic particles that make up an atom and the mass of the atom itself is known as the *mass defect*. This "missing" matter is converted into energy according to $\Delta E = \Delta mc^2$, and the energy is used to bind together the particles in the nucleus.

31. Mass of electron + proton + neutron: 0.000549 amu + 1.00728 amu + 1.00867 amu = 2.01650 amu; Electron: (0.000549 amu ÷ 2.01650 amu) × 100 = 0.0272%; Proton: (1.00728 amu ÷ 2.01650 amu) × 100 = 49.9519%; Neutron: (1.00867 amu ÷ 2.01650 amu) × 100 = 50.02%; The nucleus, containing the protons and neutrons, accounts for 49.9519% + 50.0208% = 99.9727% of the mass of the atom. *Note*: The same result can be obtained by calculating in grams rather than in amu. You may have calculated in total mass rather than unit mass, in which case all masses above would be multiplied by 6. Unit masses could not have been used if there were different numbers of electrons, protons, or neutrons.

The data are from Table 5.1, Subatomic Particles, on Page 117.

Chapter 6

Chemical Nomenclature

1. Hydrogen, H_2; nitrogen, N_2; oxygen, O_2; fluorine, F_2; chlorine, Cl_2; bromine, Br_2; iodine, I_2.

   Formulas of Elements are discussed in Section 6.2 on Pages 136-138. Figure 6.1 on Page 137 shows the seven diatomic elements and their positions in the periodic table.

2. C, I_2, Zn, Ar

   See Section 6.2, Formulas of Elements, on Pages 136-138. The formula of most elements is the elemental symbol. Seven elements exist as diatomic molecules.

3. Oxygen, calcium, barium, silver

   See Section 6.2, Formulas of Elements, on Pages 136-138. The formula of most elements is the elemental symbol. Seven elements exist as diatomic molecules.

4. H_2, Pb, Si, Na

   See Section 6.2, Formulas of Elements, on Pages 136-138. The formula of most elements is the elemental symbol. Seven elements exist as diatomic molecules.

5. Sulfur dioxide, dinitrogen oxide (or dinitrogen monoxide), PBr_3, HI(g)

   See Section 6.3, Compounds Made from Two Nonmetals, Pages 138-139. Prefixes used in naming binary molecular compounds are given in Table 6.1, Page 139. We used the symbol (g) after the formula of HI to distinguish between gaseous hydrogen iodide, HI(g), and a solution of that compound dissolved in water, hydroiodic acid, HI(aq).

6. When an atom loses one, two, or three electrons, the particle that remains is a monatomic cation. The ion has a positive charge because it has more protons than electrons.

   The formation of ions is discussed at the beginning of Section 6.4, Names and Formulas of Ions Formed by One Element, on Page 139.

7. Calcium ion, chromium(III) ion, zinc ion, phosphide ion, bromide ion

   See Section 6.4, Names and Formulas of Ions Formed by One Element, on Pages 139-143. When chromium atoms form ions, there is more than one possible charge, so the magnitude of the ionic charge must be given in the name of the ion. Zinc commonly forms only one ion, so its charge is not included in its name.

8. Li^+, NH_4^+, N^{3-}, F^- Hg^{2+}

   See Section 6.4, Names and Formulas of Ions Formed by One Element, on Pages 139-143. The name and formula of the ammonium ion are given in Section 6.7 on Page 151.

9. The element hydrogen is present in all acids discussed in this chapter.

   See the discussion at the beginning of Section 6.5, Acids and the Anions Derived from Their Total Ionization, on Page 143.

10. The suffix *-protic* refers to protons, or hydrogen ions. A polyprotic acid is an acid that has two or more ionizable hydrogens.

    See the first paragraph on Page 144.

11.

Acid Name	Acid Formula	Ion Name	Ion Formula
Hydrochloric	HCl	Chloride	Cl^-
Nitric	HNO_3	Nitrate	NO_3^-
Phosphoric	H_3PO_4	Phosphate	PO_4^{3-}
Hydrosulfuric	H_2S	Sulfide	S^{2-}
Nitrous	HNO_2	Nitrite	NO_2^-
Iodic	HIO_3	Iodate	IO_3^-
Hydrotelluric	H_2Te	Telluride	Te^{2-}
Hypochlorous	$HClO$	Hypochlorite	ClO^-
Iodous	HIO_2	Iodite	IO_2^-
Hydroselenic	H_2Se	Selenide	Se^{2-}
Perchloric	$HClO_4$	Perchlorate	ClO_4^-
Hydroiodic	HI	Iodide	I^-
Chlorous	$HClO_2$	Chlorite	ClO_2^-
Selenic	H_2SeO_4	Selenate	SeO_4^{2-}
Bromous	$HBrO_2$	Bromite	BrO_2^-

The information needed to complete this table is found in Section 6.5, Acids and the Anions Derived from Their Total Ionization, on Pages 143–150. Table 6.2, Acids and Anions, on Page 145, summarizes the five *-ic* acids that are to be memorized. Table 6.3, Prefixes and Suffixes in Acid and Anion Nomenclature, also on Page 145, summaries the system by which the five memorized *-ic* acids become the basis for naming over 1000 compounds. Memorize the acids and learn the system.

12. Hydrogen ions leave a polyprotic acid one at a time, yielding one or more intermediate anions that still contain ionizable hydrogen. This is step-by-step ionization. An equation that shows all of the ionizable hydrogen ions removed at once is total ionization.

    Section 6.6, Names and Formulas of Other Acids and Ions, explains step-by-step ionization. See Pages 150–151.

13. HSe^-, $H_2PO_4^-$

    Group 6A/16 elements form 2– ions, so selenium, Se, becomes selenide ion, Se^{2-}. Adding a H^+ to that yields hydrogen selenide ion, HSe^-. The memorized phosphoric acid, H_3PO_4, gives the formula of the phosphate ion, PO_4^{3-}. Addition of two H^+ gives the formula of the dihydrogen phosphate ion, $H_2PO_4^-$.

14. Hydrogen sulfate ion, hydrogen phosphate ion (or monohydrogen phosphate ion)

The memorized sulfuric acid, H_2SO_4, gives the formula of the sulfate ion, SO_4^{2-}. Addition of a hydrogen ion, H^+, yields the formula of the hydrogen sulfate ion, HSO_4^-. The memorized phosphoric acid, H_3PO_4, gives the formula of the phosphate ion, PO_4^{3-}. Addition of a H^+ gives the formula of the hydrogen phosphate ion, HPO_4^{2-}. The prefix *mono-* is assumed when not written, so it is optional.

15. Acetic acid, gallium ion

Your instructor will probably let you know if you are responsible for learning the names and formulas of the species in Section 6.7, Names and Formulas of Other Acids and Ion, Pages 151-152. The formula of acetic acid is memorized. Cations are named by giving the name of the element followed by the word *ion.*

16. $C_2H_3O_2^-$, Rb^+

The total ionization of an acid that ends in *-ic* yields an anion that ends in *-ate.* Acetic acid, memorized as $HC_2H_3O_2$, gives up an H^+ to yield $C_2H_3O_2^-$, acetate ion. Group 1A/1 elements form 1+ ions, so the formula of rubidium ion is Rb^+.

17. Li_2CO_3, $Ba_3(PO_4)_2$, $Al(NO_3)_3$

Lithium ion, Li^+, is from a Group 1A/1 element, which form 1+ ions. Carbonate ion, CO_3^{2-}, is derived from the memorized *-ic* acid, carbonic acid, H_2CO_3. Loss of two H^+ yields its formula. Barium ion, Ba^{2+}, is from a Group 2A/2 element, which form 2+ ions. Phosphate ion, PO_4^{3-}, is derived from the memorized *-ic* acid, phosphoric acid, H_3PO_4. Loss of three H^+ yields its formula. Aluminum ion, Al^{3+}, is from a Group 3A/13 element, which form 3+ ions. Nitrate ion, NO_3^-, is derived from the memorized *-ic* acid, nitric acid, HNO_3. Loss of one H^+ yields its formula.

18. $LiCl$, NH_4NO_3, $BaBr_2$, $Mg_3(PO_4)_2$

Lithium ion, Li^+, is from a Group 1A/1 element, which form 1+ ions. Chloride ion, Cl^-, is from a Group 7A/17 element, which form 1- ions. Ammonium ion, NH_4^+, should be memorized (Goal 7, Page 151). Nitrate ion, NO_3^-, is derived from the memorized *-ic* acid, nitric acid, HNO_3. Loss of one H^+ yields its formula. Barium ion, Ba^{2+}, is from a Group 2A/2 element, which form 2+ ions. Bromide ion, Br^-, is from a Group 7A/17 element, which form 1- ions. Magnesium ion, Mg^{2+}, is from a Group 2A/2 element, which form 2+ ions. Phosphate ion, PO_4^{3-}, is derived from the memorized *-ic* acid, phosphoric acid, H_3PO_4. Loss of three H^+ yields its formula.

19. $Ba(IO)_2$, $Cu(NO_3)_2$, $Ba(HCO_3)_2$

Barium ion, Ba^{2+}, is from a Group 2A/2 element, which form 2+ ions. Hypoiodite ion, IO^-, is based on the memorized *-ic* acid from Group 7A/17, which is chloric acid, $HClO_3$. Substitution of iodine for chlorine gives iodic acid, HIO_3. Two fewer oxygens than the *-ic* acid adds a *hypo-* prefix and changes *-ic* to *-ous,* yielding hypoiodous acid, HIO. Removing the H^+ changes *-ous* to *-ite*, giving IO^-, hypoiodite ion. The 1- anion in $Cu(NO_3)_2$ occurs twice in the formula unit, so the cation must be 2+, copper(II) ion, Cu^{2+}. Nitrate ion, NO_3^-, is derived from the memorized *-ic* acid, nitric acid, HNO_3. Loss of one H^+ yields its formula. Carbonate ion, CO_3^{2-} is derived from the memorized *-ic* acid, carbonic acid, H_2CO_3. Loss of two H^+ yields its formula. Addition of one H^+ yields the hydrogen carbonate ion, HCO_3^-.

20. Ammonium chloride, potassium hydroxide, sodium sulfate

Ammonium ion, NH_4^+, should be memorized (Goal 7, Page 151). Chloride ion, Cl^-, is from a Group 7A/17 element, which form 1– ions. Potassium ion, K^+, is from a Group 1A/1 element, which form 1+ ions. Hydroxide ion, OH^-, should be memorized (Goal 7, Page 151). Sodium ion, Na^+, is from a Group 1A/1 element, which form 1+ ions. Sulfate ion, SO_4^{2-}, is derived from the memorized *-ic* acid, sulfuric acid, H_2SO_4. Loss of two H^+ yields its formula.

21. Calcium sulfide, barium carbonate, potassium phosphate, ammonium sulfate

Calcium ion, Ca^{2+}, is from a Group 2A/2 element, which form 2+ ions. Sulfide ion, S^{2-}, is from a Group 6A/16 element, which form 2– ions. Barium ion, Ba^{2+}, is from a Group 2A/2 element, which form 2+ ions. Carbonate ion, CO_3^{2-}, is derived from the memorized *-ic* acid, carbonic acid, H_2CO_3. Loss of two H^+ yields its formula. Potassium ion, K^+, is from a Group 1A/1 element, which form 1+ ions. Phosphate ion, PO_4^{3-}, is derived from the memorized *-ic* acid, phosphoric acid, H_3PO_4. Loss of three H^+ yields its formula. Ammonium ion, NH_4^+, should be memorized (Goal 7, Page 151). Sulfate ion, SO_4^{2-}, is derived from the memorized *-ic* acid, sulfuric acid, H_2SO_4. Loss of two H^+ yields its formula.

22. Magnesium sulfite, aluminum bromate, lead(II) carbonate

Magnesium ion, Mg^{2+}, is from a Group 2A/2 element, which form 2+ ions. Sulfite ion, SO_3^{2-}, is based on the memorized *-ic* acid from Group 6A/16, which is sulfuric acid, H_2SO_4. One less oxygen than the *-ic* acid changes *-ic* to *-ous*, yielding sulfurous acid, H_2SO_3. Removing two H^+ changes *-ous* to *-ite*, giving SO_3^{2-}, sulfite ion. Aluminum ion, Al^{3+}, is from a Group 3A/13 element, which form 3+ ions. Bromate ion, BrO_3^-, is based on the memorized *-ic* acid from Group 7A/17, which is chloric acid, $HClO_3$. Substitution of bromine for chlorine gives bromic acid, $HBrO_3$. Removing the H^+ changes *-ic* to *-ite*, yielding bromate ion, BrO_3^-. Lead(II) ion is Pb^{2+}. The charge of the cation must be equal to the magnitude of the charge on the anion. Carbonate ion, CO_3^{2-}, is derived from the memorized *-ic* acid, carbonic acid, H_2CO_3. Loss of two H^+ yields its formula.

23. A hydrate is a compound that contains water molecules as part of its crystal structure. An anhydrous compound is one from which all water has been removed or one without water.

See Section 6.10, Hydrates, on Page 157.

24. Two; calcium chloride dihydrate

Calcium ion, Ca^{2+}, is from a Group 2A/2 element, which form 2+ ions. Chloride ion, Cl^-, is from a Group 7A/17 element, which form 1– ions. The prefix for 2 is *di-* (Table 6.1, Page 139).

25. $Ba(OH)_2 \cdot 8\ H_2O$, barium hydroxide octahydrate

Barium ion, Ba^{2+}, is from a Group 2A/2 element, which form 2+ ions. Hydroxide ion, OH^-, should be memorized (Goal 7, Page 151). The prefix for 8 is *octa-* (Table 6.1, Page 139).

26. HSO_3^-, KNO_3, manganese(II) sulfate, sulfur trioxide

Sulfite ion, SO_3^{2-}, is based on the memorized *-ic* acid from Group 6A/16, which is sulfuric acid, H_2SO_4. One less oxygen than the *-ic* acid changes *-ic* to *-ous*, yielding sulfurous acid, H_2SO_3. Removing two H^+ changes *-ous* to *-ite*, giving SO_3^{2-}, sulfite ion. Addition of a hydrogen ion, H^+, yields the formula of the hydrogen sulfite ion, HSO_3^-. Potassium ion, K^+, is from a Group 1A/1 element, which form 1+ ions. Nitrate ion, NO_3^-, is derived from the memorized *-ic* acid, nitric acid, HNO_3. Loss of one H^+ yields its formula. Manganese(II) ion is Mn^{2+}. The charge of the cation must be equal to the magnitude of the charge on the anion. Sulfate ion, SO_4^{2-}, is derived from the memorized *-ic* acid, sulfuric acid, H_2SO_4. Loss of two H^+ yields its formula. Sulfur trioxide is a binary molecular compound: One sulfur (*mono-* is omitted) and three (*tri-*) oxygen atoms.

27. Bromate ion, nickel hydroxide, AgCl, SiF_6

Bromate ion, BrO_3^-, is based on the memorized *-ic* acid from Group 7A/17, which is chloric acid, $HClO_3$. Substitution of bromine for chlorine gives bromic acid, $HBrO_3$. Removing the H^+ changes *-ic* to *-ite*, yielding bromate ion, BrO_3^-. Nickel ion, Ni^{2+}, is one of the transition elements that forms only one ion. Its charge must be memorized. Hydroxide ion, OH^-, should be memorized (Goal 7, Page 151). Silver ion, Ag^+, is one of the transition elements that forms only one ion. Its charge must be memorized. Chloride ion, Cl^-, is from a Group 7A/17 element, which form 1- ions. Silicon hexafluoride is a binary molecular compound: One silicon (*mono-* is omitted) and six (*hexa-*) fluorine atoms.

28. TeO_4^{2-}, $MnPO_4$, sodium acetate, dihydrogen sulfide

Tellurate ion, TeO_4^{2-}, is based on the memorized *-ic* acid from Group 6A/16, which is sulfuric acid, H_2SO_4. Substitution of tellurium for sulfur gives telluric acid, H_2TeO_4. Removing the two H^+ changes *-ic* to *-ate*, yielding tellurate ion, TeO_4^{2-}. Manganese(III) ion is Mn^{3+}. The charge of the ion is given in the name. Phosphate ion, PO_4^{3-}, is derived from the memorized *-ic* acid, phosphoric acid, H_3PO_4. Loss of three H^+ yields its formula. Sodium ion, Na^+, is from a Group 1A/1 element, which form 1+ ions. The total ionization of an acid that ends in *-ic* yields an anion that ends in *-ate*. Acetic acid, memorized as $HC_2H_3O_2$, gives up an H^+ to yield $C_2H_3O_2^-$, acetate ion. $H_2S(g)$ is a binary molecular compound: Two (*bi-*) hydrogen and one (*mono-* is omitted) sulfur atoms.

29. Hydrogen phosphate ion, copper(II) oxide, $Na_2C_2O_4$, NH_3

The memorized phosphoric acid, H_3PO_4, gives the formula of the phosphate ion, PO_4^{3-}. Addition of a H^+ gives the formula of the hydrogen phosphate ion, HPO_4^{2-}. Copper(II) ion is Cu^{2+}. The charge of the cation must be equal to the magnitude of the charge on the anion. Oxide ion, O^{2-}, is from a Group 6A/16 element, which form 2- ions. Sodium ion, Na^+, is from a Group 1A/1 element, which form 1+ ions. The oxalate ion, $C_2O_4^{2-}$, is given in Table 6.8 (Page 153). Ammonia, NH_3, is a binary molecular compound that is called by its traditional name (Goal 3, see Page 139).

30. $HClO$, $CrBr_2$, potassium hydrogen carbonate, sodium dichromate

Hypochlorous acid, $HClO$, is based on the memorized *-ic* acid from Group 7A/17, which is chloric acid, $HClO_3$. Two fewer oxygens than the *-ic* acid adds a *hypo-* prefix and changes *-ic* to *-ous*, yielding hypochlorous acid, $HClO$. Chromium(II) ion is Cr^{2+}. The magnitude of the charge is given in the name. Bromide ion, Br^-, is from a Group 7A/17 element, which form 1- ions. Potassium ion, K^+, is from a Group 1A/1 element, which form 1+ ions. Carbonate ion, CO_3^{2-}, is derived from the memorized *-ic* acid, carbonic acid, H_2CO_3. Loss of two H^+ yields its formula. Addition of one H^+ yields the hydrogen carbonate ion, HCO_3^-. Sodium ion, Na^+, is from a Group 1A/1 element, which form 1+ ions. Dichromate ion, $Cr_2O_7^{2-}$, is discussed in the *Polyatomic Anions from Transition Elements* subsection on Page 152. It also appears in Table 6.8 on Page 153.

31. Cobalt(III) oxide, sodium sulfite, HgI_2, $Al(OH)_3$

Cobalt(III) ion is Co^{3+}. The charge of the cations must be equal to the magnitude of the charge on the anions; three 2- are balanced by two 3+. Oxide ion, O^{2-}, is from a Group 6A/16 element, which form 2- ions. Sodium ion, Na^+, is from a Group 1A/1 element, which form 1+ ions. Sulfite ion, SO_3^{2-}, is based on the memorized *-ic* acid from Group 6A/16, which is sulfuric acid, H_2SO_4. One less oxygen than the *-ic* acid changes *-ic* to *-ous*, yielding sulfurous acid, H_2SO_3. Removing two H^+ changes *-ous* to *-ite*, giving SO_3^{2-}, sulfite ion. Mercury(II) ion is Hg^{2+}. The magnitude of the charge is given in the name. Iodide ion, I–, is from a Group 7A/17 element, which form 1- ions. Aluminum ion, Al^{3+}, is from a Group 3A/13 element, which form 3+ ions. Hydroxide ion, OH^-, should be memorized (Goal 7, Page 151).

32. $Ca(H_2PO_4)_2$, $KMnO_4$, ammonium iodate, selenic acid

Calcium ion, Ca^{2+}, is from a Group 2A/2 element, which form 2+ ions. The memorized phosphoric acid, H_3PO_4, gives the formula of the phosphate ion, PO_4^{3-}. Addition of two H^+ gives the formula of the dihydrogen phosphate ion, $H_2PO_4^-$. Potassium ion, K^+, is from a Group 1A/1 element, which form 1+ ions. The permanganate ion, MnO_4^-, is discussed in the *Polyatomic Anions from Transition Elements* subsection on Page 152. It also appears in Table 6.8 on Page 153. Ammonium ion, NH_4^+, should be memorized (Goal 7, Page 151). Iodate ion, IO_3^-, is based on the memorized *-ic* acid from Group 7A/17, which is chloric acid, $HClO_3$. Substitution of iodine for chlorine gives iodic acid, HIO_3. Removing the H^+ changes *-ic* to *-ate*, yielding iodate ion, IO_3^-. Selenic acid, H_2SeO_4, is based on the memorized *-ic* acid from Group 6A/16, which is sulfuric acid, H_2SO_4. Substitution of selenium for sulfur gives selenic acid, H_2SeO_4.

33. Mercury(I) chloride, periodic acid, $CoSO_4$, $Pb(NO_3)_2$

Mercury(I) ion is Hg_2^{2+}. See the discussion of this diatomic ion on Page 142. Chloride ion, Cl^-, is from a Group 7A/17 element, which form 1- ions. Periodic acid, HIO_4, is based on the memorized *-ic* acid from Group 7A/17, which is chloric acid, $HClO_3$. Substitution of iodine for chlorine gives iodic acid, HIO_3. One more oxygen than the *-ic* acid adds a *per-* prefix, yielding periodic acid, HIO_4. Cobalt(II) ion is Co^{2+}. The magnitude of the charge is given in the name. Sulfate ion, SO_4^{2-}, is derived from the memorized *-ic* acid, sulfuric acid, H_2SO_4. Loss of two H^+ yields its formula. Lead(II) ion is Pb^{2+}. The magnitude of the charge is given in the name. Nitrate ion, NO_3^-, is derived from the memorized *-ic* acid, nitric acid, HNO_3. Loss of one H^+ yields its formula.

34. P_4S_7, BaO_2, manganese(II) chloride, sodium chlorite

Tetraphosphorus heptasulfide is a binary molecular compound: Four (*tetra-*) phosphorus and seven (*hepta-*) sulfur atoms. Table 6.1 on Page 139 gives the prefixes used in naming these compounds. Barium ion, Ba^{2+}, is from a Group 2A/2 element, which form 2+ ions. Peroxide ion is O_2^{2-}. It is given in Table 6.8 on Page 153. Manganese(II) ion is Mn^{2+}. The charge of the cation must be equal to the magnitude of the charge on the anions. Chloride ion, Cl^-, is from a Group 7A/17 element, which form 1- ions. Sodium ion, Na^+, is from a Group 1A/1 element, which form 1+ ions. Chlorite ion, ClO_2^-, is based on the memorized *-ic* acid from Group 7A/17, which is chloric acid, $HClO_3$. One less oxygen than the *-ic* acid changes *-ic* to *-ous*, yielding chlorous acid, $HClO_2$. Removing the H^+ changes *-ous* to *-ite*, giving ClO_2^-, chlorite ion.

35. Potassium tellurate, zinc carbonate, $CrCl_2$, $HC_2H_3O_2$

Potassium ion, K^+, is from a Group 1A/1 element, which form 1+ ions. Tellurate ion, TeO_4^{2-}, is based on the memorized *-ic* acid from Group 6A/16, which is sulfuric acid, H_2SO_4. Substitution of tellurium for sulfur gives telluric acid, H_2TeO_4. Removing the two H^+ changes *-ic* to *-ate*, yielding tellurate ion, TeO_4^{2-}. Zinc ion, Zn^{2+}, is one of the transition elements that forms only one ion. Its charge must be memorized. Carbonate ion, CO_3^{2-}, is derived from the memorized *-ic* acid, carbonic acid, H_2CO_3. Loss of two H^+ yields its formula. Chromium(II) ion is Cr^{2+}. The magnitude of the charge is given in the name. Chloride ion, Cl^-, is from a Group 7A/17 element, which form 1- ions. The formula of acetic acid is memorized. See Section 6.7, Names and Formulas of Other Acids and Ion, Pages 151-152.

36. $BaCrO_4$, $CaSO_3$, copper(I) chloride, silver nitrate

Barium ion, Ba^{2+}, is from a Group 2A/2 element, which form 2+ ions. Chromate ion, CrO_4^{2-}, is discussed in the *Polyatomic Anions from Transition Elements* subsection on Page 152. It also appears in Table 6.8 on Page 153. Calcium ion, Ca^{2+}, is from a Group 2A/2 element, which form 2+ ions. Sulfite ion, SO_3^{2-}, is based on the memorized *-ic* acid from Group 6A/16, which is sulfuric acid, H_2SO_4. One less oxygen than the *-ic* acid changes *-ic* to *-ous*, yielding sulfurous acid, H_2SO_3. Removing two H^+ changes *-ous* to *-ite*, giving SO_3^{2-}, sulfite ion. Copper(I) ion is Cu^+. The charge of the cation must be equal to the magnitude of the charge on the anion. Chloride ion, Cl^-, is from a Group 7A/17 element, which form 1- ions. Silver ion, Ag^+, is one of the transition elements that forms only one ion. Its charge must be memorized. Nitrate ion, NO_3^-, is derived from the memorized *-ic* acid, nitric acid, HNO_3. Loss of one H^+ yields its formula.

37. Sodium peroxide, nickel carbonate, FeO, $H_2S(aq)$ [The state designation (aq) distinguishes hydrosulfuric acid from dihydrogen sulfide, $H_2S(g)$]

Sodium ion, Na^+, is from a Group 1A/1 element, which form 1+ ions. Peroxide ion is O_2^{2-}. It is given in Table 6.8 on Page 153. Nickel ion, Ni^{2+}, is one of the transition elements that forms only one ion. Its charge must be memorized. Carbonate ion, CO_3^{2-}, is derived from the memorized *-ic* acid, carbonic acid, H_2CO_3. Loss of two H^+ yields its formula. Iron(II) ion is Fe^{2+}. The magnitude of the charge is given in the name. Oxide ion, O^{2-}, is from a Group 6A/16 element, which form 2- ions. Hydrosulfuric acid, $H_2S(aq)$, has no oxygen. Its name therefore has the *hydro-* prefix and *-ic* suffix. The 2- charge of the sulfide ion is balanced by two 1+ charges on the hydrogen ion.

38. Zn_3P_2, $CsNO_3$, ammonium cyanide, disulfur decafluoride

Zinc ion, Zn^{2+}, is one of the transition elements that forms only one ion. Its charge must be memorized. Phosphide ion, P^{3-}, is from a Group 5A/5 element, which form 3– ions. Cesium ion, Cs^+, is from a Group 1A/1 element, which form 1+ ions. Nitrate ion, NO_3^-, is derived from the memorized *-ic* acid, nitric acid, HNO_3. Loss of one H^+ yields its formula. Ammonium ion, NH_4^+, should be memorized (Goal 7, Page 151). Cyanide ion, CN^-, is discussed in the *Hydrocyanic Acid and the Cyanide Ion* subsection on Page 152. It is also in Table 6.8 on Page 153. Disulfur decafluoride is a binary molecular compound: Two (*di-*) sulfur and ten (*deca-*) oxygen atoms.

39. Dinitrogen trioxide, lithium permanganate, In_2Se_3, $Hg_2(SCN)_2$

Dinitrogen trioxide is a binary molecular compound: Two (*di-*) nitrogen and three (*tri-*) oxygen atoms. Lithium ion, Li^+, is from a Group 1A/1 element, which form 1+ ions. The permanganate ion, MnO_4^-, is discussed in the *Polyatomic Anions from Transition Elements* subsection on Page 152. It also appears in Table 6.8 on Page 153. Indium ion, In^{3+}, is from a Group 3A/13 element, which form 3+ ions. Selenide ion, Se^{2-}, is from a Group 6A/16 element, which form 2– ions. Mercury(I) ion is Hg_2^{2+}. See the discussion of this diatomic ion on Page 142. Thiocyanate ion, SCN^-, is found in Table 6.8 on Page 153.

40. Cadmium chloride, nickel chlorate, $CoPO_4$, $Ca(IO_4)_2$

Cadmium ion, Cd^{2+}, is analogous to zinc ion, Zn^{2+}. Cadmium is immediately below zinc on the periodic table. Chloride ion, Cl^-, is from a Group 7A/17 element, which form 1– ions. Nickel ion, Ni^{2+}, is one of the transition elements that forms only one ion. Its charge must be memorized. Chlorate ion, ClO_3^-, is based on the memorized *-ic* acid from Group 7A/17, which is chloric acid, $HClO_3$. Removing the H^+ changes *-ic* to *-ate*, yielding chlorate ion, ClO_3^-. Cobalt(III) ion is Co^{3+}. The magnitude of the charge is given in the name. Phosphate ion, PO_4^{3-}, is derived from the memorized *-ic* acid, phosphoric acid, H_3PO_4. Loss of three H^+ yields its formula. Calcium ion, Ca^{2+}, is from a Group 2A/2 element, which form 2+ ions. Periodate ion, IO_4^-, is based on the memorized *-ic* acid in Group 7A/17 which is chloric acid, $HClO_3$. Substitution of iodine for chlorine gives iodic acid, HIO_3. Adding one oxygen adds the *per-* prefix, yielding periodic acid, HIO_4. Removing the H^+ to form the anion changes *-ic* to *-ate*, giving periodate ion, IO_4^-.

Chapter 7

Chemical Formula Relationships

1. $Ba_3(PO_4)_2$: 3 barium atoms, 2 phosphorus atoms, and 8 oxygen atoms

   See Section 7.1, The Number of Atoms in a Formula, on Pages 170-171. The formula of the barium ion is Ba^{2+}, and the formula of the phosphate ion is PO_4^{3-}.

2. Atomic mass is the mass of one atom; molecular mass is the mass of one molecule; formula mass is the mass of one formula unit.

   Atomic mass is defined in Section 5.5, Atomic Mass, on Pages 121-123. As its name implies, it is the mass of an atom. Molecular mass and formula mass are defined in Section 7.2, Pages 171-174.

3. Atomic, molecular, and formula masses are expressed in atomic mass units, amu. An amu is exactly 1/12 the mass of a carbon-12 atom.

   The atomic mass unit is defined in Section 5.5, Atomic Mass, on Pages 121-123.

4. a) 39.10 amu K + 126.9 amu I = 166.0 amu KI
   b) 22.99 amu Na + 14.01 amu N + 3(16.00 amu O) = 85.00 amu $NaNO_3$
   c) 3(24.31 amu Mg) + 2(30.97 amu P) + 8(16.00 amu O) = 262.87 amu $Mg_3(PO_4)_2$
   d) 3(12.01 amu C + 8(1.008 amu H) + 16.00 amu O = 60.09 amu C_3H_7OH
   e) 63.55 amu Cu + 32.07 amu S + 4(16.00 amu O) = 159.62 amu $CuSO_4$
   f) 52.00 amu Cr + 3(35.45 amu Cl) = 158.35 amu $CrCl_3$
   g) 22.99 amu Na + 2(12.01 amu C) + 3(1.008 amu H) + 2(16.00 amu O) = 82.03 amu $NaC_2H_3O_2$

   The procedure for finding the formula mass of a substance is given in Section 7.2, Pages 171-174.

5. The mole is the amount of any substance that contains the same number of units as the number of atoms in exactly 12 grams of carbon-12. It is convenient to think of the mole as a number: 6.02×10^{23}. The mole is used to count the huge number of atoms and molecules in macroscopic samples.

   The procedure for finding the formula mass of a substance is given in Section 7.2, Pages 171-174.

6. Avogadro's number, 6.02×10^{23}

   Avogadro's number, N, is described in Section 7.3, The Mole Concept, on Pages 175-176.

7. a) $0.818 \text{ mol K} \times \frac{6.02 \times 10^{23} \text{ atoms K}}{\text{mol K}} = 4.92 \times 10^{23} \text{ atoms K}$

b) $0.629 \text{ mol Al} \times \frac{6.02 \times 10^{23} \text{ atoms Al}}{\text{mol Al}} = 3.79 \times 10^{23} \text{ atoms Al}$

c) $1.84 \text{ mol } CS_2 \times \frac{6.02 \times 10^{23} \text{ molecules } CS_2}{\text{mol } CS_2} = 1.11 \times 10^{24} \text{ molecules } CS_2$

Avogadro's number, 6.02×10^{23}/mol, is the conversion between moles and number of particles. Conceptually, the mole is similar to the dozen. When you are given the number of moles of a substance, it is similar to being given the number of dozens. To convert to number of objects, multiply by 12.

8. a) $1.84 \times 10^{22} \text{ atoms Ar} \times \frac{1 \text{ mol Ar}}{6.02 \times 10^{23} \text{ atoms Ar}} = 0.0306 \text{ mol Ar}$

b) $9.24 \times 10^{24} \text{ molecules } H_2O \times \frac{1 \text{ mol } H_2O}{6.02 \times 10^{23} \text{ molecules } H_2O} = 15.3 \text{ mol } H_2O$

Avogadro's number, 1 mol/6.02×10^{23}, is the conversion between number of particles and number of moles. If you have 36 objects, you divide by (or multiply by the inverse) the number in a dozen, 12, to determine the number of dozens. This is similar.

9. Molar mass is the mass of one mole of molecules or formula units, expressed in grams per mole. Molecular mass is the mass of a single molecule, expressed in amu.

See the discussion of molar mass in Section 7.4, Pages 176-177.

10. a) 2(12.01 g/mol C) + 6(1.008 g/mol H) + 16.00 g/mol O = 46.07 g/mol C_2H_5OH
b) 7(12.01 g/mol C) + 5(1.008 g/mol H) + 3(14.01 g/mol N) + 6(16.00 g/mol O) = 227.14 g/mol $C_7H_5(NO_2)_3$
c) 39.10 g/mol K + 14.01 g/mol N + 2(16.00 g/mol O) = 85.11 g/mol KNO_2
d) 58.93 g/mol Co + 3(35.45 g/mol Cl) = 165.28 g/mol $CoCl_3$

Molar mass is calculated using the same procedure as for molecular mass. Review Section 7.2, Molecular Mass and Formula Mass, Pages 171-174, if necessary. If you want molecular mass, use amu units; if you want molar mass, use g/mol units.

11. a) $53.8 \text{ g Be} \times \frac{1 \text{ mol Be}}{9.012 \text{ g Be}} = 5.97 \text{ mol Be}$

b) $781 \text{ g } C_3H_4Cl_4 \times \frac{1 \text{ mol } C_3H_4Cl_4}{181.86 \text{ g } C_3H_4Cl_4} = 4.29 \text{ mol } C_3H_4Cl_4$

c) $0.756 \text{ g } Ca(OH)_2 \times \frac{1 \text{ mol } Ca(OH)_2}{74.10 \text{ g } Ca(OH)_2} = 0.0102 \text{ mol } Ca(OH)_2$

d) $9.94 \text{ g } CoBr_3 \times \frac{1 \text{ mol } CoBr_3}{298.63 \text{ g } CoBr_3} = 0.0333 \text{ mol } CoBr_3$

e) $8.80 \text{ g } (NH_4)_2Cr_2O_7 \times \frac{1 \text{ mol } (NH_4)_2Cr_2O_7}{252.08 \text{ g } (NH_4)_2Cr_2O_7} = 0.0349 \text{ mol } (NH_4)_2Cr_2O_7$

f) $28.3 \text{ g } Mg(ClO_4)_2 \times \frac{1 \text{ mol } Mg(ClO_4)_2}{223.21 \text{ g } Mg(ClO_4)_2} = 0.127 \text{ g } Mg(ClO_4)_2$

To convert from grams to moles, you use the molar mass of the substance. Molar mass is the connecting link between grams and moles. You may need to review nomenclature (Chapter 6) and/or molar mass (Section 7.4), if necessary.

12. a) $91.9 \text{ g NaClO} \times \frac{1 \text{ mol NaClO}}{74.44 \text{ g NaClO}} = 1.23 \text{ g NaClO}$

b) $881 \text{ g Al}(C_2H_3O_2)_3 \times \frac{1 \text{ mol Al}(C_2H_3O_2)_3}{204.11 \text{ g Al}(C_2H_3O_2)_3} = 4.32 \text{ mol Al}(C_2H_3O_2)_3$

c) $0.586 \text{ g } Hg_2Cl_2 \times \frac{1 \text{ mol } Hg_2Cl_2}{472.10 \text{ g } Hg_2Cl_2} = 0.00124 \text{ mol } Hg_2Cl_2$

To convert from grams to moles, you use the molar mass of the substance. Molar mass is the connecting link between grams and moles. You may need to review nomenclature (Chapter 6) and/or molar mass (Section 7.4), if necessary.

13. a) $0.542 \text{ mol } NaHCO_3 \times \frac{84.01 \text{ g } NaHCO_3}{1 \text{ mol } NaHCO_3} = 45.53 \text{ g } NaHCO_3$

b) $0.0789 \text{ mol } AgNO_3 \times \frac{169.91 \text{ g } AgNO_3}{1 \text{ mol } AgNO_3} = 13.4 \text{ g } AgNO_3$

c) $9.61 \text{ mol } Na_2HPO_4 \times \frac{141.96 \text{ g } Na_2HPO_4}{1 \text{ mol } Na_2HPO_4} = 1.36 \times 10^3 \text{ g } Na_2HPO_4$

d) $0.903 \text{ mol Ca}(BrO_3)_2 \times \frac{295.88 \text{ g Ca}(BrO_3)_2}{1 \text{ mol Ca}(BrO_3)_2} = 267 \text{ g Ca}(BrO_3)_2$

e) $1.14 \text{ mol } (NH_4)_2SO_3 \times \frac{116.15 \text{ g } (NH_4)_2SO_3}{1 \text{ mol } (NH_4)_2SO_3} = 132 \text{ g } (NH_4)_2SO_3$

To convert from moles to grams, you use the molar mass of the substance. Molar mass is the connecting link between moles and grams. You may need to review nomenclature (Chapter 6) and/or molar mass (Section 7.4), if necessary.

14. a) $0.819 \text{ mol } MnO_2 \times \frac{86.94 \text{ g } MnO_2}{1 \text{ mol } MnO_2} = 71.2 \text{ g } MnO_2$

b) $8.48 \text{ mol Al}(ClO_3)_3 \times \frac{277.33 \text{ g Al}(ClO_3)_3}{1 \text{ mol Al}(ClO_3)_3} = 2.35 \times 10^3 \text{ g Al}(ClO_3)_3$

c) $0.926 \text{ mol } CrCl_2 \times \frac{122.90 \text{ g } CrCl_2}{1 \text{ mol } CrCl_2} = 114 \text{ g } CrCl_2$

To convert from moles to grams, you use the molar mass of the substance. Molar mass is the connecting link between moles and grams. You may need to review nomenclature (Chapter 6) and/or molar mass (Section 7.4), if necessary.

15. a) $85.5 \text{ g Be(NO}_3)_2 \times \frac{1 \text{ mol Be(NO}_3)_2}{133.03 \text{ g Be(NO}_3)_2} \times \frac{6.02 \times 10^{23} \text{ Be(NO}_3)_2 \text{ units}}{\text{mol Be(NO}_3)_2} =$

$3.87 \times 10^{23} \text{ Be(NO}_3)_2 \text{ units}$

b) $9.42 \text{ g Mn} \times \frac{1 \text{ mol Mn}}{54.94 \text{ g Mn}} \times \frac{6.02 \times 10^{23} \text{ Mn atoms}}{\text{mol Mn}} = 1.03 \times 10^{23} \text{ Mn atoms}$

c) $0.0948 \text{ g C}_3\text{H}_7\text{OH} \times \frac{1 \text{ mol C}_3\text{H}_7\text{OH}}{60.09 \text{ g C}_3\text{H}_7\text{OH}} \times \frac{6.02 \times 10^{23} \text{ C}_3\text{H}_7\text{OH molecules}}{\text{mol C}_3\text{H}_7\text{OH}} =$

$9.50 \times 10^{20} \text{ C}_3\text{H}_7\text{OH molecules}$

Changing from mass to number of particles is a two-step conversion. Conversion from grams to moles requires the molar mass of the substance. Conversion from moles to number of particles is accomplished with Avogadro's number. You may need to review nomenclature (Chapter 6), the mole concept (Section 7.3), and/or molar mass (Section 7.4), if necessary.

16. a) $7.70 \text{ g I} \times \frac{1 \text{ mol I}}{126.9 \text{ g I}} \times \frac{6.02 \times 10^{23} \text{ I atoms}}{\text{mol I}} = 3.65 \times 10^{22} \text{ I atoms}$

b) $0.447 \text{ g C}_9\text{H}_{20} \times \frac{1 \text{ mol C}_9\text{H}_{20}}{128.25 \text{ g C}_9\text{H}_{20}} \times \frac{6.02 \times 10^{23} \text{ C}_9\text{H}_{20} \text{ molecules}}{\text{mol C}_9\text{H}_{20}} =$

$2.10 \times 10^{21} \text{ C}_9\text{H}_{20} \text{ molecules}$

c) $72.6 \text{ g MnCO}_3 \times \frac{1 \text{ mol MnCO}_3}{114.95 \text{ g MnCO}_3} \times \frac{6.02 \times 10^{23} \text{ MnCO}_3 \text{ units}}{\text{mol MnCO}_3} = 3.80 \times 10^{23} \text{ MnCO}_3 \text{ units}$

Changing from mass to number of particles is a two-step conversion. Conversion from grams to moles requires the molar mass of the substance. Conversion from moles to number of particles is accomplished with Avogadro's number. You may need to review nomenclature (Chapter 6), the mole concept (Section 7.3), and/or molar mass (Section 7.4), if necessary.

17. a) $2.58 \times 10^{23} \text{ FeO units} \times \frac{1 \text{ mol FeO}}{6.02 \times 10^{23} \text{ FeO units}} \times \frac{71.85 \text{ g FeO}}{\text{mol FeO}} = 30.8 \text{ g FeO}$

b) $8.67 \times 10^{24} \text{ F}_2 \text{ molecules} \times \frac{1 \text{ mol F}_2}{6.02 \times 10^{23} \text{ F}_2 \text{ molecules}} \times \frac{38.00 \text{ g F}_2}{\text{mol F}_2} = 547 \text{ g F}_2$

c) $7.36 \times 10^{23} \text{ Au atoms} \times \frac{1 \text{ mol Au}}{6.02 \times 10^{23} \text{ Au atoms}} \times \frac{197.0 \text{ g Au}}{\text{mol Au}} = 241 \text{ g Au}$

Changing from number of particles to mass is a two-step conversion. Conversion from number of particles to moles is accomplished with Avogadro's number. Conversion from moles to grams requires the molar mass of the substance. You may need to review nomenclature (Chapter 6), the mole concept (Section 7.3), and/or molar mass (Section 7.4), if necessary.

18. *GIVEN:* 0.500 carat *WANTED:* C atoms

PER/PATH: carat $\xrightarrow{200 \text{ mg C/carat}}$ mg C $\xrightarrow{1000 \text{ mg C/g C}}$ g C $\xrightarrow{12.01 \text{ g C/mol C}}$ mol C $\xrightarrow{6.02 \times 10^{23} \text{ C atoms/mol C}}$ C atoms

$0.500 \text{ carat} \times \frac{200 \text{ mg C}}{\text{carat}} \times \frac{1 \text{ g C}}{1000 \text{ mg C}} \times \frac{1 \text{ mol C}}{12.01 \text{ g C}} \times \frac{6.02 \times 10^{23} \text{ C atoms}}{\text{mol C}} =$

$5.01 \times 10^{21} \text{ C atoms}$

A key problem-solving strategy used in this problem is to think backwards. You must realize that to get to number of atoms, you need to get to moles. To get to moles, you need mass. You can convert from one mass unit, carats, to another, grams.

19. *GIVEN:* 2.7 kg C_8H_{18} *WANTED:* C_8H_{18} molecules

PER/PATH: $\text{kg } C_8H_{18} \xrightarrow{1000 \text{ g } C_8H_{18}/\text{kg } C_8H_{18}} \text{g } C_8H_{18} \xrightarrow{114.22 \text{ g } C_8H_{18}/\text{mol } C_8H_{18}}$

$\text{mol } C_8H_{18} \xrightarrow{6.02 \times 10^{23} C_8H_{18} \text{ molecules/mol } C_8H_{18}} C_8H_{18} \text{ molecules}$

$$2.7 \text{ kg } C_8H_{18} \times \frac{1000 \text{ g } C_8H_{18}}{\text{kg } C_8H_{18}} \times \frac{1 \text{ mol } C_8H_{18}}{114.22 \text{ g } C_8H_{18}} \times \frac{6.02 \times 10^{23} C_8H_{18} \text{ molecules}}{\text{mol } C_8H_{18}} = 1.42 \times 10^{25} C_8H_{18} \text{ molecules}$$

The volume of the gasoline is irrelevant in this problem. The essence of the question asks you to convert from mass to number of particles. Mass is converted to grams, molar mass takes you from grams to moles, and Avogadro's number allows you to convert from moles to number of molecules.

20. a) $$3.61 \text{ g } F_2 \times \frac{1 \text{ mol } F_2}{38.00 \text{ g } F_2} \times \frac{6.02 \times 10^{23} F_2 \text{ molecules}}{\text{mol } F_2} = 5.72 \times 10^{22} F_2 \text{ molecules}$$

b) $$3.61 \text{ g } F_2 \times \frac{1 \text{ mol } F_2}{38.00 \text{ g } F_2} \times \frac{6.02 \times 10^{23} F_2 \text{ molecules}}{\text{mol } F_2} \times \frac{2 \text{ F atoms}}{F_2 \text{ molecule}} = 1.14 \times 10^{23} \text{ F atoms}$$

c) $$3.61 \text{ g F} \times \frac{1 \text{ mol F}}{19.00 \text{ g F}} \times \frac{6.02 \times 10^{23} F_2 \text{ molecules}}{\text{mol } F_2} = 1.14 \times 10^{23} \text{ F atoms}$$

d) $$3.61 \times 10^{23} \text{ F atoms} \times \frac{1 \text{ mol F}}{6.02 \times 10^{23} \text{ F atoms}} \times \frac{19.00 \text{ g F}}{\text{mol F}} = 11.4 \text{ g F}$$

e) $$3.61 \times 10^{23} F_2 \text{ molecules} \times \frac{1 \text{ mol } F_2}{6.02 \times 10^{23} F_2 \text{ molecules}} \times \frac{38.00 \text{ g } F_2}{\text{mol } F_2} = 22.8 \text{ g } F_2$$

All parts of this question are solved with a grams-to-moles-to-number-of-particles strategy or vice versa. Molar mass takes you from grams to moles, and Avogadro's number allows you to convert from moles to number of atoms or molecules.

21. a) $Mg(NO_3)_2$
$$\frac{24.31 \text{ g Mg}}{148.33 \text{ g } Mg(NO_3)_2} \times 100 = 16.39\% \text{ Mg}$$

$$\frac{2(14.01 \text{ g N})}{148.33 \text{ g } Mg(NO_3)_2} \times 100 = 18.89\% \text{ N} \qquad \frac{6(16.00 \text{ g O})}{148.33 \text{ g } Mg(NO_3)_2} \times 100 = 64.72\% \text{ O}$$

16.39% + 18.89% + 64.72% = 100.00%

b) Na_3PO_4
$$\frac{3(22.99 \text{ g Na})}{163.94 \text{ g } Na_3PO_4} \times 100 = 42.07\% \text{ Na}$$

$$\frac{30.97 \text{ g P}}{163.94 \text{ g } Na_3PO_4} \times 100 = 18.89\% \text{ P} \qquad \frac{4(16.00 \text{ g O})}{163.94 \text{ g } Na_3PO_4} \times 100 = 39.04\% \text{ O}$$

42.07% + 18.89% + 39.04% = 100.00%

c) $CuCl_2$
$$\frac{63.55 \text{ g Cu}}{134.45 \text{ g } CuCl_2} \times 100 = 47.27\% \text{ Cu}$$

$$\frac{2(35.45 \text{ g Cl})}{134.45 \text{ g } CuCl_2} \times 100 = 52.73\% \text{ Cl}$$

47.27% + 52.73% = 100.00%

d) $Cr_2(SO_4)_3$ $\quad \frac{2(52.00\ g\ Cr)}{392.21\ g\ Cr_2(SO_4)_3} \times 100 = 26.52\%\ Cr$

$\frac{3(32.07\ g\ S)}{392.21\ g\ Cr_2(SO_4)_3} \times 100 = 24.53\%\ S$ $\quad \frac{12(16.00\ g\ O)}{392.21\ g\ Cr_2(SO_4)_3} \times 100 = 48.95\%\ O$

$26.52\% + 24.53\% + 48.95\% = 100.00\%$

e) Ag_2CO_3 $\quad \frac{2(107.9\ g\ Ag)}{275.81\ g\ Ag_2CO_3} \times 100 = 78.24\%\ Ag$

$\frac{12.01\ g\ C}{275.81\ g\ Ag_2CO_3} \times 100 = 4.354\%\ C$ $\quad \frac{3(16.00\ g\ O)}{275.81\ g\ Ag_2CO_3} \times 100 = 17.40\%\ O$

$78.24\% + 4.354\% + 17.40\% = 99.99\%$

Percentage composition is discussed in Section 7.6, Pages 180–183. Equation 7.2 is applied to each part of this question. If you had trouble with the name-to-formula conversions, review Chapter 6, Chemical Nomenclature.

22. *GIVEN:* 7.50 g NH_4Br *WANTED:* mass Br (assume g)

PER/PATH: g NH_4Br $\xrightarrow{79.90\ g\ Br/97.94\ g\ NH_4Br}$ g Br

$$7.50\ g\ NH_4Br \times \frac{79.90\ g\ Br}{97.94\ g\ NH_4Br} = 6.12\ g\ Br$$

The molar mass of NH_4Br is [14.01 g N + 4(1.008 g H) + 79.90 g Br] = 97.94 g NH_4Br. The *PER* expression is based on the mass of Br and the mass of the compound. See Page 182, starting after Example 7.11, to the end of the section.

23. *GIVEN:* 1.82 kg MgO *WANTED:* mass Mg (assume kg)

PER/PATH: kg MgO $\xrightarrow{24.31\ kg\ Mg/40.31\ kg\ MgO}$ kg Mg

$$1.82\ kg\ MgO \times \frac{24.31\ kg\ Mg}{40.31\ kg\ MgO} = 1.10\ kg\ Mg$$

The molar mass of MgO is (24.31 g Mg + 16.00 g O) = 40.31 g MgO. The ratio 24.31 g Mg/ 40.31 g MgO can be multiplied by 1000 on the top and bottom to arrive at the *PER* expression, 24.31 kg Mg/40.31 kg MgO. See Page 182, starting after Example 7.11, to the end of the section.

24. *GIVEN:* 445 g $C_{12}H_{22}O_{11}$ *WANTED:* g O

PER/PATH: g $C_{12}H_{22}O_{11}$ $\xrightarrow{11(16.00\ g\ O)/342.30\ g\ C_{12}H_{22}O_{11}}$ g O

$$445\ g\ C_{12}H_{22}O_{11} \times \frac{11(16.00\ g\ O)}{342.30\ g\ C_{12}H_{22}O_{11}} = 229\ g\ O$$

The molar mass of $C_{12}H_{22}O_{11}$ is [12(12.01 g C) + 22(1.008 g H) + 11(16.00 g O)] = 342.30 g $C_{12}H_{22}O_{11}$. The *PER* expression is based on the mass of O and the mass of the compound. See Page 182, starting after Example 7.11, to the end of the section.

25. *GIVEN:* 87.1 g CH_3COCH_3 *WANTED:* g H

PER/PATH: g CH_3COCH_3 $\xrightarrow{6(1.008\ g\ H)/58.08\ g\ CH_3COCH_3}$ g H

$$87.1\ g\ CH_3COCH_3 \times \frac{6(1.008\ g\ H)}{58.08\ g\ CH_3COCH_3} = 9.07\ g\ H$$

The molar mass of CH_3COCH_3 is [3(12.01 g C) + 6(1.008 g H) + 16.00 g O] = 58.08 g CH_3COCH_3. The *PER* expression is based on the mass of H and the mass of the compound. See Page 182, starting after Example 7.11, to the end of the section.

26. *GIVEN:* 7.86 g N *WANTED:* g $Sr(NO_3)_2$

PER/PATH: g N $\xrightarrow{2(14.01 \text{ g N})/211.64 \text{ g } Sr(NO_3)_2}$ g $Sr(NO_3)_2$

$$7.86 \text{ g N} \times \frac{211.64 \text{ g } Sr(NO_3)_2}{2(14.01 \text{ g N})} = 59.4 \text{ g } Sr(NO_3)_2$$

The molar mass of $Sr(NO_3)_2$ is [87.62 g Sr + 2(14.01 g N) + 6(16.00 g O)] = 211.64 g $Sr(NO_3)_2$. The *PER* expression is based on the mass of N and the mass of the compound. See Page 182, starting after Example 7.11, to the end of the section.

27. C_2H_6O and N_2O_5 are empirical formulas. The empirical formula of Na_2O_2 is NaO; the empirical formula of $C_2H_4O_2$ is CH_2O.

See the Empirical formulas and Molecular Formulas subsection on Page 185.

	Element	Grams	Moles	Mole Ratio	Formula Ratio	Empirical Formula	Molecular Formula
28.	Na	29.1	1.27	1.01	2		
	S	40.5	1.26	1.00	2	$Na_2S_2O_3$	
	O	30.4	1.90	1.51	3		
29.	N	1.69	0.121	1.00	2		
	O	4.80	0.300	2.48	5	N_2O_5	
30.	Na	19.2	0.835	1.00	1		
	H	1.7	1.7	2.04	2		
	P	25.8	0.833	1.00	1	NaH_2PO_4	
	O	53.3	3.33	4.00	4		
31.	C	54.6	4.55	2.00	2		
	H	9.0	8.9	3.9	4	C_2H_4O	$\frac{88}{44} = 2$
	O	36.4	2.28	1.00	1		$C_4H_8O_2$
32.	Al	23.1	0.856	1.00	2		$\frac{234}{234} = 1$
	C	15.4	1.28	1.50	3	$Al_2C_3O_9$	$Al_2C_3O_9$
	O	61.5	3.84	4.49	9		$[Al_2(CO_3)_3]$

Grams column: If percentage composition is given, use those figures because they represent the grams of each element in a 100-g sample of the compound. Question 29: (6.49 g N + O) − 4.80 g O = 1.69 g N
Moles column: Grams ÷ Molar mass = Moles
Mole ratio column: Moles ÷ Smallest number of moles = Mole ratio
Formula ratio column: Multiply Mole ratio numbers by 2, 3, or 4 to get whole numbers for the Empirical formula column.

Molecular formula column: Ratio of molar mass of compound to molar mass of empirical formula = Number of empirical formula units in the compound.
See Section 7.8, Empirical Formula of a Compound, Pages 185–190, and Section 7.9, Determination of a Molecular Formula, Pages 190–191.

34. All statements are false.

a) The term *molecular mass* applies mostly to molecular compounds. The term formula mass applies mostly to ionic compounds.
b) Molar mass is measured in grams per mole.
c) Grams are larger than atomic mass units. Molar mass and atomic mass are numerically the same.
d) The molar mass of monatomic hydrogen is read directly from the periodic table. The molar mass of the H_2 molecule is 2 times the mass of a hydrogen atom.
e) A molecular formula may or may not be an empirical formula and vice versa.

35. Hardly—the mass of 10^{25} atoms of copper is 2 pounds:
GIVEN: 10^{25} Cu atoms *WANTED:* lb Cu

PER/PATH: Cu atoms $\xrightarrow{6.02 \times 10^{23} \text{ Cu atoms/mol Cu}}$ mol Cu $\xrightarrow{63.55 \text{ g Cu/mol Cu}}$ g Cu $\xrightarrow{454 \text{ g Cu/lb Cu}}$ lb Cu

$$10^{25} \text{ Cu atoms} \times \frac{1 \text{ mol Cu}}{6.02 \times 10^{23} \text{ Cu atoms}} \times \frac{63.55 \text{ g Cu}}{\text{mol Cu}} \times \frac{1 \text{ lb Cu}}{454 \text{ g Cu}} = 2 \text{ lb}$$

There are many ways to answer this question. You need to come up with a way of expressing 10^{25} atoms of copper in terms that are more familiar so that it is easy to judge whether transportation by truck is reasonable. We converted to pounds, familiar to those who live in the U.S. If you are comfortable with other mass units, your answer in those units is an acceptable alternative.

36. *GIVEN:* 2.95×10^{22} "air" molecules *WANTED:* g "air"

PER/PATH: "air" molecules $\xrightarrow{6.02 \times 10^{23} \text{ "air" molecules/mol "air"}}$ mol "air" $\xrightarrow{29 \text{ g "air"/mol "air"}}$ g "air"

$$2.95 \times 10^{22} \text{ "air" molecules} \times \frac{1 \text{ mol "air"}}{6.02 \times 10^{23} \text{ "air" molecules}} \times \frac{29 \text{ g "air"}}{\text{mol "air"}} = 1.42 \text{ g "air"}$$

This is a molecules-to-mass conversion, similar to those in Question 17. Changing from number of molecules to mass is a two-step conversion. Conversion from number of particles to moles is accomplished with Avogadro's number. Conversion from moles to grams requires the molar mass of the substance. You may need to review the mole concept (Section 7.3), and/or molar mass (Section 7.4), if necessary.

37. *GIVEN:* 85.9 g CO_2 *WANTED:* g C

PER/PATH: g CO_2 $\xrightarrow{12.01 \text{ g C}/44.01 \text{ g CO}_2}$ g C

$$85.9 \text{ g CO}_2 \times \frac{12.01 \text{ g C}}{44.01 \text{ g CO}_2} = 23.4 \text{ g C}$$

GIVEN: 35.5 g H_2O *WANTED:* g H

PER/PATH: g H_2O $\xrightarrow{2(1.008 \text{ g H})/18.02 \text{ g H}_2\text{O}}$ g H

$$35.5 \text{ g H}_2\text{O} \times \frac{2(1.008 \text{ g H})}{18.02 \text{ g H}_2\text{O}} = 3.97 \text{ g H}$$

Element	Grams	Moles	Mole Ratio	Formula Ratio	Empirical Formula
C	23.4	1.95	1.00	1	
H	3.97	3.94	2.02	2	CH_2

The molar mass of CO_2 is [12.01 g C + 2(16.00 g O)] = 44.01 g CO_2. The *PER* expression is based on the mass of C and the mass of the compound. The molar mass of H_2O is [2(1.008 g H) + 16.00 g O] = 18.02 g H_2O. The *PER* expression is based on the mass of H and the mass of the compound. See Page 182, starting after Example 7.11, to the end of the section. Also see Section 7.8, Empirical Formula of a Compound, Pages 185-190, if necessary.

Chapter 8

Reactions and Equations

1. a) $CO + H_2O \rightarrow CO_2 + H_2$
b) Particulate: One molecule of CO reacts with one molecule of H_2O to form one molecule of CO_2 and one molecule of H_2. Molar: One mole of CO reacts with one mole of H_2O to form one mole of CO_2 and one mole of H_2.

a) Black + red = CO; 2 whites + red = H_2O; 2 reds + black = CO_2; 2 whites = H_2
b) The particulate interpretation comes from thinking of the models and symbols as representations of molecules. The molar interpretation considers the coefficients in the equation, all of which are one in this equation, to be numbers of moles.

2. $2\ Ca(s) + O_2(g) \rightarrow 2\ CaO(s)$

Unbalanced equation: $Ca + O_2 \rightarrow CaO$
Balance O: $Ca + O_2 \rightarrow 2\ CaO$
Balance Ca: $2\ Ca + O_2 \rightarrow 2\ CaO$

3. $4\ P(s) + 5\ O_2(g) \rightarrow P_4O_{10}(s)$ *or* $P_4(s) + 5\ O_2(g) \rightarrow P_4O_{10}(s)$

Unbalanced equation: $P + O_2 \rightarrow P_4O_{10}$
Balance O: $P + 5\ O_2 \rightarrow P_4O_{10}$
Balance P: $4\ P + 5\ O_2 \rightarrow P_4O_{10}$
If your instructor requires using P_4 as the formula of phosphorus, the second equation is correct. It is written by following logic similar to that outlined above.

4. $2\ K(s) + F_2(g) \rightarrow 2\ KF(s)$

Unbalanced equation: $K + F_2 \rightarrow KF$
Balance F: $K + F_2 \rightarrow 2\ KF$
Balance K: $2\ K + F_2 \rightarrow 2\ KF$

5. $2\ NaCl(\ell) \rightarrow 2\ Na(s) + Cl_2(g)$

Unbalanced equation: $NaCl \rightarrow Na + Cl_2$
Balance Cl: $2\ NaCl \rightarrow Na + Cl_2$
Balance Na: $2\ NaCl \rightarrow 2\ Na + Cl_2$

6. $2\ HClO(aq) \rightarrow H_2O(\ell) + Cl_2O(g)$

Unbalanced equation: $HClO \rightarrow H_2O + Cl_2O$
Balance hydrogen: $2\ HClO \rightarrow H_2O + Cl_2O$
Chlorine and oxygen are balanced

7. $HC_2H_3O_2(\ell) + 2\ O_2(g) \rightarrow 2\ CO_2(g) + 2\ H_2O(g)$

Unbalanced equation: $HC_2H_3O_2 + O_2 \rightarrow CO_2 + H_2O$
Balance C: $HC_2H_3O_2 + O_2 \rightarrow 2\ CO_2 + H_2O$
Balance H: $HC_2H_3O_2 + O_2 \rightarrow 2\ CO_2 + 2\ H_2O$
Balance O: $HC_2H_3O_2 + 2\ O_2 \rightarrow 2\ CO_2 + 2\ H_2O$

8. $2\ C_2H_2(g) + 5\ O_2(g) \rightarrow 4\ CO_2(g) + 2\ H_2O(g)$

Unbalanced equation: $C_2H_2 + O_2 \rightarrow CO_2 + H_2O$
Balance C: $C_2H_2 + O_2 \rightarrow 2\ CO_2 + H_2O$
H is balanced
Balance O: $C_2H_2 + 5/2\ O_2 \rightarrow 2\ CO_2 + H_2O$
Clear fraction: $2\ C_2H_2 + 5\ O_2 \rightarrow 4\ CO_2 + 2\ H_2O$

9. $Mg(s) + H_2SO_4(aq) \rightarrow H_2(g) + MgSO_4(aq)$

Unbalanced equation: $Mg + H_2SO_4 \rightarrow H_2 + MgSO_4$
All elements are balanced

10. $Br_2(\ell) + 2\ NaI(aq) \rightarrow I_2(aq) + 2\ NaBr(aq)$

Unbalanced equation: $Br_2 + NaI \rightarrow I_2 + NaBr$
Balance I: $Br_2 + 2\ NaI \rightarrow I_2 + NaBr$
Balance Br: $Br_2 + 2\ NaI \rightarrow I_2 + 2\ NaBr$
Na is balanced

11. $MgCl_2(aq) + 2\ NaF(aq) \rightarrow MgF_2(s) + 2\ NaCl(aq)$

Unbalanced equation: $MgCl_2 + NaF \rightarrow MgF_2 + NaCl$
Balance Cl: $MgCl_2 + NaF \rightarrow MgF2 + 2\ NaCl$
Balance F: $MgCl_2 + 2\ NaF \rightarrow MgF2 + 2\ NaCl$
Mg and Na are balanced

12. $BaCl_2(aq) + Na_2SO_4(aq) \rightarrow BaSO_4(s) + 2\ NaCl(aq)$

Unbalanced equation: $BaCl_2 + Na_2SO_4 \rightarrow BaSO_4 + NaCl$
Balance Cl: $BaCl_2 + Na_2SO_4 \rightarrow BaSO_4 + 2\ NaCl$
Ba, Na, and SO_4 are balanced

13. $KOH(aq) + HNO_3(aq) \rightarrow KNO_3(aq) + H_2O(\ell)$

Unbalanced equation: $KOH + HNO_3 \rightarrow KNO_3 + HOH$
All elements are balanced

14. $Mg(OH)_2(s) + 2\ HCl(aq) \rightarrow MgCl_2(aq) + 2\ H_2O(\ell)$

Unbalanced equation: $Mg(OH)_2 + HCl \rightarrow MgCl_2 + HOH$
Balance OH: $Mg(OH)_2 + HCl \rightarrow MgCl_2 + 2\ HOH$
Balance H: $Mg(OH)_2 + 2\ HCl \rightarrow MgCl_2 + 2\ HOH$
Mg and Cl are balanced

15. $AgNO_3(aq) + KBr(aq) \rightarrow AgBr(s) + KNO_3(aq)$

Unbalanced equation: $AgNO_3 + KBr \rightarrow AgBr + KNO_3$
All elements are balanced

16. $2\ CH_3CHO(\ell) + 5\ O_2(g) \rightarrow 4\ CO_2(g) + 4\ H_2O(g)$

Unbalanced equation: $CH_3CHO + O_2 \rightarrow CO_2 + H_2O$
Balance C: $CH_3CHO + O_2 \rightarrow 2\ CO_2 + H_2O$
Balance H: $CH_3CHO + O_2 \rightarrow 2\ CO_2 + 2\ H_2O$
Balance O: $CH_3CHO + 5/2\ O_2 \rightarrow 2\ CO_2 + 2\ H_2O$
Clear fraction: $2\ CH_3CHO + 5\ O_2 \rightarrow 4\ CO_2 + 4\ H_2O$

17. $H_2CO_3(aq) \rightarrow H_2O(\ell) + CO_2(g)$

Unbalanced equation: $H_2CO_3 \rightarrow H_2O + CO_2$
All elements are balanced

18. $Ba(s) + 2\ H_2O(\ell) \rightarrow Ba(OH)_2(aq) + H_2(g)$

Unbalanced equation: $Ba + HOH \rightarrow Ba(OH)_2 + H_2$
Balance OH: $Ba + 2\ HOH \rightarrow Ba(OH)_2 + H_2$
Ba and H are balanced

19. $Mg(s) + NiCl_2(aq) \rightarrow MgCl_2(aq) + Ni(s)$

Unbalanced equation: $Mg + NiCl_2 \rightarrow MgCl_2 + Ni$
All elements are balanced

20. $2\ AgNO_3(aq) + Na_2S(aq) \rightarrow Ag_2S(s) + 2\ NaNO_3(aq)$

Unbalanced equation: $AgNO_3 + Na_2S \rightarrow Ag_2S + NaNO_3$
Balance Na: $AgNO_3 + Na_2S \rightarrow Ag_2S + 2\ NaNO_3$
Balance NO3: $2\ AgNO_3 + Na_2S \rightarrow Ag_2S + 2\ NaNO_3$
Ag and S are balanced

21. $Si(s) + 2\ Cl_2(g) \rightarrow SiCl_4(\ell)$

Unbalanced equation: $Si + Cl_2 \rightarrow SiCl_4$
Balance Cl: $Si + 2\ Cl_2 \rightarrow SiCl_4$
Si is balanced

22. $2\ H_2O_2(\ell) \rightarrow 2\ H_2O(\ell) + O_2(g)$

Unbalanced equation: $H_2O_2 \rightarrow H_2O + O_2$
Create an even number of O on the right: $H_2O_2 \rightarrow 2\ H_2O + O_2$
Balance O: $2\ H_2O_2 \rightarrow 2\ H_2O + O_2$
H is balanced

23. $C_{12}H_{22}O_{11}(s) + 12\ O_2(g) \rightarrow 12\ CO_2(g) + 11\ H_2O(\ell)$

Unbalanced equation: $C_{12}H_{22}O_{11} + O_2 \rightarrow CO_2 + H_2O$
Balance C: $C_{12}H_{22}O_{11} + O_2 \rightarrow 12\ CO_2 + H_2O$
Balance H: $C_{12}H_{22}O_{11} + O_2 \rightarrow 12\ CO_2 + 11\ H_2O$
Balance O: $C_{12}H_{22}O_{11} + 12\ O_2 \rightarrow 12\ CO_2 + 11\ H_2O$

24. $2\ F_2(g) + O_2(g) \rightarrow 2\ OF_2(g)$

Unbalanced equation: $F_2 + O_2 \rightarrow OF_2$
Balance O: $F_2 + O_2 \rightarrow 2\ OF_2$
Balance F: $2\ F_2 + O_2 \rightarrow 2\ OF_2$

25. $AlCl_3(aq) + 3\ NaC_2H_3O_2(aq) \rightarrow Al(C_2H_3O_2)_3(s) + 3\ NaCl(aq)$

Unbalanced equation: $AlCl_3 + NaC_2H_3O_2 \rightarrow Al(C_2H_3O_2)_3 + NaCl$
Balance Cl: $AlCl_3 + NaC_2H_3O_2 \rightarrow Al(C_2H_3O_2)_3 + 3\ NaCl$
Balance Na: $AlCl_3 + 3\ NaC_2H_3O_2 \rightarrow Al(C_2H_3O_2)_3 + NaCl$
Al and $C_2H_3O_2$ are balanced

26. $3\ Mg(s) + N_2(g) \rightarrow Mg_3N_2(s)$

Unbalanced equation: $Mg + N_2 \rightarrow Mg_3N_2$
Balance Mg: $3\ Mg + N_2 \rightarrow Mg_3N_2$
N is balanced

27. $NiCl_2(aq) + Na_2CO_3(aq) \rightarrow NiCO_3(s) + 2\ NaCl(aq)$

Unbalanced equation: $NiCl_2 + Na_2CO_3 \rightarrow NiCO_3 + NaCl$
Balance Cl: $NiCl_2 + Na_2CO_3 \rightarrow NiCO_3 + 2\ NaCl$
Ni, Na, and CO_3 are balanced

28. $2\ Li(s) + MnCl_2(aq) \rightarrow 2\ LiCl(aq) + Mn(s)$

Unbalanced equation: $Li + MnCl_2 \rightarrow LiCl + Mn$
Balance Cl: $Li + MnCl_2 \rightarrow 2\ LiCl + Mn$
Balance Li: $2\ Li + MnCl_2 \rightarrow 2\ LiCl + Mn$
Mn is balanced

29. $3\ FeO(s) + 2\ Al(s) \rightarrow Al_2O_3(s) + 3\ Fe(s)$

Unbalanced equation: $FeO + Al \rightarrow Al_2O_3 + Fe$
Balance O: $3\ FeO + Al \rightarrow Al_2O_3 + Fe$
Balance Al: $3\ FeO + 2\ Al \rightarrow Al_2O_3 + Fe$
Balance Fe: $3\ FeO + 2\ Al \rightarrow Al_2O_3 + 3\ Fe$

30. $Zn(s) + H_2O(g) \rightarrow ZnO(s) + H_2(g)$

Unbalanced equation: $Zn + H_2O \rightarrow ZnO + H_2$
All elements are balanced

31. $BaO(s) + H_2O(\ell) \rightarrow Ba(OH)_2(aq)$

Unbalanced equation: $BaO + H_2O \rightarrow Ba(OH)_2$
All elements are balanced

32. $Fe_2O_3(s) + 3\ CO(g) \rightarrow 2\ Fe(s) + 3\ CO_2(g)$

Unbalanced equation: $Fe_2O_3 + CO \rightarrow Fe + CO_2$
Oxygens on left must be an even number; therefore coefficient on CO must be an odd number (odd number of O in Fe_2O_3; odd + odd = even). Coefficients on CO and CO_2 must be the same to keep C in balance. Coefficient of 1 on CO doesn't work (4 O on left, 2 O on right), so try 3: $Fe_2O_3 + 3\ CO \rightarrow Fe + 3\ CO_2$. Oxygens and carbons are balanced.
Balance Fe: $Fe_2O_3 + 3\ CO \rightarrow 2\ Fe + 3\ CO_2$

34. True: a, b, c, d, e. False: f. Note: Statement d is true, but it is not "the whole truth." The products of a decomposition reaction may be an element and a compound or two or more compounds.

f) A nonmetal replaces a nonmetal in a single-replacement reaction.

35.

a)	Single replacement redox	$Pb + Cu(NO_3)_2 \rightarrow Pb(NO_3)_2 + Cu$
b)	Double replacement neutralization	$Mg(OH)_2 + 2\ HBr \rightarrow MgBr_2 + 2\ H_2O$
c)	Burning	$C_5H_{10}O + 7\ O_2 \rightarrow 5\ CO_2 + 5\ H_2O$
d)	Double replacement precipitation	$Na_2CO_3 + CaSO_4 \rightarrow Na_2SO_4 + CaCO_3$
e)	Decomposition	$2\ LiBr \rightarrow 2\ Li + Br_2$

Review the appropriate section in the chapter, as necessary.

36. Under the circumstances discussed in this book, you may *never* write the formula of a diatomic molecule without a subscript 2 in its formula.

See Section 6.2, Formulas of Elements, Pages 136-138.

37. $H_2SO_4 + CaCO_3 \rightarrow CaSO_4 + H_2CO_3$; $H_2SO_4 + CaCO_3 \rightarrow CaSO_4 + H_2O + CO_2$

The reactions are of the double replacement type. Under common conditions, carbonic acid decomposes quickly to carbon dioxide and water: $H_2CO_3(aq) \rightarrow CO_2(g) + H_2O(\ell)$.

38. $S + 3\ F_2 \rightarrow SF_6$ *or* $S_8 + 24\ F_2 \rightarrow 8\ SF_6$
$S + Cl_2 \rightarrow SCl_2$ *or* $S_8 + 8\ Cl_2 \rightarrow 8\ SCl_2$
$2\ S + Br_2 \rightarrow S_2Br_2$ *or* $S_8 + 4\ Br_2 \rightarrow 4\ S_2Br_2$

If your instructor requires using S_8 as the formula of sulfur, the second equation in each case is correct.

39. $2\ LiOH(aq) + H_2SO_4(aq) \rightarrow Li_2SO_4(aq) + 2\ H_2O(\ell)$

The reactants in a neutralization reaction have the general form HX + MOH. The cation in the product Li_2SO_4 is Li^+, so it is combined with OH^- as one of the reactants, LiOH. The anion in the product is SO_4^{2-}, and it is therefore combined with H^+ as a reactant, H_2SO_4. The second product of a neutralization reaction comes from the combination of hydrogen ion and hydroxide ion, $H^+ + OH^- \rightarrow HOH$.

40. $H_2C_2O_4(aq) + 2\ NaOH(aq) \rightarrow Na_2C_2O_4(aq) + 2\ HOH(\ell)$

Oxalic acid, $H_2C_2O_4$, yields the oxalate ion, $C_2O_4^{2-}$, when its hydrogen ions are removed.

Chapter 9

Quantity Relationships in Chemical Reactions

1. a) *GIVEN:* 3.40 mol C_4H_{10} *WANTED:* mol O_2

PER/PATH: mol C_4H_{10} $\xrightarrow{13 \text{ mol } O_2/2 \text{ mol } C_4H_{10}}$ mol O_2

$$3.40 \text{ mol } C_4H_{10} \times \frac{13 \text{ mol } O_2}{2 \text{ mol } C_4H_{10}} = 22.1 \text{ mol } O_2$$

b) *GIVEN:* 4.68 mol C_4H_{10} *WANTED:* mol CO_2

PER/PATH: mol C_4H_{10} $\xrightarrow{8 \text{ mol } CO_2/2 \text{ mol } C_4H_{10}}$ mol CO_2

$$4.68 \text{ mol } C_4H_{10} \times \frac{8 \text{ mol } CO_2}{2 \text{ mol } C_4H_{10}} = 18.7 \text{ mol } CO_2$$

c) *GIVEN:* 0.568 mol CO_2 *WANTED:* mol H_2O

PER/PATH: mol CO_2 $\xrightarrow{10 \text{ mol } H_2O/8 \text{ mol } CO_2}$ mol H_2O

$$0.568 \text{ mol } CO_2 \times \frac{10 \text{ mol } H_2O}{8 \text{ mol } CO_2} = 0.710 \text{ mol } H_2O$$

See Section 9.1, Conversion Factors from a Chemical Equation, Pages 224-225.

2. $PCl_5 + 4\ H_2O \rightarrow H_3PO_4 + 5\ HCl$

GIVEN: 0.239 mol H_2O *WANTED:* mol HCl

PER/PATH: mol H_2O $\xrightarrow{5 \text{ mol HCl}/4 \text{ mol } H_2O}$ mol HCl

$$0.239 \text{ mol } H_2O \times \frac{5 \text{ mol HCl}}{4 \text{ mol } H_2O} = 0.299 \text{ mol HCl}$$

See Section 9.1, Conversion Factors from a Chemical Equation, Pages 224-225.

3. $Zn + 2\ HCl \rightarrow H_2 + ZnCl_2$

GIVEN: 0.0837 mol HCl *WANTED:* mol H_2

PER/PATH: mol HCl $\xrightarrow{1 \text{ mol } H_2/2 \text{ mol HCl}}$ mol H_2

$$0.0837 \text{ mol HCl} \times \frac{1 \text{ mol } H_2}{2 \text{ mol HCl}} = 0.0419 \text{ mol } H_2$$

See Section 9.1, Conversion Factors from a Chemical Equation, Pages 224-225.

4. a) *GIVEN:* 1.42 mol O_2 *WANTED:* g C_4H_{10}

PER/PATH: mol O_2 $\xrightarrow{2 \text{ mol } C_4H_{10}/13 \text{ mol } O_2}$ mol C_4H_{10} $\xrightarrow{58.12 \text{ g } C_4H_{10}/\text{mol } C_4H_{10}}$ g C_4H_{10}

$$1.42 \text{ mol } O_2 \times \frac{2 \text{ mol } C_4H_{10}}{13 \text{ mol } O_2} \times \frac{58.12 \text{ g } C_4H_{10}}{1 \text{ mol } C_4H_{10}} = 12.7 \text{ g } C_4H_{10}$$

b) *GIVEN:* 9.43 g O_2 *WANTED:* mol H_2O

PER/PATH: g O_2 $\xrightarrow{32.00 \text{ g } O_2/\text{mol } O_2}$ mol O_2 $\xrightarrow{10 \text{ mol } H_2O/13 \text{ mol } O_2}$ mol H_2O

$$9.43 \text{ g } O_2 \times \frac{1 \text{ mol } O_2}{32.00 \text{ g } O_2} \times \frac{10 \text{ mol } H_2O}{13 \text{ mol } O_2} = 0.227 \text{ mol } H_2O$$

c) *GIVEN:* 78.4 g C_4H_{10} *WANTED:* g CO_2

PER/PATH: g C_4H_{10} $\xrightarrow{58.12 \text{ g } C_4H_{10}/\text{mol } C_4H_{10}}$ mol C_4H_{10} $\xrightarrow{8 \text{ mol } CO_2/2 \text{ mol } C_4H_{10}}$ mol CO_2 $\xrightarrow{44.01 \text{ g } CO_2/\text{mol } CO_2}$ g CO_2

$$78.4 \text{ g } C_4H_{10} \times \frac{1 \text{ mol } C_4H_{10}}{58.12 \text{ g } C_4H_{10}} \times \frac{8 \text{ mol } CO_2}{2 \text{ mol } C_4H_{10}} \times \frac{44.01 \text{ g } CO_2}{\text{mol } CO_2} = 237 \text{ g } CO_2$$

d) *GIVEN:* 43.8 g H_2O *WANTED:* g O_2

PER/PATH: g H_2O $\xrightarrow{18.02 \text{ g } H_2O/\text{mol } H_2O}$ mol H_2O $\xrightarrow{13 \text{ mol } O_2/10 \text{ mol } H_2O}$ mol O_2 $\xrightarrow{32.00 \text{ g } O_2/\text{mol } O_2}$ g O_2

$$43.8 \text{ g } H_2O \times \frac{1 \text{ mol } H_2O}{18.02 \text{ g } H_2O} \times \frac{13 \text{ mol } O_2}{10 \text{ mol } H_2O} \times \frac{32.00 \text{ g } O_2}{\text{mol } O_2} = 101 \text{ g } O_2$$

See Section 9.2, Mass Calculations, Pages 226-230. A summary of the stoichiometry path is given in a *PROCEDURE* box on Page 226.

5. *GIVEN:* 105 g $C_{600}H_{1000}O_{500}$ *WANTED:* g $C_{12}H_{22}O_{11}$

PER/PATH: g $C_{600}H_{1000}O_{500}$ $\xrightarrow{16{,}214.00 \text{ g } C_{600}H_{1000}O_{500}/\text{mol } C_{600}H_{1000}O_{500}}$ mol $C_{600}H_{1000}O_{500}$ $\xrightarrow{50 \text{ mol } C_{12}H_{22}O_{11}/1 \text{ mol } C_{600}H_{1000}O_{500}}$ mol $C_{12}H_{22}O_{11}$ $\xrightarrow{342.30 \text{ g } C_{12}H_{22}O_{11}/\text{mol } C_{12}H_{22}O_{11}}$ g $C_{12}H_{22}O_{11}$

$$105 \text{ g } C_{600}H_{1000}O_{500} \times \frac{1 \text{ mol } C_{600}H_{1000}O_{500}}{16{,}214.00 \text{ g } C_{600}H_{1000}O_{500}} \times \frac{50 \text{ mol } C_{12}H_{22}O_{11}}{1 \text{ mol } C_{600}H_{1000}O_{500}} \times \frac{342.30 \text{ g } C_{12}H_{22}O_{11}}{1 \text{ mol } C_{12}H_{22}O_{11}} = 111 \text{ g } C_{12}H_{22}O_{11}$$

See Section 9.2, Mass Calculations, Pages 226-230. A summary of the stoichiometry path is given in a *PROCEDURE* box on Page 226.

6. *GIVEN:* 596 g $NaHCO_3$ *WANTED:* g H_2SO_4

PER/PATH: g $NaHCO_3$ $\xrightarrow{84.01 \text{ g } NaHCO_3/\text{mol } NaHCO_3}$ mol $NaHCO_3$ $\xrightarrow{1 \text{ mol } H_2SO_4/2 \text{ mol } NaHCO_3}$ mol H_2SO_4 $\xrightarrow{98.09 \text{ g } H_2SO_4/\text{mol } H_2SO_4}$ g H_2SO_4

$$596 \text{ g } NaHCO_3 \times \frac{1 \text{ mol } NaHCO_3}{84.01 \text{ g } NaHCO_3} \times \frac{1 \text{ mol } H_2SO_4}{2 \text{ mol } NaHCO_3} \times \frac{98.09 \text{ g } H_2SO_4}{\text{mol } H_2SO_4} = 348 \text{ g } H_2SO_4$$

See Section 9.2, Mass Calculations, Pages 226-230. A summary of the stoichiometry path is given in a *PROCEDURE* box on Page 226.

7. *GIVEN:* 5.00 kg Na_2SO_4 *WANTED:* kg NaCl

PER/PATH: kg Na_2SO_4 $\xrightarrow{142.05 \text{ kg } Na_2SO_4/\text{kmol } Na_2SO_4}$ kmol Na_2SO_4 $\xrightarrow{4 \text{ kmol NaCl}/2 \text{ kmol } Na_2SO_4}$ kmol NaCl $\xrightarrow{58.44 \text{ kg NaCl/kmol NaCl}}$ kg NaCl

$$5.00 \text{ kg } Na_2SO_4 \times \frac{1 \text{ kmol } Na_2SO_4}{142.05 \text{ kg } Na_2SO_4} \times \frac{4 \text{ kmol NaCl}}{2 \text{ kmol } Na_2SO_4} \times \frac{58.44 \text{ kg NaCl}}{\text{kmol NaCl}} = 4.11 \text{ kg NaCl}$$

See Section 9.2, Mass Calculations, Pages 226-230. A summary of the stoichiometry path is given in a *PROCEDURE* box on Page 226.

8. a) *GIVEN:* 1.90 kg C_7H_8 *WANTED:* quantity HNO_3 (assume kg)

PER/PATH: kg C_7H_8 $\xrightarrow{92.13 \text{ kg } C_7H_8/\text{kmol } C_7H_8}$ kmol C_7H_8 $\xrightarrow{3 \text{ kmol } HNO_3/1 \text{ kmol } C_7H_8}$ kmol HNO_3 $\xrightarrow{63.02 \text{ kg } HNO_3/\text{kmol } HNO_3}$ kg HNO_3

$$1.90 \text{ kg } C_7H_8 \times \frac{1 \text{ kmol } C_7H_8}{92.13 \text{ kg } C_7H_8} \times \frac{3 \text{ kmol } HNO_3}{1 \text{ kmol } C_7H_8} \times \frac{63.02 \text{ kg } HNO_3}{1 \text{ kmol } HNO_3} = 3.90 \text{ kg } HNO_3$$

b) *GIVEN:* 1.90 kg C_7H_8 *WANTED:* kg $C_7H_5N_3O_6$

PER/PATH: kg C_7H_8 $\xrightarrow{92.13 \text{ kg } C_7H_8/\text{kmol } C_7H_8}$ kmol C_7H_8 $\xrightarrow{1 \text{ kmol } C_7H_5N_3O_6/1 \text{ mol } C_7H_8}$ kmol $C_7H_5N_3O_6$ $\xrightarrow{227.14 \text{ kg } C_7H_5N_3O_6/\text{kmol } C_7H_5N_3O_6}$ kg $C_7H_5N_3O_6$

$$1.90 \text{ kg } C_7H_8 \times \frac{1 \text{ kmol } C_7H_8}{92.13 \text{ kg } C_7H_8} \times \frac{1 \text{ kmol } C_7H_5N_3O_6}{1 \text{ kmol } C_7H_8} \times \frac{227.14 \text{ kg } C_7H_5N_3O_6}{1 \text{ kmol } C_7H_5N_3O_6} = 4.68 \text{ kg } C_7H_5N_3O_6$$

See Section 9.2, Mass Calculations, Pages 226-230. A summary of the stoichiometry path is given in a *PROCEDURE* box on Page 226.

9. *GIVEN:* 778 kg Fe_2O_3 *WANTED:* quantity Fe (assume kg)

PER/PATH: kg Fe_2O_3 $\xrightarrow{159.70 \text{ kg } Fe_2O_3/\text{kmol } Fe_2O_3}$ kmol Fe_2O_3 $\xrightarrow{2 \text{ kmol Fe}/1 \text{ kmol } Fe_2O_3}$ kmol Fe $\xrightarrow{55.85 \text{ kg Fe/kmol Fe}}$ kg Fe

$$778 \text{ kg } Fe_2O_3 \times \frac{1 \text{ kmol } Fe_2O_3}{159.70 \text{ kg } Fe_2O_3} \times \frac{2 \text{ kmol Fe}}{1 \text{ kmol } Fe_2O_3} \times \frac{55.85 \text{ kg Fe}}{1 \text{ kmol Fe}} = 544 \text{ kg Fe}$$

See Section 9.2, Mass Calculations, Pages 226-230. A summary of the stoichiometry path is given in a *PROCEDURE* box on Page 226.

10. *GIVEN:* 81.2 g NH_3 *WANTED:* g NH_4HCO_3

PER/PATH: g NH_3 $\xrightarrow{17.03 \text{ g } NH_3/\text{mol } NH_3}$ mol NH_3 $\xrightarrow{1 \text{ mol } NH_4HCO_3/1 \text{ mol } NH_3}$ mol NH_4HCO_3 $\xrightarrow{79.06 \text{ g } NH_4HCO_3/\text{mol } NH_4HCO_3}$ g NH_4HCO_3

$$81.2 \text{ g } NH_3 \times \frac{1 \text{ mol } NH_3}{17.03 \text{ g } NH_3} \times \frac{1 \text{ mol } NH_4HCO_3}{1 \text{ mol } NH_3} \times \frac{79.06 \text{ g } NH_4HCO_3}{1 \text{ mol } NH_4HCO_3} = 377 \text{ g } NH_4HCO_3$$

See Section 9.2, Mass Calculations, Pages 226-230. A summary of the stoichiometry path is given in a *PROCEDURE* box on Page 226.

11. *GIVEN:* 448 g Na_2CO_3 *WANTED:* mass $NaHCO_3$ (assume g)

PER/PATH: g Na_2CO_3 $\xrightarrow{105.99 \text{ g } Na_2CO_3/\text{mol } Na_2CO_3}$ mol Na_2CO_3 $\xrightarrow{2 \text{ mol } NaHCO_3/1 \text{ mol } Na_2CO_3}$ mol $NaHCO_3$ $\xrightarrow{84.01 \text{ g } NaHCO_3/\text{mol } NaHCO_3}$ g $NaHCO_3$

$$448 \text{ g } Na_2CO_3 \times \frac{1 \text{ mol } Na_2CO_3}{105.99 \text{ g } Na_2CO_3} \times \frac{2 \text{ mol } NaHCO_3}{1 \text{ mol } Na_2CO_3} \times \frac{84.01 \text{ g } NaHCO_3}{\text{mol } NaHCO_3} = 7.10 \times 10^2 \text{ g } NaHCO_3$$

See Section 9.2, Mass Calculations, Pages 226-230. A summary of the stoichiometry path is given in a *PROCEDURE* box on Page 226.

12. $Ca(OH)_2 + 2\ HCl \rightarrow CaCl_2 + 2\ H_2O$

GIVEN: 6.34 g HCl *WANTED:* g $Ca(OH)_2$

PER/PATH: $\text{g HCl} \xrightarrow{36.46\text{ g HCl/mol HCl}} \text{mol HCl} \xrightarrow{1\text{ mol Ca(OH)}_2/2\text{ mol HCl}}$ $\text{mol Ca(OH)}_2 \xrightarrow{74.10\text{ g Ca(OH)}_2/\text{mol Ca(OH)}_2} \text{g Ca(OH)}_2$

$$6.34\text{ g HCl} \times \frac{1\text{ mol HCl}}{36.46\text{ g HCl}} \times \frac{1\text{ mol Ca(OH)}_2}{2\text{ mol HCl}} \times \frac{74.10\text{ g Ca(OH)}_2}{1\text{ mol Ca(OH)}_2} = 6.44\text{ g Ca(OH)}_2$$

See Section 9.2, Mass Calculations, Pages 226-230. A summary of the stoichiometry path is given in a *PROCEDURE* box on Page 226.

13. $Na_2CO_3 + CaCl_2 \rightarrow 2\ NaCl + CaCO_3$

GIVEN: 0.523 g $CaCl_2$ *WANTED:* mass $CaCO_3$ (assume g)

PER/PATH: $\text{g CaCl}_2 \xrightarrow{110.98\text{ g CaCl}_2/\text{mol CaCl}_2} \text{mol CaCl}_2$ $\xrightarrow{1\text{ mol CaCO}_3/1\text{ mol CaCl}_2} \text{mol CaCO}_3 \xrightarrow{100.09\text{ g CaCO}_3/\text{mol CaCO}_3} \text{g CaCO}_3$

$$0.523\text{ g CaCl}_2 \times \frac{1\text{ mol CaCl}_2}{110.98\text{ g CaCl}_2} \times \frac{1\text{ mol CaCO}_3}{1\text{ mol CaCl}_2} \times \frac{100.09\text{ g CaCO}_3}{1\text{ mol CaCO}_3} = 0.472\text{ g CaCO}_3$$

See Section 9.2, Mass Calculations, Pages 226-230. A summary of the stoichiometry path is given in a *PROCEDURE* box on Page 226.

14. $K_2CO_3 + 2\ LiCl \rightarrow 2\ KCl + Li_2CO_3$

GIVEN: 3.36 g Li_2CO_3 *WANTED:* g LiCl

PER/PATH: $\text{g Li}_2\text{CO}_3 \xrightarrow{73.98\text{ g Li}_2\text{CO}_3/\text{mol Li}_2\text{CO}_3} \text{mol Li}_2\text{CO}_3$ $\xrightarrow{2\text{ mol LiCl}/1\text{ mol Li}_2\text{CO}_3} \text{mol LiCl} \xrightarrow{42.39\text{ g LiCl/mol LiCl}} \text{g LiCl}$

$$3.36\text{ g Li}_2\text{CO}_3 \times \frac{1\text{ mol Li}_2\text{CO}_3}{73.89\text{ g Li}_2\text{CO}_3} \times \frac{2\text{ mol LiCl}}{1\text{ mol Li}_2\text{CO}_3} \times \frac{42.39\text{ g LiCl}}{\text{mol LiCl}} = 3.86\text{ g LiCl}$$

See Section 9.2, Mass Calculations, Pages 226-230. A summary of the stoichiometry path is given in a *PROCEDURE* box on Page 226.

15. $Fe_2O_3 + 2\ Al \rightarrow Al_2O_3 + 2\ Fe$

GIVEN: 47.1 g Al *WANTED:* g Fe_2O_3

PER/PATH: $\text{g Al} \xrightarrow{26.98\text{ g Al/mol Al}} \text{mol Al} \xrightarrow{1\text{ mol Fe}_2\text{O}_3/2\text{ mol Al}}$ $\text{mol Fe}_2\text{O}_3 \xrightarrow{159.70\text{ g Fe}_2\text{O}_3/\text{mol Fe}_2\text{O}_3} \text{g Fe}_2\text{O}_3$

$$47.1\text{ g Al} \times \frac{1\text{ mol Al}}{26.98\text{ g Al}} \times \frac{1\text{ mol Fe}_2\text{O}_3}{2\text{ mol Al}} \times \frac{159.70\text{ g Fe}_2\text{O}_3}{\text{mol Fe}_2\text{O}_3} = 139\text{ g Fe}_2\text{O}_3$$

See Section 9.2, Mass Calculations, Pages 226-230. A summary of the stoichiometry path is given in a *PROCEDURE* box on Page 226.

16. Molar volume is the volume occupied by one mole of a gas.

Molar volume is defined on Page 230.

17. *GIVEN:* 27.2 L O_2 at STP *WANTED:* mol O_2

PER/PATH: L O_2 $\xrightarrow{22.4\ L\ O_2/mol\ O_2}$ mol O_2

$$27.2\ L\ O_2 \times \frac{1\ mol\ O_2}{22.4\ L\ O_2} = 1.21\ mol\ O_2$$

Memorize that one mole of any gas at STP occupies a volume of 22.4 liters. See Section 9.3, Gas Stoichiometry at Standard Temperature and Pressure, Pages 230–233.

18. *GIVEN:* 8.55 g $NaHCO_3$ *WANTED:* STP volume of CO_2 (assume L)

PER/PATH: g $NaHCO_3$ $\xrightarrow{84.01\ g\ NaHCO_3/mol\ NaHCO_3}$ mol $NaHCO_3$ $\xrightarrow{1\ mol\ CO_2/1\ mol\ NaHCO_3}$ mol CO_2 $\xrightarrow{22.4\ L\ CO_2/mol\ CO_2}$ L CO_2

$$8.55\ g\ NaHCO_3 \times \frac{1\ mol\ NaHCO_3}{84.01\ g\ NaHCO_3} \times \frac{1\ mol\ CO_2}{1\ mol\ NaHCO_3} \times \frac{22.4\ L\ CO_2}{mol\ CO_2} = 2.28\ L\ CO_2$$

Memorize that one mole of any gas at STP occupies a volume of 22.4 liters. See Section 9.3, Gas Stoichiometry at Standard Temperature and Pressure, Pages 230–233.

19. $CH_4 + 2\ O_2 \rightarrow CO_2 + 2\ H_2O$

GIVEN: 35.0 L CH_4 at STP *WANTED:* g CO_2

PER/PATH: L CH_4 $\xrightarrow{22.4\ L\ CH_4/mol\ CH_4}$ mol CH_4 $\xrightarrow{1\ mol\ CO_2/1\ mol\ CH_4}$ mol CO_2 $\xrightarrow{44.01\ g\ CO_2/mol\ CO_2}$ g CO_2

$$35.0\ L\ CH_4 \times \frac{1\ mol\ CH_4}{22.4\ L\ CH_4} \times \frac{1\ mol\ CO_2}{1\ mol\ CH_4} \times \frac{44.01\ g\ CO_2}{mol\ CO_2} = 68.8\ g\ CO_2$$

Memorize that one mole of any gas at STP occupies a volume of 22.4 liters. See Section 9.3, Gas Stoichiometry at Standard Temperature and Pressure, Pages 230–233.

20. *GIVEN:* 155 L O_2 at STP *WANTED:* L NO_2 at STP

PER/PATH: L O_2 $\xrightarrow{2\ L\ NO_2/1\ L\ O_2}$ L NO_2

$$155\ L\ O_2 \times \frac{2\ L\ NO_2}{1\ L\ O_2} = 3.10 \times 10^2\ L\ NO_2$$

The ratio of volumes of gases in a reaction is the same as the ratio of moles if the volumes are measured at the same temperature and pressure. See the Volume-Volume Stoichiometry subsection on Pages 232–233.

21. *GIVEN:* 4.83 g $NaHCO_3$ *WANTED:* volume CO_2 (assume L)

PER/PATH: g $NaHCO_3$ $\xrightarrow{84.01\ g\ NaHCO_3/mol\ NaHCO_3}$ mol $NaHCO_3$ $\xrightarrow{1\ mol\ CO_2/1\ mol\ NaHCO_3}$ mol CO_2 $\xrightarrow{22.4\ L\ CO_2/mol\ CO_2}$ L CO_2

$$4.83\ g\ NaHCO_3 \times \frac{1\ mol\ NaHCO_3}{84.01\ g\ NaHCO_3} \times \frac{1\ mol\ CO_2}{1\ mol\ NaHCO_3} \times \frac{22.4\ L\ CO_2}{mol\ CO_2} = 1.29\ L\ CO_2$$

	Volume	Temperature	Pressure
Initial Value (1)	1.29 L	0°C; 273 K	1 atm
Final Value (2)	V_2	343°C; 616 K	1.04 atm

$$1.29\ L \times \frac{1\ atm}{1.04\ atm} \times \frac{616\ K}{273\ K} = 2.80\ L$$

Our first step was to find the volume of gas produced at STP. We then used the Combined Gas Laws to find the volume occupied at the given temperature and pressure. See Section 9.4, Gas Stoichiometry at Nonstandard Conditions, Pages 233-236.

22. $CH_4 + 2\,O_2 \rightarrow CO_2 + 2\,H_2O$

	Volume	Temperature	Pressure
Initial Value (1)	19.2 L	26°C; 299 K	0.813 atm
Final Value (2)	V_2	0°C; 273 K	1 atm

$$19.2\text{ L} \times \frac{273\text{ K}}{299\text{ K}} \times \frac{0.813\text{ atm}}{1\text{ atm}} = 14.3\text{ L } CH_4$$

GIVEN: 14.3 L CH_4 *WANTED:* g CO_2

PER/PATH: $\text{L } CH_4 \xrightarrow{22.4\text{ L }CH_4/\text{mol }CH_4} \text{mol } CH_4 \xrightarrow{1\text{ mol }CO_2/1\text{ mol }CH_4} \text{mol } CO_2 \xrightarrow{44.01\text{ g }CO_2/\text{mol }CO_2} \text{g } CO_2$

$$14.3\text{ L } CH_4 \times \frac{1\text{ mol } CH_4}{22.4\text{ L } CH_4} \times \frac{1\text{ mol } CO_2}{1\text{ mol } CH_4} \times \frac{44.01\text{ g } CO_2}{\text{mol } CO_2} = 28.1\text{ g } CO_2$$

Our first step was to adjust the given temperature and pressure to STP by applying the Combined Gas Laws. We then converted from the given STP volume to the wanted mass. See Section 9.4, Gas Stoichiometry at Nonstandard Conditions, Pages 233-236.

23. *GIVEN:* 1 kg $CaCO_3 \cdot MgCO_3$ *WANTED:* L CO_2

PER/PATH: $\text{kg } CaCO_3 \cdot MgCO_3 \xrightarrow{184.41\text{ kg }CaCO_3 \cdot MgCO_3/\text{kmol }CaCO_3 \cdot MgCO_3} \text{kmol } CaCO_3 \cdot MgCO_3 \xrightarrow{2\text{ kmol }CO_2/1\text{ kmol }CaCO_3 \cdot MgCO_3} \text{kmol } CO_2 \xrightarrow{22.4\text{ kL }CO_2/\text{kmol }CO_2} \text{kL } CO_2$

$$1\text{ kg } CaCO_3 \cdot MgCO_3 \times \frac{1\text{ kmol } CaCO_3 \cdot MgCO_3}{184.41\text{ kg } CaCO_3 \cdot MgCO_3} \times \frac{2\text{ kmol } CO_2}{1\text{ kmol } CaCO_3 \cdot MgCO_3} \times \frac{22.4\text{ kL } CO_2}{\text{kmol } CO_2} = 0.243\text{ kL } CO_2$$

	Volume	Temperature	Pressure
Initial Value (1)	0.243 kL	0°C; 273 K	1 atm
Final Value (2)	V_2	264°C; 537 K	1.09 atm

$$0.243\text{ kL } CO_2 \times \frac{537\text{ K}}{273\text{ K}} \times \frac{1\text{ atm}}{1.09\text{ atm}} \times \frac{1000\text{ L}}{\text{kL}} = 439\text{ L } CO_2$$

Our first step was to convert from the given mass of solid to the wanted STP volume of gas. We then converted from the STP volume to the volume at the wanted temperature and pressure, using the Combined Gas Laws. See Section 9.4, Gas Stoichiometry at Nonstandard Conditions, Pages 233-236.

24. (a)

	Volume	Temperature	Pressure
Initial Value (1)	1.94 L	19°C; 292 K	691 torr
Final Value (2)	V_2	0°C; 273 K	760 torr

$$1.94\text{ L} \times \frac{691\text{ torr}}{760\text{ torr}} \times \frac{273\text{ K}}{292\text{ K}} = 1.65\text{ L } NO_2$$

GIVEN: 1.65 L NO_2 *WANTED:* g Sn

PER/PATH: $\text{L } NO_2 \xrightarrow{22.4\text{ L }NO_2/\text{mol }NO_2} \text{mol } NO_2 \xrightarrow{1\text{ mol Sn}/4\text{ mol }NO_2} \text{mol Sn} \xrightarrow{118.7\text{ g Sn/mol Sn}} \text{g Sn}$

$$1.65 \text{ L } NO_2 \times \frac{1 \text{ mol } NO_2}{22.4 \text{ L } NO_2} \times \frac{1 \text{ mol Sn}}{4 \text{ mol } NO_2} \times \frac{118.7 \text{ g Sn}}{\text{mol Sn}} = 2.19 \text{ g Sn}$$

(b) *GIVEN:* 2.19 g Sn; 4.77 g solder *WANTED:* % Sn

EQUATION: $\frac{\text{g Sn}}{\text{total mass}} \times 100 = \frac{2.19 \text{ g}}{4.77 \text{ g}} \times 100 = 45.9\% \text{ Sn}$

(a) Our first step was to convert the volume of gas at the given temperature and pressure to STP volume using the Combined Gas Laws. We then converted the STP volume of gas to the wanted mass of solid. See Section 9.4, *Gas Stoichiometry at Nonstandard Conditions*, Pages 233-236.
(b) The percent concept is discussed on Page 180.

25. *GIVEN:* 41.9 g Cu_2S *WANTED:* g Cu

PER/PATH: g Cu_2S $\xrightarrow{159.17 \text{ g } Cu_2S/\text{mol } Cu_2S}$ mol Cu_2S $\xrightarrow{2 \text{ mol Cu}/1 \text{ mol } Cu_2S}$ mol Cu $\xrightarrow{63.55 \text{ g Cu/mol Cu}}$ g Cu

$$41.9 \text{ g } Cu_2S \times \frac{1 \text{ mol } Cu_2S}{159.17 \text{ g } Cu_2S} \times \frac{2 \text{ mol Cu}}{1 \text{ mol } Cu_2S} \times \frac{63.55 \text{ g Cu}}{1 \text{ mol Cu}} = 33.5 \text{ g Cu}$$

$$\% \text{ yield} = \frac{\text{actual yield}}{\text{theoretical yield}} \times 100 = \frac{29.2 \text{ g Cu (act)}}{33.5 \text{ g Cu (theo)}} \times 100 = 87.2\% \text{ yield}$$

The first step is to find the theoretical yield, using the standard stoichiometry path. You then use the percent yield definition to determine its value. See Section 9.5, *Percent Yield*, Pages 236-241.

26. *GIVEN:* 557 kg NaCl; 82.6% yield *WANTED:* quantity HCl (assume kg)

PER/PATH: kg NaCl $\xrightarrow{58.44 \text{ kg NaCl/kmol NaCl}}$ kmol NaCl $\xrightarrow{2 \text{ kmol HCl}/2 \text{ kmol NaCl}}$ kmol HCl $\xrightarrow{36.46 \text{ kg HCl (theo)/kmol HCl}}$ kg HCl (theo) $\xrightarrow{82.6 \text{ kg HCl (act)}/100 \text{ kg HCl (theo)}}$ kg HCl (act)

$$557 \text{ kg NaCl} \times \frac{1 \text{ kmol NaCl}}{58.44 \text{ kg NaCl}} \times \frac{2 \text{ kmol HCl}}{2 \text{ kmol NaCl}} \times \frac{36.46 \text{ kg HCl (theo)}}{1 \text{ kmol HCl}} \times \frac{82.6 \text{ kg HCl (act)}}{100 \text{ kg HCl (theo)}} = 287 \text{ kg HCl (act)}$$

In this problem, you are given mass of reactant and asked for actual mass of product. Recall that *yield* applies to product. Since you are starting with quantity of reactant, the yield is not applied yet. The first three steps are the standard stoichiometry path: mass given → moles given → moles wanted → mass wanted. That mass wanted is the theoretical yield. It is what you get under perfect conditions. The 82.6% yield tells you that only 82.6 kg are actually produced under the real conditions of this reaction for every 100 kg you expect theoretically. Therefore, the last step is to adjust for the percent yield. See Section 9.5, *Percent Yield*, Pages 236-241.

27. *GIVEN:* 38.4 g CCl_4; 85% yield *WANTED:* g S_2Cl_2

PER/PATH: g CCl_4 (act) $\xrightarrow{85 \text{ g } CCl_4 \text{ (act)}/100 \text{ g } CCl_4 \text{ (theo)}}$ g CCl_4 (theo) $\xrightarrow{153.81 \text{ g } CCl_4/\text{mol } CCl_4}$ mol CCl_4 $\xrightarrow{2 \text{ mol } S_2Cl_2/1 \text{ mol } CCl_4}$ mol S_2Cl_2 $\xrightarrow{135.04 \text{ g } S_2Cl_2/\text{mol } S_2Cl_2}$ g S_2Cl_2

$$38.4 \text{ g } CCl_4 \text{ (act)} \times \frac{100 \text{ g } CCl_4 \text{ (theo)}}{85 \text{ g } CCl_4 \text{ (act)}} \times \frac{1 \text{ mol } CCl_4}{153.81 \text{ g } CCl_4} \times \frac{2 \text{ mol } S_2Cl_2}{1 \text{ mol } CCl_4} \times \frac{135.04 \text{ g } S_2Cl_2}{\text{mol } S_2Cl_2} = 79 \text{ g } S_2Cl_2$$

The given in this problem is the mass of product. This is the amount that you actually want to produce, so we label it g (act). Since this is a product, yield is applied here. The percent yield says that 85 g are actually produced for every 100 g theoretically produced, and this is the basis of the next conversion factor. The problem then follows the standard stoichiometry path. The two significant figure percent yield limits the final answer to two significant figures. See Section 9.5, Percent Yield, Pages 236-241.

28. *GIVEN:* 5.95 g MnO_2; 4.22 g Cl_2 (act) *WANTED:* % yield

PER/PATH: $\text{g } MnO_2 \xrightarrow{86.94 \text{ g } MnO_2/\text{mol } MnO_2} \text{mol } MnO_2 \xrightarrow{1 \text{ mol } Cl_2/1 \text{ mol } MnO_2} \text{mol } Cl_2 \xrightarrow{70.90 \text{ g } Cl_2/\text{mol } Cl_2} \text{g } Cl_2 \text{ (theo)}$

$$5.95 \text{ g } MnO_2 \times \frac{1 \text{ mol } MnO_2}{86.94 \text{ g } MnO_2} \times \frac{1 \text{ mol } Cl_2}{1 \text{ mol } MnO_2} \times \frac{70.90 \text{ g } Cl_2}{\text{mol } Cl_2} = 4.85 \text{ g } Cl_2 \text{ (theo)}$$

$$\% \text{ yield} = \frac{\text{actual yield}}{\text{theoretical yield}} \times 100 = \frac{4.22 \text{ g } Cl_2 \text{ (act)}}{4.85 \text{ g } Cl_2 \text{ (theo)}} \times 100 = 87.0\% \text{ yield}$$

The first step is to find the theoretical yield, using the standard stoichiometry path. You then use the percent yield definition to determine its value. See Section 9.5, Percent Yield, Pages 236-241.

29. *GIVEN:* 397 kg CH_3OH; 84.9% yield *WANTED:* quantity HCHO (assume kg)

PER/PATH: $\text{kg } CH_3OH \xrightarrow{32.04 \text{ kg } CH_3OH/\text{kmol } CH_3OH} \text{kmol } CH_3OH \xrightarrow{2 \text{ kmol HCHO}/2 \text{ kmol } CH_3OH} \text{kmol HCHO} \xrightarrow{30.03 \text{ kg HCHO (theo)/mol HCHO}} \text{kg HCHO (theo)} \xrightarrow{84.9 \text{ kg HCHO (act)}/100 \text{ kg HCHO (theo)}} \text{kg HCHO (act)}$

$$397 \text{ kg } CH_3OH \times \frac{1 \text{ kmol } CH_3OH}{32.04 \text{ kg } CH_3OH} \times \frac{2 \text{ kmol HCHO}}{2 \text{ kmol } CH_3OH} \times \frac{30.03 \text{ kg HCHO (theo)}}{1 \text{ kmol HCHO}} \times \frac{84.9 \text{ kg HCHO (act)}}{100 \text{ kg HCHO (theo)}} = 316 \text{ HCHO kg (act)}$$

In this problem, you are given mass of reactant and asked for actual mass of product. Recall that *yield* applies to product. Since you are starting with quantity of reactant, the yield is not applied yet. The first three steps are the standard stoichiometry path: mass given → moles given → moles wanted → mass wanted. This problem uses kilounits. That mass wanted is the theoretical yield. It is what you get under perfect conditions. The 84.9% yield tells you that only 84.9 kg are actually produced under the real conditions of this reaction for every 100 kg you expect theoretically. Therefore, the last step is to adjust for the percent yield. See Section 9.5, Percent Yield, Pages 236-241.

Questions 30–32: Both the comparison-of-moles and smaller-amount methods are shown. Differences in answers between the methods are caused by round-offs in calculations.

30.

Comparison-of-moles method:	NH_3	+	HNO_3	→	NH_4NO_3
Grams at start	74.4		159		0
Molar mass	17.03		63.02		80.05
Moles at start	4.37		2.52		0
Moles used (–), produced (+)	– 2.52		– 2.52		+ 2.52
Moles at end	1.85		0		2.52
Grams at end	31.5		0		202

Grams at start line: The given masses are listed.
Molar mass line: The molar mass of each substance is calculated from its formula.

Moles at start line: $74.4 \text{ g } NH_3 \times \frac{1 \text{ mol } NH_3}{17.03 \text{ g } NH_3} = 4.37 \text{ mol } NH_3$

$159 \text{ g } HNO_3 \times \frac{1 \text{ mol } HNO_3}{63.02 \text{ g } HNO_3} = 2.52 \text{ mol } HNO_3$

Moles used (-), produced (+) line: There is a 1:1 reaction stoichiometry, so the smaller number of moles, 2.52 mol HNO_3, limits the reaction and the moles of NH_3 that react is the same. All reaction coefficients are one, so the moles of NH_4NO_3 produced are also the same, with a + sign to indicate that they are formed.
Moles at end line: Moles at start - Moles used or + Moles produced.

Grams at end line: $1.85 \text{ mol } NH_3 \times \frac{17.03 \text{ g } NH_3}{\text{mol } NH_3} = 31.5 \text{ g } NH_3$

$2.52 \text{ mol } NH_4NO_3 \times \frac{80.05 \text{ g } NH_4NO_3}{\text{mol } NH_4NO_3} = 202 \text{ g } NH_4NO_3$

Smaller-amount method:

GIVEN: 74.4 g NH_3 *WANTED:* g NH_4NO_3

PER/PATH: $\text{g } NH_3 \xrightarrow{17.03 \text{ g } NH_3/\text{mol } NH_3} \text{mol } NH_3 \xrightarrow{1 \text{ mol } NH_4NO_3/1 \text{ mol } NH_3} \text{mol } NH_4NO_3 \xrightarrow{80.05 \text{ g } NH_4NO_3/\text{mol } NH_4NO_3} \text{g } NH_4NO_3$

$74.4 \text{ g } NH_3 \times \frac{1 \text{ mol } NH_3}{17.03 \text{ g } NH_3} \times \frac{1 \text{ mol } NH_4NO_3}{1 \text{ mol } NH_3} \times \frac{80.05 \text{ g } NH_4NO_3}{\text{mol } NH_4NO_3} = 3.50 \times 10^2 \text{ g } NH_4NO_3$

GIVEN: 159 g HNO_3 *WANTED:* g NH_4NO_3

PER/PATH: $\text{g } HNO_3 \xrightarrow{63.02 \text{ g } HNO_3/\text{mol } HNO_3} \text{mol } HNO_3 \xrightarrow{1 \text{ mol } NH_4NO_3/1 \text{ mol } HNO_3} \text{mol } NH_4NO_3 \xrightarrow{80.05 \text{ g } NH_4NO_3/\text{mol } NH_4NO_3} \text{g } NH_4NO_3$

$159 \text{ g } HNO_3 \times \frac{1 \text{ mol } HNO_3}{63.02 \text{ g } HNO_3} \times \frac{1 \text{ mol } NH_4NO_3}{1 \text{ mol } HNO_3} \times \frac{80.05 \text{ g } NH_4NO_3}{\text{mol } NH_4NO_3} = 202 \text{ g } NH_4NO_3$

HNO_3 is the limiting reactant. The yield is the smaller amount, 202 g NH_4NO_3.

GIVEN: 159 g HNO_3 *WANTED:* g NH_3

PER/PATH: $\text{g } HNO_3 \xrightarrow{63.02 \text{ g } HNO_3/\text{mol } HNO_3} \text{mol } HNO_3 \xrightarrow{1 \text{ mol } NH_3/1 \text{ mol } HNO_3} \text{mol } NH_3 \xrightarrow{17.03 \text{ g } NH_3/\text{mol } NH_3} \text{g } NH_3$

$159 \text{ g } HNO_3 \times \frac{1 \text{ mol } HNO_3}{63.02 \text{ g } HNO_3} \times \frac{1 \text{ mol } NH_3}{1 \text{ mol } HNO_3} \times \frac{17.03 \text{ g } NH_3}{\text{mol } NH_3} = 43.0 \text{ g } NH_3$

74.4 g NH_3 (initial) – 43.0 g NH_3 (used) = 31.4 g NH_3 left

The first step is to follow the standard stoichiometry pattern to convert the mass of each of the reactants into mass of product. We found that 74.4 g NH_3 can form 3.50×10^2 g NH_4NO_3 if there is more than enough HNO_3 to react. We then found that 159 g HNO_3 can form 202 g NH_4NO_3 if there is more than enough NH_3 to react. The smaller amount of product, 202 g NH_4NO_3, is the quantity formed. This also tells us that HNO_3 is the limiting reactant. We then use the mass of limiting reactant to calculate the mass of excess reactant consumed in the reaction. The initial mass of the excess reactant minus the amount used gives the mass that remains.

31.

Comparison-of-moles method:	CCl_3CHO +	2 C_6H_5Cl →	$(ClC_6H_4)_2CHCCl_3 + H_2O$
Grams at start	3.19	4.54	0
Molar mass	147.38	112.55	354.46
Moles at start	0.0216	0.0403	0
Moles used (–), produced (+)	– 0.0202	– 0.0403	+ 0.0202
Moles at end	0.0014	0	0.0202
Grams at end	0.21	0	7.16

Grams at start line: The given masses are listed.

Molar mass line: The molar mass of each substance is calculated from its formula.

Moles at start line: $3.19 \text{ g } CCl_3CHO \times \frac{1 \text{ mol } CCl_3CHO}{147.38 \text{ g } CCl_3CHO} = 0.0216 \text{ mol } CCl_3CHO$

$4.54 \text{ g } C_6H_5Cl \times \frac{1 \text{ mol } C_6H_5Cl}{112.55 \text{ g } C_6H_5Cl} = 0.0403 \text{ mol } C_6H_5Cl$

Moles used (-), produced (+) line: There are 2 moles of C_6H_5Cl for every 1 mole of CCl_3CHO. If we compare two times the moles of CCl_3CHO, $2 \times 0.0216 = 0.0432$, to the moles of C_6H_5Cl, 0.0403, we see that moles of C_6H_5Cl are stoichiometrically smaller and therefore limiting. The moles of CCl_3CHO used will be: $0.0403 \text{ mol } C_6H_5Cl \times \frac{1 \text{ mol } CCl_3CHO}{2 \text{ mol } C_6H_5Cl} = 0.0202 \text{ mol } CCl_3CHO$.

There is a 1:1 stoichiometry ratio between CCl_3CHO and $(ClC_6H_4)_2CHCCl_3$, so the moles of $(ClC_6H_4)_2CHCCl_3$ produced is also 0.0202 mol.

Moles at end line: Moles at start - Moles used or + Moles produced.

Grams at end line: $0.0014 \text{ mol } CCl_3CHO \times \frac{147.38 \text{ g } CCl_3CHO}{\text{mol } CCl_3CHO} = 0.21 \text{ g } CCl_3CHO$

$0.0202 \text{ mol } (ClC_6H_4)_2CHCCl_3 \times \frac{354.46 \text{ g } (ClC_6H_4)_2CHCCl_3}{\text{mol } (ClC_6H_4)_2CHCCl_3} = 7.16 \text{ g } (ClC_6H_4)_2CHCCl_3$

Smaller-amount method:

GIVEN: 3.19 g CCl_3CHO *WANTED:* g $(ClC_6H_4)_2CHCCl_3$

PER/PATH: g $CCl_3CHO \xrightarrow{147.38 \text{ g } CCl_3CHO/\text{mol } CCl_3CHO}$ mol CCl_3CHO $\xrightarrow{1 \text{ mol } (ClC_6H_4)_2CHCCl_3/1 \text{ mol } CCl_3CHO}$ mol $(ClC_6H_4)_2CHCCl_3$ $\xrightarrow{354.46 \text{ g } (ClC_6H_4)_2CHCCl_3/\text{mol } (ClC_6H_4)_2CHCCl_3}$ g $(ClC_6H_4)_2CHCCl_3$

$$3.19 \text{ g } CCl_3CHO \times \frac{1 \text{ mol } CCl_3CHO}{147.38 \text{ g } CCl_3CHO} \times \frac{1 \text{ mol } (ClC_6H_4)_2CHCCl_3}{1 \text{ mol } CCl_3CHO} \times \frac{354.46 \text{ g } (ClC_6H_4)_2CHCCl_3}{\text{mol } (ClC_6H_4)_2CHCCl_3} = 7.67 \text{ g } (ClC_6H_4)_2CHCCl_3$$

GIVEN: 4.54 g C_6H_5Cl *WANTED:* g $(ClC_6H_4)_2CHCCl_3$

PER/PATH: g C_6H_5Cl $\xrightarrow{112.55\ g\ C_6H_5Cl/mol\ C_6H_5Cl}$ mol C_6H_5Cl $\xrightarrow{1\ mol\ (ClC_6H_4)_2CHCCl_3/2\ mol\ C_6H_5Cl}$ mol $(ClC_6H_4)_2CHCCl_3$ $\xrightarrow{354.46\ g\ (ClC_6H_4)_2CHCCl_3/mol\ (ClC_6H_4)_2CHCCl_3}$ g $(ClC_6H_4)_2CHCCl_3$

$$4.54\ g\ C_6H_5Cl \times \frac{1\ mol\ C_6H_5Cl}{112.55\ g\ C_6H_5Cl} \times \frac{1\ mol\ (ClC_6H_4)_2CHCCl_3}{2\ mol\ C_6H_5Cl} \times \frac{354.46\ g\ (ClC_6H_4)_2CHCCl_3}{mol\ (ClC_6H_4)_2CHCCl_3} = 7.15\ g\ (ClC_6H_4)_2CHCCl_3$$

C_6H_5Cl is the limiting reactant. The yield is the smaller amount, 7.15 g $(ClC_6H_4)_2CHCCl_3$

GIVEN: 4.54 g C_6H_5Cl *WANTED:* g CCl_3CHO

PER/PATH: g C_6H_5Cl $\xrightarrow{112.55\ g\ C_6H_5Cl/mol\ C_6H_5Cl}$ mol C_6H_5Cl $\xrightarrow{1\ mol\ CCl_3CHO/2\ mol\ C_6H_5Cl}$ mol CCl_3CHO $\xrightarrow{147.38\ g\ CCl_3CHO/mol\ CCl_3CHO}$ g CCl_3CHO

$$4.54\ g\ C_6H_5Cl \times \frac{1\ mol\ C_6H_5Cl}{112.55\ g\ C_6H_5Cl} \times \frac{1\ mol\ CCl_3CHO}{2\ mol\ C_6H_5Cl} \times \frac{147.38\ g\ CCl_3CHO}{mol\ CCl_3CHO} = 2.97\ g\ CCl_3CHO$$

3.19 g CCl_3CHO (initial) − 2.97 g CCl_3CHO (used) = 0.22 g CCl_3CHO left

The first step is to follow the standard stoichiometry pattern to convert the mass of each of the reactants into mass of product. We found that 3.19 g CCl_3CHO can form 7.67 g $(ClC_6H_4)_2CHCCl_3$ if there is more than enough C_6H_5Cl to react. We then found that 4.54 g C_6H_5Cl can form 7.15 g $(ClC_6H_4)_2CHCCl_3$ if there is more than enough CCl_3CHO to react. The smaller amount of product, 7.15 g $(ClC_6H_4)_2CHCCl_3$, is the quantity formed. This also tells us that C_6H_5Cl is the limiting reactant. We then use the mass of limiting reactant to calculate the mass of excess reactant consumed in the reaction. The initial mass of the excess reactant minus the amount used gives the mass that remains.

32.

Comparison-of-moles method:	2 HNO_3 +	Na_2CO_3 →	CO_2 + others
Grams at start	188	135	0
Molar mass	63.02	105.99	44.01
Moles at start	2.98	1.27	0
Moles used (–), produced (+)	– 2.54	– 1.27	+ 1.27
Moles at end	0.44	0	1.27
Grams at end			55.9

Na_2CO_3 is the limiting reactant, so 135 g Na_2CO_3 is not enough to neutralize 188 g HNO_3. 55.9 grams of CO_2 is released.

Grams at start line: The given masses are listed.

Molar mass line: The molar mass of each substance is calculated from its formula.

Moles at start line: $188\ g\ HNO_3 \times \frac{1\ mol\ HNO_3}{63.02\ g\ HNO_3} = 2.98\ mol\ HNO_3$

$$135\ g\ Na_2CO_3 \times \frac{1\ mol\ Na_2CO_3}{105.99\ g\ Na_2CO_3} = 1.27\ mol\ Na_2CO_3$$

Moles used (-), produced (+) line: There are 2 moles of HNO_3 for every 1 mole of Na_2CO_3. If we compare two times the moles of Na_2CO_3, 2 × 1.27 = 2.54, to the moles of HNO_3, 2.98, we see that moles of Na_2CO_3 are stoichiometrically smaller and therefore limiting. The moles of HNO_3 used will be: $1.27 \text{ mol } Na_2CO_3 \times \frac{2 \text{ mol } HNO_3}{1 \text{ mol } Na_2CO_3} = 2.54 \text{ mol } HNO_3$

There is a 1:1 stoichiometry ratio between Na_2CO_3 and CO_2, so the moles of CO_2 produced is also 1.27 mol.

Moles at end line: Moles at start - Moles used or + Moles produced.

Grams at end line: $1.27 \text{ mol } CO_2 \times \frac{44.01 \text{ g } CO_2}{\text{mol } CO_2} = 55.9 \text{ g } CO_2$

Smaller-amount method:

GIVEN: 135 g Na_2CO_3 *WANTED:* g CO_2

PER/PATH: g Na_2CO_3 $\xrightarrow{105.99 \text{ g } Na_2CO_3/\text{mol } Na_2CO_3}$ mol Na_2CO_3 $\xrightarrow{1 \text{ mol } CO_2/1 \text{ mol } Na_2CO_3}$ mol CO_2 $\xrightarrow{44.01 \text{ g } CO_2/\text{mol } CO_2}$ g CO_2

$$135 \text{ g } Na_2CO_3 \times \frac{1 \text{ mol } Na_2CO_3}{105.99 \text{ g } Na_2CO_3} \times \frac{1 \text{ mol } CO_2}{1 \text{ mol } Na_2CO_3} \times \frac{44.01 \text{ g } CO_2}{\text{mol } CO_2} = 56.1 \text{ g } CO_2$$

GIVEN: 188 g HNO_3 *WANTED:* g CO_2

PER/PATH: g HNO_3 $\xrightarrow{63.02 \text{ g } HNO_3/\text{mol } HNO_3}$ mol HNO_3 $\xrightarrow{1 \text{ mol } CO_2/2 \text{ mol } HNO_3}$ mol CO_2 $\xrightarrow{44.01 \text{ g } CO_2/\text{mol } CO_2}$ g CO_2

$$188 \text{ g } HNO_3 \times \frac{1 \text{ mol } HNO_3}{63.02 \text{ g } HNO_3} \times \frac{1 \text{ mol } CO_2}{2 \text{ mol } HNO_3} \times \frac{44.01 \text{ g } CO_2}{\text{mol CO2}} = 65.6 \text{ g } CO_2$$

Na_2CO_3 is the limiting reactant, so 135 g Na_2CO_3 is not enough to neutralize 188 g HNO_3. 56.1 grams of CO_2 is released.

The first step is to follow the standard stoichiometry pattern to convert the mass of each of the reactants into mass of product. We found that 135 g Na_2CO_3 can form 56.1 g CO_2 if there is more than enough HNO_3 to react. We then found that 188 g HNO_3 can form 65.6 g CO_2 if there is more than enough Na_2CO_3 to react. The smaller amount of product, 56.1 g CO_2, is the quantity formed.

33. a) *GIVEN:* 60.5 cal *WANTED:* J

PER/PATH: cal $\xrightarrow{4.184 \text{ J/cal}}$ J

$$60.5 \text{ cal} \times \frac{4.184 \text{ J}}{\text{cal}} = 253 \text{ J}$$

b) *GIVEN:* 8.32 kJ *WANTED:* cal

PER/PATH: kJ $\xrightarrow{1000 \text{ J/kJ}}$ J $\xrightarrow{4.184 \text{ J/cal}}$ cal

$$8.32 \text{ kJ} \times \frac{1000 \text{ J}}{\text{kJ}} \times \frac{1 \text{ cal}}{4.184 \text{ J}} = 1.99 \times 10^3 \text{ cal}$$

c) *GIVEN:* 0.753 kJ *WANTED:* kcal

PER/PATH: kJ $\xrightarrow{4.184 \text{ kJ/kcal}}$ kcal

$$0.753 \text{ kJ} \times \frac{1 \text{ kcal}}{4.184 \text{ kJ}} = 0.180 \text{ kcal}$$

The key *PER* expression in all parts of this question is the relationship between joules and calories, 4.184 J = 1 cal. Unit-kilounit conversions are also applied. See Section 9.9, Energy, on Pages 249–250.

34. *GIVEN:* 912 cal *WANTED:* J, kJ

PER/PATH: $\text{cal} \xrightarrow{4.184\ \text{J/cal}} \text{J} \xrightarrow{1000\ \text{J/kJ}} \text{kJ}$

$$912\ \text{cal} \times \frac{4.184\ \text{J}}{\text{cal}} = 3.82 \times 10^3\ \text{J} \qquad 3.82 \times 10^3\ \text{J} \times \frac{1\ \text{kJ}}{1000\ \text{J}} = 3.82\ \text{kJ}$$

The key *PER* expression in this question is the relationship between joules and calories, 4.184 J = 1 cal. Unit-kilounit conversions are also applied. See Section 9.9, Energy, on Pages 249-250.

35. *GIVEN:* 56.7 g sucrose; 3.94×10^3 cal/g *WANTED:* kJ

PER/PATH: $\text{g sucrose} \xrightarrow{3.94 \times 10^3\ \text{cal/g sucrose}} \text{cal} \xrightarrow{4.184\ \text{J/cal}} \text{J} \xrightarrow{1000\ \text{J/kJ}} \text{kJ}$

$$56.7\ \text{g sucrose} \times \frac{3.94 \times 10^3\ \text{cal}}{\text{g sucrose}} \times \frac{4.184\ \text{J}}{\text{cal}} \times \frac{1\ \text{kJ}}{1000\ \text{J}} = 935\ \text{kJ}$$

The key *PER* expression in this question is the relationship between joules and calories, 4.184 J = 1 cal. Unit-kilounit conversions are also applied. See Section 9.9, Energy, on Pages 249-250.

36. $C_3H_8(g) + 5\ O_2(g) \rightarrow 3\ CO_2(g) + 4\ H_2O(\ell) + 2.22 \times 10^3\ \text{kJ}$

$C_3H_8(g) + 5\ O_2(g) \rightarrow 3\ CO_2(g) + 4\ H_2O(\ell) \quad \Delta H = -2.22 \times 10^3\ \text{kJ}$

The problem statement says that heat is released, so the reaction is exothermic. The energy term therefore appears as a product in one equation, and the sign of ΔH is negative in the other equation. See Section 9.10, Thermochemical Equations, Pages 250-252.

37. $CaO(s) + H_2O(\ell) \rightarrow Ca(OH)_2(s) + 65.3\ \text{kJ}$

$CaO(s) + H_2O(\ell) \rightarrow Ca(OH)_2(s) \quad \Delta H = -65.3\ \text{kJ}$

The problem statement says that heat energy is released, so the reaction is exothermic. The energy term therefore appears as a product in one equation, and the sign of ΔH is negative in the other equation. See Section 9.10, Thermochemical Equations, Pages 250-252.

38. $5.4 \times 10^2\ \text{kJ} + \frac{1}{2}\ Al_2O_3(s) + \frac{3}{4}\ C(s) \rightarrow Al(s) + \frac{3}{4}\ CO_2(g)$

$\frac{1}{2}\ Al_2O_3(s) + \frac{3}{4}\ C(s) \rightarrow Al(s) + \frac{3}{4}\ CO_2(g) \quad \Delta H = +5.4 \times 10^2\ \text{kJ}$ *or*

$2.2 \times 10^3\ \text{kJ} + 2\ Al_2O_3(s) + 3\ C(s) \rightarrow 4\ Al(s) + 3\ CO_2(g)$

$2\ Al_2O_3(s) + 3\ C(s) \rightarrow 4\ Al(s) + 3\ CO_2(g) \quad \Delta H = +2.2 \times 10^3\ \text{kJ}$

The problem statement says that energy is required, so the reaction is endothermic. The energy term therefore appears as a reactant in one equation, and the sign of ΔH is positive in the other equation. Our cautioning statement was given because of the energy requirement of 5.4×10^2 kJ/mol Al. If you balance the equation with the smallest whole-number coefficients possible, you have 4 mol Al. The energy requirement becomes $4 \times 5.4 \times 10^2$ kJ/4 mol Al. Alternatively, you can balance the equation so that the coefficient of Al is one and use the given energy. See Section 9.10, Thermochemical Equations, Pages 250-252.

39. *GIVEN:* 356 kJ *WANTED:* mL H_2O

PER/PATH: kJ $\xrightarrow{572 \text{ kJ}/2 \text{ mol } H_2O}$ mol H_2O $\xrightarrow{18.02 \text{ g } H_2O/\text{mol } H_2O}$ g H_2O $\xrightarrow{1.00 \text{ g } H_2O/\text{mL } H_2O}$ mL H_2O

$$356 \text{ kJ} \times \frac{2 \text{ mol } H_2O}{572 \text{ kJ}} \times \frac{18.02 \text{ g } H_2O}{\text{mol } H_2O} \times \frac{1 \text{ mL } H_2O}{1.00 \text{ g } H_2O} = 22.4 \text{ mL } H_2O$$

The given thermochemical equation provides the *PER* expression that relates kJ and mol H_2O. See Section 9.11, Thermochemical Stoichiometry, Pages 252-253.

40. *GIVEN:* 454 g $C_6H_{12}O_6$ *WANTED:* quantity of energy (assume kJ)

PER/PATH: g $C_6H_{12}O_6$ $\xrightarrow{180.16 \text{ g } C_6H_{12}O_6/\text{mol } C_6H_{12}O_6}$ mol $C_6H_{12}O_6$ $\xrightarrow{2.82 \times 10^3 \text{ kJ/mol } C_6H_{12}O_6}$ kJ

$$454 \text{ g } C_6H_{12}O_6 \times \frac{1 \text{ mol } C_6H_{12}O_6}{180.16 \text{ g } C_6H_{12}O_6} \times \frac{2.82 \times 10^3 \text{ kJ}}{\text{mol } C_6H_{12}O_6} = 7.11 \times 10^3 \text{ kJ}$$

The given thermochemical equation provides the *PER* expression that relates kJ and mol $C_6H_{12}O_6$. See Section 9.11, Thermochemical Stoichiometry, Pages 252-253.

41. *GIVEN:* 1.50 kg C_4H_{10} *WANTED:* quantity of energy (assume kJ)

PER/PATH: kg C_4H_{10} $\xrightarrow{1000 \text{ g } C_4H_{10}/\text{kg } C_4H_{10}}$ g C_4H_{10} $\xrightarrow{58.12 \text{ g } C_4H_{10}/\text{mol } C_4H_{10}}$ mol C_4H_{10} $\xrightarrow{5.77 \times 10^3 \text{ kJ}/2 \text{ mol } C_4H_{10}}$ kJ

$$1.50 \text{ kg } C_4H_{10} \times \frac{1000 \text{ g } C_4H_{10}}{\text{kg } C_4H_{10}} \times \frac{1 \text{ mol } C_4H_{10}}{58.12 \text{ g } C_4H_{10}} \times \frac{5.77 \times 10^3 \text{ kJ}}{2 \text{ mol } C_4H_{10}} = 7.45 \times 10^4 \text{ kJ}$$

The given thermochemical equation provides the *PER* expression that relates kJ and mol C_4H_{10}. See Section 9.11, Thermochemical Stoichiometry, Pages 252-253.

43. True: a, d, e, f. False: b, c.

b) Stoichiometry problems require balanced equations for their solution in order to find the mole ratio of the given and wanted.
c) In solving a stoichiometry problem, quantity of given substance is changed to moles of given substance, which is then changed to moles of wanted substance, and finally to quantity of wanted substance. The mass-to-moles change and vice versa is based on molar masses, and the mole-to-mole change is based on the balanced chemical equation.

44. *GIVEN:* 35 g N_2 *WANTED:* g Na_2CO_3

PER/PATH: g N_2 $\xrightarrow{28.02 \text{ g } N_2/\text{mol } N_2}$ mol N_2 $\xrightarrow{1 \text{ mol } Na_2CO_3/1 \text{ mol } N_2}$ mol Na_2CO_3 $\xrightarrow{105.99 \text{ g } Na_2CO_3/\text{mol } Na_2CO_3}$ g Na_2CO_3

$$35 \text{ g } N_2 \times \frac{1 \text{ mol } N_2}{28.02 \text{ g } N_2} \times \frac{1 \text{ mol } Na_2CO_3}{1 \text{ mol } N_2} \times \frac{105.99 \text{ g } Na_2CO_3}{\text{mol } Na_2CO_3} = 1.3 \times 10^2 \text{ g } Na_2CO_3$$

See Section 9.2, Mass Calculations, Pages 226-230. A summary of the stoichiometry path is given in a *PROCEDURE* box on Page 226.

45. *GIVEN:* 125 g KO_2 *WANTED:* g O_2

PER/PATH: g KO_2 $\xrightarrow{71.10\text{ g }KO_2/\text{mol }KO_2}$ mol KO_2 $\xrightarrow{3\text{ mol }O_2/4\text{ mol }KO_2}$ mol O_2 $\xrightarrow{32.00\text{ g }O_2/\text{mol }O_2}$ g O_2

$$125\text{ g }KO_2 \times \frac{1\text{ mol }KO_2}{71.10\text{ g }KO_2} \times \frac{3\text{ mol }O_2}{4\text{ mol }KO_2} \times \frac{32.00\text{ g }O_2}{\text{mol }O_2} = 42.2\text{ g }O_2$$

See Section 9.2, Mass Calculations, Pages 226-230. A summary of the stoichiometry path is given in a *PROCEDURE* box on Page 226.

46. *GIVEN:* 1.68 g Al *WANTED:* g Al_2O_3

PER/PATH: g Al $\xrightarrow{26.98\text{ g Al/mol Al}}$ mol Al $\xrightarrow{2\text{ mol }Al_2O_3/4\text{ mol Al}}$ mol Al_2O_3 $\xrightarrow{101.96\text{ g }Al_2O_3/\text{mol }Al_2O_3}$ g Al_2O_3

$$1.68\text{ g Al} \times \frac{1\text{ mol Al}}{26.98\text{ g Al}} \times \frac{2\text{ mol }Al_2O_3}{4\text{ mol Al}} \times \frac{101.96\text{ g }Al_2O_3}{\text{mol }Al_2O_3} = 3.17\text{ g }Al_2O_3$$

GIVEN: 3.17 g Al_2O_3; 12.8 g ore Wanted: % Al_2O_3

EQUATION: $\%\ Al_2O_3 = \frac{\text{g }Al_2O_3}{\text{g ore}} \times 100 = \frac{3.17\text{ g }Al_2O_3}{12.8\text{ g ore}} \times 100 = 24.8\%\ Al_2O_3$ in the ore

The question asks for % Al_2O_3 in the ore. This means that we need the ratio of mass of Al_2O_3 to mass of ore, the fraction of Al_2O_3 in the ore. The percent concept is discussed in Section 7.6 on Page 180. We have applied Equation 7.2 in this problem. Mass of ore is given, but mass of Al_2O_3 is not, so we must find mass of Al_2O_3. We are given mass of Al, and we can use stoichiometry to find mass of Al_2O_3. See Section 9.2, Mass Calculations, Pages 226-230. A summary of the stoichiometry path is given in a *PROCEDURE* box on Page 226.

47. *GIVEN:* 0.500 ton $Ca(H_2PO_4)_2$ *WANTED:* kg rock

PER/PATH: ton $Ca(H_2PO_4)_2$ $\xrightarrow{2000\text{ lb }Ca(H_2PO_4)_2/\text{ton }Ca(H_2PO_4)_2}$ lb $Ca(H_2PO_4)_2$ $\xrightarrow{2.20\text{ lb }Ca(H_2PO_4)_2/\text{kg }Ca(H_2PO_4)_2}$ kg $Ca(H_2PO_4)_2$ $\xrightarrow{234.05\text{ kg }Ca(H_2PO_4)_2/\text{kmol }Ca(H_2PO_4)_2}$ kmol $Ca(H_2PO_4)_2$ $\xrightarrow{1\text{ kmol }Ca_3(PO_4)_2/1\text{ kmol }Ca(H_2PO_4)_2}$ kmol $Ca_3(PO_4)_2$ $\xrightarrow{310.18\text{ kg }Ca_3(PO_4)_2/\text{kmol }Ca_3(PO_4)_2}$ kg $Ca_3(PO_4)_2$ $\xrightarrow{79.4\text{ kg }Ca_3(PO_4)_2/100\text{ kg rock}}$ kg rock

$$0.500\text{ ton }Ca(H_2PO_4)_2 \times \frac{2000\text{ lb }Ca(H_2PO_4)_2}{\text{ton }Ca(H_2PO_4)_2} \times \frac{1\text{ kg }Ca(H_2PO_4)_2}{2.20\text{ lb }Ca(H_2PO_4)_2} \times \frac{1\text{ kmol }Ca(H_2PO_4)_2}{234.05\text{ kg }Ca(H_2PO_4)_2} \times \frac{1\text{ kmol }Ca_3(PO_4)_2}{1\text{ kmol }Ca(H_2PO_4)_2} \times \frac{310.18\text{ kg }Ca_3(PO_4)_2}{\text{kmol }Ca_3(PO_4)_2} \times \frac{100\text{ kg rock}}{79.4\text{ kg }Ca_3(PO_4)_2} = 759\text{ kg rock}$$

The first step in the solution requires knowledge of the USCS conversion 2000 lb = 1 ton. The relationship 2.20 lb = 1 kg is found in Table 3.3, Metric-USCS Conversion Factors, on Page 71. The three steps in the middle of the solution are the standard stoichiometry pattern, using kilounits. A summary of the stoichiometry path is given in a *PROCEDURE* box on Page 226. The final step uses the percent concept (see Section 7.6 on Page 180).

48. *GIVEN:* 40.1 kg sludge *WANTED:* amount NaCN (assume kg)

PER/PATH: kg sludge $\xrightarrow{23.1\text{ kg AgCl}/100\text{ kg sludge}}$ kg AgCl $\xrightarrow{143.4\text{ kg AgCl/kmol AgCl}}$ kmol AgCl $\xrightarrow{4\text{ kmol NaCN}/2\text{ kmol AgCl}}$ kmol NaCN $\xrightarrow{49.01\text{ kg NaCN/kmol NaCN}}$ kg NaCN

$$40.1\text{ kg sludge} \times \frac{23.1\text{ kg AgCl}}{100\text{ kg sludge}} \times \frac{1\text{ kmol AgCl}}{143.4\text{ kg AgCl}} \times \frac{4\text{ kmol NaCN}}{2\text{ kmol AgCl}} \times \frac{49.01\text{ kg NaCN}}{\text{kmol NaCN}} = 6.33\text{ kg NaCN}$$

The first step is an application of the percent concept (see Section 7.6 on Page 180). The next three steps are the standard stoichiometry pattern, using kilounits. A summary of the stoichiometry path is given in a *PROCEDURE* box on Page 226.

49. *GIVEN:* 105 kg Cl_2 (act) *WANTED:* mass of solution (assume kg)

PER/PATH: kg Cl_2 (act) $\xrightarrow{61\text{ kg }Cl_2\text{ (act)}/100\text{ kg }Cl_2\text{ (theo)}}$ kg Cl_2 (theo) $\xrightarrow{70.90\text{ kg }Cl_2/\text{kmol }Cl_2}$ kmol Cl_2 $\xrightarrow{2\text{ kmol NaCl}/1\text{ kmol }Cl_2}$ kmol NaCl $\xrightarrow{58.44\text{ kg NaCl/kmol NaCl}}$ kg NaCl $\xrightarrow{9.6\text{ kg NaCl}/100\text{ kg solution}}$ kg solution

$$105\text{ kg }Cl_2\text{ (act)} \times \frac{100\text{ kg }Cl_2\text{ (theo)}}{61\text{ kg }Cl_2\text{ (act)}} \times \frac{1\text{ kmol }Cl_2}{70.90\text{ kg }Cl_2} \times \frac{2\text{ kmol NaCl}}{1\text{ kmol }Cl_2} \times \frac{100\text{ kg solution}}{9.6\text{ kg NaCl}} = 3.0 \times 10^3\text{ kg water}$$

In the first step, we apply the percent yield concept (Section 9.5, Pages 236-241). The next three steps are the standard stoichiometry pattern, using kilounits. A summary of the stoichiometry path is given in a *PROCEDURE* box on Page 226. Finally, the percentage composition of the solution is used to convert from kg NaCl to kg solution. The percent concept is discussed in Section 7.6 on Page 180.

50. *GIVEN:* 239 mg $Ca(OH)_2$ *WANTED:* mg SnF_2

PER/PATH: mg $Ca(OH)_2$ $\xrightarrow{74.10\text{ mg }Ca(OH)_2/\text{mmol }Ca(OH)_2}$ mmol $Ca(OH)_2$ $\xrightarrow{1\text{ mmol }SnF_2/1\text{ mmol }Ca(OH)_2}$ mmol SnF_2 $\xrightarrow{156.7\text{ mg }SnF_2/\text{mmol }SnF_2}$ mg SnF_2

$$239\text{ mg }Ca(OH)_2 \times \frac{1\text{ mmol }Ca(OH)_2}{74.10\text{ mg }Ca(OH)_2} \times \frac{1\text{ mmol }SnF_2}{1\text{ mmol }Ca(OH)_2} \times \frac{156.7\text{ mg }SnF_2}{1\text{ mmol }SnF_2} =$$

505 mg SnF_2 is needed to treat 239 mg $Ca(OH)_2$

505 mg SnF_2 needed – 305 mg used = 2.00×10^2 mg SnF_2 additional should be used

The first question in the problem statement asks if there is enough SnF_2 to convert 239 mg $Ca(OH)_2$. Our approach to answer the question was to determine the mass of SnF_2 needed to convert 239 mg $Ca(OH)_2$. We followed the standard stoichiometry pattern, using milliunits. A summary of the stoichiometry path is given in a *PROCEDURE* box on Page 226. After we found that more SnF_2 was needed, the difference between the needed quantity and the used quantity gives the additional amount that should have been used.

51. *GIVEN:* 454 g Al *WANTED:* kw-hr

PER/PATH: g Al $\xrightarrow{26.98\text{ g Al/mol Al}}$ mol Al $\xrightarrow{1.97 \times 10^3\text{ kJ}/4\text{ mol Al}}$ kJ $\xrightarrow{1\text{ kw-hr}/3.60 \times 10^3\text{ kJ}}$ kw-hr

$$454 \text{ g Al} \times \frac{1 \text{ mol Al}}{26.98 \text{ g Al}} \times \frac{1.97 \times 10^3 \text{ kJ}}{4 \text{ mol Al}} \times \frac{1 \text{ kw-hr}}{3.60 \times 10^3 \text{ kJ}} = 2.30 \text{ kw-hr}$$

The first two steps in the solution are the standard steps in a thermochemical stoichiometry problem. See Section 9.11, Pages 252-253. The conversion factor 1.97×10^3 kJ/4 mol Al comes from the thermochemical equation. The conversion factor to change from kJ to kw-hr is given in the problem statement.

52. *GIVEN:* 2.056 g AgCl *WANTED:* g NaCl

$NaCl + AgNO_3 \rightarrow NaNO_3 + AgCl$

PER/PATH: g AgCl $\xrightarrow{143.4 \text{ g AgCl/mol AgCl}}$ mol AgCl $\xrightarrow{1 \text{ mol NaCl/1 mol AgCl}}$ mol NaCl $\xrightarrow{58.44 \text{ g NaCl/mol NaCl}}$ g NaCl

$$2.056 \text{ g AgCl} \times \frac{1 \text{ mol AgCl}}{143.4 \text{ g AgCl}} \times \frac{1 \text{ mol NaCl}}{1 \text{ mol AgCl}} \times \frac{58.44 \text{ g NaCl}}{\text{mol NaCl}} = 0.8379 \text{ g NaCl}$$

1.6240 g mixture of NaCl and $NaNO_3$ – 0.8379 g NaCl = 0.7861 g $NaNO_3$

EQUATION: $\% \text{ NaCl} = \frac{\text{g NaCl}}{\text{g mixture}} = \frac{0.8379 \text{ g NaCl}}{1.6240 \text{ g mixture}} \times 100 = 51.59\% \text{ NaCl}$

EQUATION: $\% \text{ NaNO}_3 = \frac{\text{g NaNO}_3}{\text{g mixture}} = \frac{0.7861 \text{ g NaNO}_3}{1.6240 \text{ g mixture}} \times 100 = 48.41\% \text{ NaNO}_3$

The mass of NaCl is found from the mass of AgCl by applying the standard stoichiometry pattern. A summary of the stoichiometry path is given in a *PROCEDURE* box on Page 226. Since there are two compounds in the mixture, the mass of the second compound is the total mass minus the mass of the first compound. Finally, we apply the percent concept (see Section 7.6 on Page 180) to answer the question.

53. *GIVEN:* 50.0 mL; 17.0% NaOH; 1.19 g/mL *WANTED:* g $Mg(NO_3)_2$

PER/PATH: mL solution $\xrightarrow{1.19 \text{ g solution/mL solution}}$ g solution $\xrightarrow{17.0 \text{ g NaOH/100 g solution}}$ g NaOH $\xrightarrow{40.00 \text{ g NaOH/mol NaOH}}$ mol NaOH $\xrightarrow{1 \text{ mol Mg(NO}_3)_2\text{/2 mol NaOH}}$ mol $Mg(NO_3)_2$ $\xrightarrow{148.33 \text{ g Mg(NO}_3)_2\text{/mol Mg(NO}_3)_2}$ g $Mg(NO_3)_2$

$$50.0 \text{ mL} \times \frac{1.19 \text{ g soln}}{\text{mL}} \times \frac{17.0 \text{ g NaOH}}{100 \text{ g soln}} \times \frac{1 \text{ mol NaOH}}{40.00 \text{ g NaOH}} \times \frac{1 \text{ mol Mg(NO}_3)_2}{2 \text{ mol NaOH}} \times \frac{148.33 \text{ g Mg(NO}_3)_2}{\text{mol Mg(NO}_3)_2} = 18.8 \text{ g Mg(NO}_3)_2$$

The problem statement gives two ratios. If you break these ratios down to their component parts, it is easier to see the *PER/PATH*. 17.0% NaOH is 17.0 g NaOH/100 g solution. The 1.19 g/mL density given for the solution is 1.19 g solution/mL solution. The given quantity is 50.0 mL of solution. From there, you can convert to g solution and then to g NaOH. From this point forward, you apply the standard stoichiometry pattern. A summary of the stoichiometry path is given in a *PROCEDURE* box on Page 226.

54. The molar volume of a gas is 22.4 L/mol at STP. If temperature and pressure are both increased, the pressure change will reduce the volume, whereas the temperature change will increase the volume. If the increase and decrease in volume offset each other exactly, the molar volume remains at 22.4 L/mol. If both temperature and pressure go down, they again have opposite effects on molar volume; if the changes are balanced, molar volume remains at 22.4 L/mol. There are many temperature–pressure combinations can that produce a molar volume of 22.4 L/mol, but the probability of their natural occurrence is remote, except near sea level. One combination is 676 torr and –30°C—possible, perhaps, on a very high mountain in the depths of winter.

Molar volume is defined on Page 230. The STP molar volume of a gas is also discussed on that page.

Chapter 10

Atomic Theory: The Quantum Model of the Atom

1. A discrete line spectrum is one that consists of discrete, or separate, lines of color. Spectra of white light do not have discrete lines. Rather, the colors blend into each other to form a continuous spectrum. A rainbow is an example of a spectrum that does not have discrete lines. Discrete line spectra are not normally encountered outside of chemistry and physics laboratories.

   See Section 10.1, The Bohr Model of the Hydrogen Atom, Pages 264-267. Line spectra, discrete and continuous, are discussed as item number 3 in the list of well-known facts upon which Bohr's model of the atom was based (Page 265). Also review Figure 10.1 on Page 264, which shows a continuous spectrum and Figure 10.3 on Page 265, which shows line spectra.

2. Both are forms of electromagnetic radiation.

   The electromagnetic spectrum is illustrated in Figure 10.1(b) on Page 264. Visible light is found in the middle of this spectrum. Radio waves are in the Hertzian waves box. Also see item number 1 in the list of well-known facts upon which Bohr's model of the atom was based (Page 264).

3. (b) and (c) are quantized.

   The term *quantized* is defined on Page 265. It means that an amount is limited to specific values. Also see Figure 10.4 on Page 266. The elevation of the woman on the stairs is quantized.

4. Electron energies can have one of several possible values, but it may never have an energy between them.

   The term *quantized* is defined on Page 265. It means that an amount is limited to specific values. Also see Figure 10.4 on Page 266. The elevation of the woman on the stairs is quantized.

5. Discrete line spectra are evidence of quantized electron energy levels. Each discrete line represents one of the possible energy transitions for an electron.

   Figure 10.3 on Page 265 shows how the light produced from heating a gaseous element yields a line spectrum.

6. An atom is in its ground state when all electrons are at their lowest possible energies.

   There is a discussion of ground (and excited) states in the last paragraph on Page 266.

7. See Figure 10.5.

   Review Section 10.1, The Bohr Model of the Hydrogen Atom, on Pages 264-267. Figure 10.5 on Page 266 is a summary of the Bohr model.

8. Hydrogen is the only atom that fits the Bohr model. Additionally, the Bohr electron should lose energy and crash into the nucleus.

   The shortcomings are discussed in the next-to-last paragraph of Section 10.1 on Page 267.

9. The principal energy levels of an atom are the main electron energy levels.

   See the Principal Energy Levels subsection of Section 10.2 on Page 268.

10. Each principal energy level is divided into one or more sublevels. They are identified by the letters *s*, *p*, *d*, and *f*.

    See the Sublevels subsection of Section 10.2 on Page 268.

11. An orbital is a mathematically described region in space within an atom in which there is a high probability that an electron will be found. An *s* orbital is spherical; a *p* orbital is dumbbell shaped. See Figure 10.7 for sketches of shapes.

    See the Electron Orbitals subsection of Section 10.2 on Pages 268-270. Figures 10.6 and 10.7 on Page 269 also are important in the development of your understanding of orbitals.

12. There is never more than one *p* sublevel at any value of n.

    At $n = 2$, there are two sublevels, 2*s* and 2*p*; at $n = 3$, there are three sublevels, 3*s*, 3*p*, and 3*d*.

13. An orbital is a region in space where there is a high probability of finding an electron. This implies correctly that there is a low but real probability of finding the electron outside that region—outside the ball.

    Figure 10.6 on Page 269 illustrates the quantum model of the atom. The r_{90} radius on this illustration shows the volume of space within with the electron is found 90% of the time. The probability of finding the electron "outside the ball" is 10%.

14. The Pauli exclusion principle says, in effect, that no more than two electrons can occupy the same orbital.

    See The Pauli Exclusion Principle subsection of Section 10.2 on Pages 270-271.

15. The quantum model of the atom gives no indication of the path of an electron. Furthermore, the orbital is three dimensional, not two as suggested by a figure 8.

    See Figure 10.6 on Page 269. An electron following a path is consistent with the Bohr model, which is incorrect. In the quantum model, the path of an electron is unknown.

16. If $n \geq 3$, the *actual* (not maximum) number of *d* orbitals is always five.

    See the Summary: The Quantum Mechanical Model of the Atom on Pages 270-271. When n = 3, 4, 5, ..., the principal energy level has *d* sublevels. The *d* sublevel has 5 orbitals.

17. Energies of principal energy levels increase in the order 1, 2, 3, ... Energies of sublevels increase in the order *s*, *p*, *d*, *f*.

See the Summary: The Quantum Mechanical Model of the Atom on Pages 270-271.

18. The models are consistent in identifying quantized energy levels. The quantum model substitutes orbitals (regions in space) for orbits (electron paths) in the Bohr model. The quantum model goes beyond the Bohr model in identifying sublevels.

See the Principal Energy Levels subsection of Section 10.2 on Page 268. The principal energy levels of the Bohr model are part of the quantum model. Beyond that, the models are very different. See Figure 10.6 on Page 269.

19. The symbol $3p^4$ means there are four electrons in the $3p$ sublevel.

The total number of electrons in any sublevel is shown by a superscript number. See Section 10.3, Pages 272-278.

20. Fluorine, which is in Period 2, Group 7A/17.

There are 2 + 2 + 5 = 9 electrons, and fluorine, F, is the element with atomic number 9.

21. The neon core, $1s^2 2s^2 2p^6$.

The neon core is discussed in the third paragraph of Section 10.3 on Page 275.

22. a) neon, b) phosphorus, c) manganese

a) 2 + 2 + 6 = 10, Z = 10 for Ne, neon
b) 2 + 2 + 6 + 2 + 3 = 15, Z = 15 for P, phosphorus
c) 2 + 2 + 6 + 2 + 6 + 2 + 5 = 25, Z = 25 for Mn, manganese

23. a) carbon, b) chlorine, c) arsenic

a) 2 (He) + 2 + 2 = 6, Z = 6 for C, carbon
b) 10 (Ne) + 2 + 5 = 17, Z = 17 for Cl, chlorine
c) 18 (Ar) + 2 + 10 + 3 = 33, Z = 33 for As, arsenic

24. N: $1s^2 2s^2 2p^3$; Ti: $1s^2 2s^2 2p^6 3s^2 3p^6 4s^2 3d^2$

See Section 10.3, Electron Configuration, Pages 272-278.

25. Ca: $1s^2 2s^2 2p^6 3s^2 3p^6 4s^2$; Cu: $1s^2 2s^2 2p^6 3s^2 3p^6 4s^1 3d^{10}$

See Section 10.3, Electron Configuration, Pages 272-278.

26. Ti: $[Ar]4s^2 3d^2$; Ca: $[Ar]4s^2$; Cu: $[Ar]4s^1 3d^{10}$

[Ar] is $1s^2 2s^2 2p^6 3s^2 3p^6$. See Section 10.3, Electron Configuration, Pages 272-278.

27. Ge: $[Ar]4s^1 3d^{10} 4p^2$

[Ar] is $1s^2 2s^2 2p^6 3s^2 3p^6$. See Section 10.3, Electron Configuration, Pages 272-278.

28. Valence electrons are the highest energy *s* and *p* electrons in an atom.

    See Section 10.4, Valence Electrons, Pages 279-280.

29. Valence electrons can be represented by their electron configuration or by Lewis symbols. Examples for each group in the periodic table are given in Table 10.1.

    See Section 10.4, Valence Electrons, Pages 279-280.

30. Group 4A/14

    The total number of valence electrons is 2 + 2 = 4, which corresponds to Group 4A/14. See Section 10.4, Valence Electrons, Pages 279-280.

31. Ionization energy (first) is the energy required to remove one electron from a neutral gaseous atom of an element. The energy required to remove a second electron from an atom is its second ionization energy.

    See the Ionization Energy subsection in Section 10.5 on Pages 280-281.

32. Atoms in the same family become larger as atomic number increases. The highest-energy electrons are farther from the nucleus and easier to remove. This appears as lower ionization energies.

    See the Ionization Energy subsection in Section 10.5 on Pages 280-281.

33. A chemical family is a group of elements having similar chemical properties because of similar valence electron configuration, appearing in the same column of the periodic table.

    See the Chemical Families subsection in Section 10.5 on Pages 281-283.

34. ns^1

    The alkali metals are the Group 1A/1 elements (except hydrogen). See the Chemical Families subsection in Section 10.5 on Pages 281-283.

35. Isoelectronic species have the same number of electrons.

    The term *isoelectronic* is defined in the Chemical Families subsection in Section 10.5 on Page 282. The prefix *iso-* refers to same, and electronic refers to electrons.

36. Iodine is a halogen; rubidium is an alkali metal.

    Iodine, I, is in Group 7A/17, the halogens. Rubidium, Rb, is in Group 1A/1, the alkali metals. See Figure 10.13 on Page 282.

37. Chemical properties of elements are often determined by the number of valence electrons. Both magnesium and calcium have two: ns^2.

    See the Chemical Families subsection in Section 10.5 on Pages 281-283.

38. Main group elements are in the A groups (Groups 1, 2, 13 to 18) of the periodic table. Elements to the left of the stair-step line are metals, and those to the right are nonmetals.

    Main group elements were introduced on Page 126 in Section 5.6, The Periodic Table. To review metals and nonmetals, see the Metals and Nonmetals subsection in Section 10.5 on Pages 285-286.

39. Atoms of antimony (Z = 51) and bismuth (Z = 83) are larger than atoms of arsenic, and atoms of phosphorus and nitrogen are smaller.

    See the Atomic Size subsection in Section 10.5 on Pages 283-285. Atomic size increases from top to bottom in a group.

40. As you go down a column in the periodic table, valence electrons occupy higher and higher principal energy levels. The electrons are farther and farther from the nucleus, so the atoms become larger.

    See the Atomic Size subsection in Section 10.5 on Pages 283-285. Note, in particular, item 1 in the list of influences on atomic size on Page 283.

41. The $3s^1$ electron in a sodium atom is in the third principal energy level. Removal of that electron to form an ion leaves the second principal energy level as the highest occupied level. Therefore, we would expect the ion to be smaller. The prediction is correct; the radius of the ion is 0.102 nm compared to 0.186 nm for the atom.

    See Figure 10.11 on Page 281.

42. Generally, an element is known as a metal if it can lose one or more electrons and become a positively charged ion.

    See the Metals and Nonmetals subsection in Section 10.5 on Pages 285-286.

43. (a) Q and M, (b) D and E

    (a) The halogens are in Group 7A/17. (b) The alkali metals are in Group 1A/1.

44. J, W, and L

    The transition elements are those in the B groups (3 to 12) of the periodic table (see Page 126 in Section 5.6, The Periodic Table).

45. E > D > G

    See the Atomic Size subsection in Section 10.5 on Pages 283-285. Atomic size increases from top to bottom in a group and from right to left across a period. E is the largest atom because is the farthest down and left. D is larger than G because it is farther left.

47. True: b, e, g, i, k, m, n, p. False: a, c, d, f, h, j, l, o.

a) Electron energies are quantized no matter whether in the ground state or an excited state.
c) Energy is released as an electron passes from an excited state to ground state.
d) The energy of an electron is never between two quantized levels.
f) The Bohr model of the atom describes orbits in which electrons travel around the nucleus; the quantum mechanical model states that it is not possible to describe such orbits.
h) All *s* sublevels have one orbital; all *p* sublevels have three orbitals; all *d* sublevels have five orbitals; all *f* sublevels have seven orbitals.
j) In the d sublevel, the maximum number of electrons is 10.
l) The dot structure of the alkaline earths is X:, where X is the symbol of any element in the family.
o) Atomic numbers 52, 35, and 18 are arranged in order of decreasing atomic size.

48. Something that behaves like a wave has properties normally associated with waves, some of which appear in Figure 10.2.

An electron behaves like a wave. Figure 10.2 is on Page 265.

49. Ba: $[Xe]6s^2$, Tc: $[Kr]5s^2 4d^5$

See Section 10.3, Electron Configuration, on Pages 272-278. Figure 10.8 on Page 274 is a guide to writing electron configurations.

50. The first negatively charged electron is removed from a neutral atom, and the second electron is removed from a positively charged ion.

See the Ionization Energy subsection in Section 10.5 on Pages 280-281.

51. Aluminum atoms have a lower nuclear charge—fewer protons in the nucleus—than chlorine atoms. It is therefore easier to remove a 3*p* electron from an aluminum atom than from a chlorine atom.

See the Ionization Energy subsection in Section 10.5 on Pages 280-281.

52. The quantum and Bohr model explanations of atomic spectra are essentially the same.

See Section 10.1, The Bohr Model of the Hydrogen Atom, on Pages 264-267. Also see the Principal Energy Levels subsection of Section 10.2 on Page 268. The principal energy levels of the Bohr model are part of the quantum model. Beyond that, the models are very different.

53. Sc^{3+} is isoelectronic with an argon atom.

The electron configuration of Sc^{3+} and Ar is $1s^2 2s^2 2p^6 3s^2 3p^6$. See Section 10.3, Electron Configuration, Pages 272-278.

54. The smaller atoms in Group 5A/15 tend to complete their octets by gaining or sharing electrons, which is a characteristic of nonmetals. Larger atoms in the group tend to lose their highest-energy *s* electrons and form positively charged ions, a characteristic of metals.

See the Metals and Nonmetals subsection in Section 10.5 on Pages 285-286. Figure 10.15 on Page 286 shows that N and P are nonmetals, As and Sb are metalloids, and Bi is a metal.

55. Iron loses two electrons from the 4*s* orbital to form Fe^{2+} and a third from a 3*d* orbital to form Fe^{3+}. This is an example of *d* electrons contributing to the chemical properties of an element.

The electron configuration of iron is [Ar]$4s^2 3d^6$. You may therefore speculate that the 2+ ion forms from the loss of the two 4*s* electrons. Since there are five d orbitals, the loss of one electron from these orbitals leaves one electron in each of the orbitals in the 3+ ion.

Chapter 11

Chemical Bonding

1. Na^+, Mg^{2+}, Al^{3+}, P^{3-}, S^{2-}, and Cl^-.

   Isoelectronic species have the same electron configurations. See Section 11.1, Monatomic Ions with Noble-Gas Electron Configurations, Pages 294-296. Sodium ion, magnesium ion, and aluminum ion are isoelectronic with neon. Phosphide ion, sulfide ion, and chloride ion are isoelectronic with argon.

2. Any two of N^{3-}, O^{2-}, F^-.

   Isoelectronic species have the same electron configurations. See Section 11.1, Monatomic Ions with Noble-Gas Electron Configurations, Pages 294-296.

3. Any two of P^{3-}, S^{2-}, K^+, Ca^{2+}, Sc^{3+}.

   Isoelectronic species have the same electron configurations. See Section 11.1, Monatomic Ions with Noble-Gas Electron Configurations, Pages 294-296.

4. *Please see the answer on textbook Page 312.*

   A sulfur atom has six valence electrons. A sodium atom has one valence electron. Two sodium atoms will therefore donate their valence electrons to a single sulfur atom to form two sodium ions and a sulfide ion. Large numbers of these ions will form a crystal, a solid with a definite geometric pattern in which ions are arranged. See Section 11.2, Ionic Bonds, on Pages 296-298.

5. A potassium atom forms a K^+ ion by losing one electron. A sulfur atom can accept two electrons to form a S^{2-} ion, so it takes two potassium atoms to donate the two electrons to a single sulfur atom.

   An ionic compound must have a net neutral charge: the total positive charge must equal the total negative charge. See Section 11.2, Ionic Bonds, on Pages 296-298.

6. An electron cloud is the area of space around an atom or between bonded atoms that is occupied by electrons.

   See Section 11.3, Covalent Bonds, on Pages 299-300. Figure 11.5 on Page 300 shows electron clouds for separate hydrogen atoms and the hydrogen molecule formed when they bond.

7. *Please see the answer on textbook Page 312.*

   Atoms of iodine and chlorine each have seven valence electrons. One electron from each atom is shared, in the form of a covalent bond, in the molecule. See Section 11.3, Covalent Bonds, on Pages 299-300.

8. There are one bonding pair and three lone pairs on each atom.

   Each pair of dots is a lone pair. The dash in between the atoms is a bonding pair. See Section 11.3, Covalent Bonds, on Pages 299-300.

9. Decrease

   See the last paragraph of Section 11.3, Covalent Bonds, on Page 300.

10. In a polar bond, the bonding electrons spend more time nearer the atoms having the higher electronegativity. This unequal sharing of electrons makes the bond polar. When two atoms of the same element are bonded, the electronegativity difference between them is zero, and the bond is completely nonpolar.

    See Section 11.4, Polar and Nonpolar Covalent Bonds, on Pages 301-303.

11. Cl—Cl; Br—Cl; I—Cl; F—Cl. You may have placed the F—Cl bond in a different position in the ordering if you weren't referring to electronegativity value table, but you should have the other three bonds in the correct order. The Cl—Cl bond is nonpolar; the other bonds are polar.

    Note that Cl is present in each bond. The other bonded atom is also in Group 7A/17. Electronegativity decreases down a group in the periodic table, generally dropping off the most between the second period and the third period, and then in relatively small increments after that. Given that, the F—Cl electronegativity difference would be expected to be large. Br should be more electronegative than I because it is closer to the top of the periodic table; therefore, the Cl—Br *difference* should be less than the Cl—I difference. See Section 11.4, Polar and Nonpolar Covalent Bonds, on Pages 301-303. Note, in particular, Figure 11.7 on Page 301 for actual electronegativity values.

12. In Br—Cl and I—Cl, chlorine is the negative pole; in F—Cl, fluorine is the negative pole.

    The negative pole is the more electronegative atom. Electronegativity is highest at the top of a group, and it decreases down the group. See Section 11.4, Polar and Nonpolar Covalent Bonds, on Pages 301-303. Note, in particular, Figure 11.7 on Page 301 for actual electronegativity values.

13. Electronegativity is the ability of an atom of that element in a molecule to attract bonding electron pairs to itself. Noble gases are sometimes omitted from electronegativity tables because there are very few compounds formed from noble-gas atoms.

    Electronegativity is defined on Page 301. The Noble Gases subsection on Page 283 states that "only a small number of compounds of noble gases are known, and none occur naturally."

14. *Multiple bond* is a general term that includes both double and triple bonds. In a single bond, one electron pair is shared between bonded atoms. In double and triple bonds, two and three electron pairs, respectively, are shared between bonded atoms.

    See Section 11.5, Multiple Bonds, on Page 303.

15. When a central atom has four electron pairs surrounding it, it has eight total electrons, which is an octet. The four pairs can bond to a maximum of four atoms. If an atom in a molecule is truly "central," it must be bonded to a minimum of two other atoms.

    See Section 11.6, Atoms That Are Bonded to Two or More Other Atoms, on Pages 303-304.

16. *Please see the answer on textbook Page 312.*

    See Section 11.6, Atoms That Are Bonded to Two or More Other Atoms, on Pages 303-304.

17. In order to conform to the octet rule, each atom in a molecule must be surrounded by eight electrons. Eight is an even number, and no combination of eights can result in an odd number.

    See Section 11.7, Exceptions to the Octet Rule, on Pages 304-305.

18. An metallic bond is more similar to a covalent bond because, in both types of bonds, electrons are shared. In an ionic bond, electrons are transferred.

    The electrons shared in a covalent bond are shared between two nuclei. Electrons shared in a metallic bond are shared among many nuclei. To review ionic bonds, see Section 11.2, Ionic Bonds, on Pages 296-298. To review covalent bonds, see Section 11.3, Covalent Bonds, on Pages 299-300. To review metallic bonds, see Section 11.8, Metallic Bonds, on Pages 306-308.

19. *Please see the answer on textbook Page 312.*

    Potassium atoms have one valence electron: $[Ar]4s^1$. The electron-sea model therefore will be the one on the left in Figure 11.9 (Page 308), with one valence electron per 1+ ion.

20. An alloy is a mixture of two or more metals. Steel, brass, and bronze are common alloys.

    Alloy is defined at the bottom of Page 308. Table 11.2 on Page 308 gives the composition of some common alloys.

22. True: a, b, c, d, e, f, h, i. False: g, j, k.

    g) Multiple bonds can form between atoms of the same element and atoms of different elements.
    j) Valence electrons are delocalized among many atoms in a metal.
    k) Alloys are mixtures.

23. Ions are formed when neutral atoms lose or gain electrons. The electron(s) that is(are) lost by one atom is(are) transferred to another atom. The attraction between the ions is an ionic bond. Covalent bonds are formed when a pair of electrons is shared by the two bonded atoms. Effectively, the electrons belong to both atoms, spending some time near each nucleus.

    To review ionic bonds, see Section 11.2, Ionic Bonds, on Pages 296-298. To review covalent bonds, see Section 11.3, Covalent Bonds, on Pages 299-300.

24. The K—Cl bond is ionic, formed by "transferring" an electron from a potassium atom to a chlorine atom. The Cl—Cl bond is covalent, formed by two chlorine atoms sharing a pair of electrons.

Potassium is a metal; chlorine is a nonmetal. A potassium atom therefore transfers an electron to a chlorine atom to form a potassium ion and a chloride ion, which are attracted to one another in the form of an ionic bond. Two nonmetal atoms bond by sharing electrons in a covalent bond. To review ionic bonds, see Section 11.2, Ionic Bonds, on Pages 296-298. To review covalent bonds, see Section 11.3, Covalent Bonds, on Pages 299-300.

25. The H^+ ion has no electrons, so it has no electron configuration.

The most common isotope of hydrogen, ^{1}H, consists of one proton and one electron. The hydrogen ion is a proton.

26. $4p$ from bromine and $2p$ from oxygen.

Br: $[Ar]4s^2 3d^{10} 4p^5$. There is one half-filled $4p$ orbital in bromine.
O: $1s^2 2s^2 2p^4$. There are two half-filled $2p$ orbitals in oxygen.

27. Electronegativities are highest at the upper right corner of the periodic table and lowest at the lower left corner. Therefore, the electronegativity of A is higher than the electronegativity of B. Because X is higher in the table than Y, the electronegativity of X should be larger than that of Y, but because Y is farther to the right, the electronegativity of X should be smaller than Y. Therefore, no prediction can be made for X and Y.

See Pages 301-302 in Section 11.4 for a discussion of electronegativity. Also review Figure 11.7 to see the periodic trends in electronegativity values.

28. (a) Kr, krypton, Z = 36; (b) The 2– ion had two electrons added to the neutral atom. The neutral atom is therefore Z = 36 – 2 = 34. Z = 34 is Se, selenium. The ion is Se^{2-}, selenide ion. (c) A 1+ ion has one electron removed from the neutral atom, therefore the neutral atom is Z = 36 + 1 = 37. Z = 37 is Rb, rubidium. The ion is Rb^+, rubidium ion.

2 + 2 + 6 + 2 + 6 + 2 + 10 + 6 = 36; Z = 36 for Kr, krypton. See Section 11.1, Monatomic Ions with Noble-Gas Electron Configurations, Pages 294-296.

29. Rb_2Se

The positive charge of two 1+ ions balances the negative charge of one 2- ion.

30. Nonmetal atoms are usually one, two, or possibly three or four short of an octet of electrons. They achieve that octet most easily by gaining the missing electrons. When two nonmetal atoms combine, the easiest way for both atoms to reach the octet is to share each other's electrons, forming a covalent bond. If the second atom is a metal, however, it has one, two, or possibly three electrons more than an octet. It reaches the octet by giving its electrons to the nonmetal, becoming a positive ion itself, and making the nonmetal atom a negative ion. The two atoms form an ionic bond.

To review ionic bonds, see Section 11.2, Ionic Bonds, on Pages 296-298. To review covalent bonds, see Section 11.3, Covalent Bonds, on Pages 299-300.

31. An F—Si bond is more polar than an O—P bond. F has a higher electronegativity than O, and Si has a lower electronegativity than P, based on their relative positions in the periodic table (high at the upper right, low at the lower left). Therefore, the *difference* in electronegativities is largest for F—Si, which makes it the more polar bond.

See Pages 301-302 in Section 11.4 for a discussion of electronegativity. Also review Figure 11.7 to see the periodic trends in electronegativity values.

32. AsI_5 can be formed if each of the lone-pair electrons forms a bonding pair with one electron from an I atom. *Please see the Lewis diagrams on textbook Page 313.*

See Section 11.7, Exceptions to the Octet Rule, on Pages 304-305.

Chapter 12

Structure and Shape

Note: Throughout this chapter, we follow the five-step *Procedure: Drawing a Lewis Diagram* from Page 317 to explain the thought process for drawing Lewis diagrams.

1. *Please see the answers on textbook Page 344.*

HF: 1) 1 + 7 = 8 valence electrons available
2) Only two atoms; no central atom
3) H-F single bond and add 3 unshared pairs to F
4) 8 valence electrons in tentative diagram match 8 available
5) H has 2 electrons; F has 8 electrons

OF_2: 1) 6 + 2(7) = 20 valence electrons available
2) O is the least electronegative atom and therefore the central atom
3) Single bond F-O-F and add 3 unshared pairs to Fs and 2 unshared pairs to O
4) 20 valence electrons in tentative diagram match 20 available
5) Each atom has 8 electrons

NF_3: 1) 5 + 3(7) = 26 valence electrons available
2) N is the least electronegative atom and therefore the central atom
3) Single bond each F to N and add 3 unshared pairs to Fs and 1 unshared pair to N
4) 26 valence electrons in tentative diagram match 26 available
5) Each atom has 8 electrons

2. *Please see the answers on textbook Page 344.*

PO_4^{3-}: 1) 5 + 4(6) + 3 = 32 valence electrons available
2) P is the least electronegative atom and therefore the central atom
3) Single bond each O to P and add 3 unshared pairs to Os
4) 32 valence electrons in tentative diagram match 26 available
5) Each atom has 8 electrons

BrO_3^-: 1) 7 + 3(6) + 1 = 26 valence electrons available
2) Br is the least electronegative atom and therefore the central atom
3) Single bond each O to Br and add 3 unshared pairs to Os and 1 unshared pair to Br
4) 26 valence electrons in tentative diagram match 26 available
5) Each atom has 8 electrons

SO_4^{2-}: 1) 6 + 4(6) + 2 = 32 valence electrons available
2) S is the least electronegative atoms and therefore the central atom
3) Single bond each O to S and add 3 unshared pairs to Os
4) 32 valence electrons in tentative diagram match 32 available
5) Each atom has 8 electrons

3. *Please see the answers on textbook Page 344.*

ClO_2^-: 1) 7 + 2(6) + 1 = 20 valence electrons available
2) Cl is the least electronegative atom and therefore the central atom
3) Single bond each O to Cl and add 3 unshared pairs to Os and 2 unshared pairs to Cl
4) 20 valence electrons in tentative diagram match 20 available
5) Each atom has 8 electrons

HIO_3: 1) 1 + 7 + 3(6) = 26 valence electrons available
2) I is the least electronegative atom and therefore the central atom (H cannot be central)
3) Single bond each O to I, single bond H to O, add 2 unshared pairs to O with H bonded, and add 3 unshared pairs to other two O atoms
4) 26 valence electrons in tentative diagram match 26 available
5) H has 2 electrons; Os and I have 8 electrons

H_3PO_4: 1) 3(1) + 5 + 4(6) = 32 valence electrons available
2) P is the least electronegative atom and therefore the central atom (H cannot be central)
3) Single bond each O to P, single bond each H to O, add 2 unshared pairs to Os with H bonded, and add 3 unshared pairs to other O atom
4) 32 valence electrons in tentative diagram match 32 available
5) Each H has 2 electrons; Os and P have 8 electrons

4. *Please see the answers on textbook Page 344.*

CH_2ClBr: 1) 4 + 2(1) + 7 + 7 = 20 valence electrons available
2) C is the least electronegative atom and therefore the central atom (H cannot be central)
3) Single bond each H, Cl, and Br to C and add 3 unshared pairs each to Cl and Br
4) 20 valence electrons in tentative diagram match 20 available
5) Each H has 2 electrons; C, Cl, and Br each have 8 electrons

CH_2F_2: 1) 4 + 2(1) + 2(7) = 20 valence electrons available
2) C is the least electronegative atom and therefore the central atom (H cannot be central)
3) Single bond each H and each F to C and add 3 unshared pairs to each F
4) 20 valence electrons in tentative diagram match 20 available
5) Each H has 2 electrons; Fs and C have 8 electrons

CBr_2I_2: 1) 4 + 2(7) + 2(7) = 32 valence electrons available
2) C is the least electronegative atom and therefore the central atom
3) Single bond each Br and each I to C and add 3 unshared pairs to each Br and I
4) 32 valence electrons in tentative diagram match 32 available
5) Each atom has 8 electrons

5. *Please see the answers on textbook Page 344.*

$C_2H_4Br_2$: 1) 2(4) + 4(1) + 2(7) = 26 valence electrons available
2) C is the least electronegative atom and therefore the central atom (H cannot be central)
3) Single bond the Cs to one another, single bond Hs and Brs to the Cs (Brs can go on the same C or on different Cs), add 3 unshared pairs to each Br
4) 26 valence electrons in tentative diagram match 26 available
5) Each H has 2 electrons; Cs and Brs each have 8 electrons

C_2H_4BrF: 1) 2(4) + 4(1) + 7 + 7 = 26 valence electrons available
2) C is the least electronegative atom and therefore the central atom (H cannot be central)
3) Single bond the Cs to one another, single bond Hs and Br and F to the Cs (Br and F can go on the same C or on different Cs), add 3 unshared pairs to Br and F
4) 26 valence electrons in tentative diagram match 26 available
5) Each H has 2 electrons; Cs, Br, and F each have 8 electrons

$C_3H_5FBr_2$: 1) 3(4) + 5(1) + 7 + 2(7) = 38 valence electrons available
2) C is the least electronegative atom and therefore the central atom (H cannot be central)
3) Single bond the Cs to one another, single bond Hs, F, and Brs to the Cs (9 different arrangements are possible), add 3 unshared pairs to F and Brs
4) 38 valence electrons in tentative diagram match 38 available
5) Each H has 2 electrons; Cs, F, and Brs each have 8 electrons

6. *Please see the answers on textbook Page 344.*

C_3H_8: 1) 3(4) + 8(1) = 20 valence electrons available
2) C is the least electronegative atom and therefore the central atom (H cannot be central)
3) Single bond the Cs to one another, single bond Hs to the Cs
4) 20 valence electrons in tentative diagram match 20 available
5) Each H has 2 electrons; each C has 8 electrons

C_3H_6: 1) 3(4) + 6(1) = 18 valence electrons available
2) C is the least electronegative atom and therefore the central atom (H cannot be central)
3) Single bond the Cs to one another, single bond Hs to the Cs, add 2 unshared pairs to Cs
4) 20 valence electrons in tentative diagram are 2 too many; erase lone pairs and double bond one pair of Cs; 18 valence electrons in revised diagram match 18 available
5) Each H has 2 electrons; each C has 8 electrons

C_3H_6O: 1) 3(4) + 6(1) + 6 = 24 valence electrons available
2) C is the least electronegative atom and therefore the central atom (H cannot be central)
3) There are many correct approaches to arrive at correct Lewis diagrams. In general, if you start by bonding the Cs to each other and then add one —O—H and 5 single-bonded Hs, you will need to add two unshared pairs to the Cs and two unshared pairs to the O. This results in 26 valence electrons in the tentative diagram, 2 too many.
4) Erasing the lone pairs and double bonding one pair of Cs gives one of the isomers. The —O—H can be moved among the carbons to find other isomers. Another set of isomers can be drawn by double-bonding the oxygen to the Cs. Another isomer (not shown the answer) can be drawn by placing a single-bonded O between two carbons and double bonding the other C=C atoms.
5) Each H has 2 electrons; each C and the O have 8 electrons.

7. *Please see the answers on textbook Pages 344–345.*

C_5H_{12}: 1) 5(4) + 12(1) = 32 valence electrons available
2) C is the least electronegative atom and therefore the central atom (H cannot be central)
3) Single bond the Cs to one another and single bond Hs to the Cs
4) 32 valence electrons in tentative diagram match 32 available
Additional isomers can be drawn by single bonding 4 carbons to one another, attaching a $-CH_3$ group to a carbon, and adding the appropriate number of Hs to the carbons; and by bonding 4 $-CH_3$ groups to a central carbon

5) Each H has 2 electrons; each C has 8 electrons

C_4H_8O: 1) 4(4) + 8(1) + 6 = 30 valence electrons
2) C is the least electronegative atom and therefore the central atom (H cannot be central)
3) There are many correct approaches to arrive at correct Lewis diagrams. In general, if you start by bonding the Cs to each other and then add one —O—H and 7 single-bonded Hs, you will need to add two unshared pairs to the Cs and two unshared pairs to the O. This results in 32 valence electrons in the tentative diagram, 2 too many.
4) Erasing the lone pairs and double bonding one pair of Cs gives one of the isomers. The —O—H can be moved among the carbons to find other isomers. Another set of isomer (not shown in the answer) can be drawn by double-bonding the oxygen to the Cs. Another set of isomers can be drawn by placing a single-bonded O between two carbons and double bonding the other C=C atoms.
5) Each H has 2 electrons; each C and the O have 8 electrons.

C_4H_8: 1) 4(4) + 8(1) = 24 valence electrons available
2) C is the least electronegative atom and therefore the central atom (H cannot be central)
3) Single bond the Cs to one another, single bond Hs to the Cs, add 2 unshared pairs to Cs
4) 26 valence electrons in tentative diagram are 2 too many; erase lone pairs and double bond one pair of Cs; 24 valence electrons in revised diagram match 24 available; one additional isomer can be drawn
5) Each H has 2 electrons; each C has 8 electrons

8. *Please see the answer on textbook Page 345.*

1) 4 + 3(1) + 4 + 6 + 6 +1 = 24 valence electrons available
2) C is the least electronegative atom and therefore the central atom (H cannot be central)
3) The line formula gives information about the structure; 3 Hs are bonded to one of the Cs and an O and an -O-H are bonded to the other C.
4) Doing this with all single bonds gives 26 valence electrons in the tentative structure. Erase a lone pair from the C bonded to the O and from the O, and add a double bond. The structure now has 24 valence electrons.
5) Each H has 2 electrons; each C and each O have 8 electrons

9. *Please see the answer on textbook Page 345.*

OH^-: 1) 6 + 1 + 1 = 8 valence electrons available
2) Only two atoms; no central atom
3) O-H single bond and add 3 unshared pairs to O
4) 8 valence electrons in tentative diagram match 8 available
5) H has 2 electrons; O has 8 electrons

H_2O: 1) 2(1) + 6 = 8 valence electrons available
2) O is the central atom (H cannot be central)
3) Single bond each H to the O and add 2 unshared pairs to O
4) 8 valence electrons in tentative diagram match 8 available
5) Hs have 2 electrons; O has 8 electrons

CH_4O: 1) 4 + 4(1) + 6 = 14 valence electrons available
2) C is the least electronegative atom and therefore the central atom (H cannot be central)
3) Single bond 3 Hs to C and single bond an —O—H to the C and add 2 unshared pairs to O

4) 14 valence electrons in the tentative diagram match 14 available
5) Each H has 2 electrons, C and O each have 8 electrons

10. *Please see the answers on textbook Page 345.*

$BeCl_2$: 1) 2 + 2(7) = 16 valence electrons available
2) Be is the least electronegative atom and therefore the central atom
3) Single bond Cls to Be; Be is an exception to the octet rule so don't add unshared pairs
4) 16 valence electrons in the tentative diagram match 16 available
5) Each Cl has 8 electrons and Be has 2 pairs of electrons (4 electrons)
Electron-pair geometry: 2 pairs is linear
Molecular geometry: 2 pairs, both bonded is linear

ICl: 1) 7 + 7 = 14 valence electrons available
2) Only two atoms; no central atom
3) Single bond I to Cl, add three unshared pairs to each
4) 14 valence electrons in the tentative diagram match 14 available
5) I and Cl each have 8 electrons
Electron-pair geometry: 2 atoms must be linear
Molecular geometry: 2 atoms must be linear

SiH_4: 1) 4 + 4(1) = 8 valence electrons available
2) Si is the central atom (H cannot be central)
3) Single bond each H to Si
4) 8 valence electrons in the tentative diagram match 8 available
5) Each H has 2 electrons and Si has 8 electrons
Electron-pair geometry: 4 pairs is tetrahedral
Molecular geometry: 4 pairs, 4 bonded is tetrahedral

11. *Please see the answers on textbook Page 345.*

ClO_4^-: 1) 7 + 4(6) + 1 = 32 valence electrons available
2) Cl is the least electronegative atom and therefore the central atom
3) Single bond each O to Cl and add 3 unshared pairs to each Cl
4) 32 valence electrons in the tentative diagram match 32 available
5) Each O and the Cl have 8 electrons
Electron-pair geometry: 4 pairs is tetrahedral
Molecular geometry: 4 pairs, 4 bonded is tetrahedral

IO_2^-: 1) 7 + 2(6) + 1 = 20 valence electrons available
2) I is the least electronegative atom and therefore the central atom
3) Single bond each O to I and add 3 unshared pairs to each O and 2 unshared pairs to I
4) 20 valence electrons in the tentative diagram match 20 available
5) Each I and O have 8 electrons
Electron-pair geometry: 4 pairs is tetrahedral
Molecular geometry: 4 pairs, 2 bonded is bent

SO_4^{2-}: 1) 6 + 4(6) + 2 = 32 valence electrons available
2) S is the least electronegative atom and therefore the central atom
3) Single bond each O to S and add 3 unshared pairs to each O
4) 32 valence electrons in the tentative diagram match 32 available

5) Each O and S have 8 electrons
Electron-pair geometry: 4 pairs is tetrahedral
Molecular geometry: 4 pairs, 4 bonded is tetrahedral

12. *Please see the answers on textbook Page 345.*

1) 3(4) + 7(1) + 6 + 1 = 26 valence electrons are available
2) C is the least electronegative atom and therefore the central atom (H cannot be central)
3) The line formula gives information about the structure; single bond the Cs to one another, single bond 7 Hs to the Cs, and bond an —O—H to the Cs; add two unshared pairs to the O
4) 26 valence electrons in the tentative diagrams match 26 available
5) Each H has 2 electrons; each C and the O have 8 electrons
Electron-pair geometry around O: 4 pairs is tetrahedral
Molecular geometry around O: 4 pairs, 2 bonded is bent

13. *Please see the answers on textbook Page 346.*

1) 2(4) + 5(1) + 5 + 2(1) = 20 valence electrons available
2) C is the least electronegative atom and therefore the central atom (H cannot be central)
3) The line formula gives information about the structure; single bond the Cs to one another, single bond 5 Hs to the Cs, and bond a $—NH_2$ to the Cs; add an unshared pair to the N
4) 20 valence electrons in the tentative diagram match 20 available
5) Each H has 2 electrons; each C and the N have 8 electrons
Electron-pair geometry around each C: 4 pairs is tetrahedral
Molecular geometry around each C: 4 pairs, 4 bonded is tetrahedral

14. *Please see the answers on textbook Page 346.*

1) 2(4) + 4(1) = 12 valence electrons available
2) C is the least electronegative atom and therefore the central atom (H cannot be central)
3) Single bond the Cs to one another, single bond 4 Hs to Cs, and add 2 unshared pairs to the Cs
4) 14 valence electrons in the tentative diagram is 2 too many; erase two lone pairs and add an additional bond between the Cs, which yields 12 valence electrons, matching the number available
5) Each H has 2 electrons and each C has 8 electrons
Electron-pair geometry around each C: 3 regions of electron density (a double or triple bond is one region of density) is trigonal planar
Molecular geometry around each C: 3 regions of electron density, 3 bonded is trigonal planar

15. *Please see the answers on textbook Page 346.*

1) 6 + 4 + 2(7) = 24 valence electrons available
2) C is the least electronegative atom and therefore the central atom
3) The line formula gives information about the structure; single bond O to C and single bond 2 Cls to C, add 3 unshared pairs to O and each Cl, and add 1 unshared pair to C
4) 26 valence electrons in the tentative diagram is 2 too many; erase two lone pairs and add an additional bond between the C and O (you could also double bond a Cl, but for reasons beyond the scope of this text, the correct diagram has a C=O)
5) Each Cl, C, and O have 8 electrons

Electron-pair geometry: 3 regions of electron density (a double or triple bond is one region of density) is trigonal planar
Molecular geometry: 3 regions of electron density, 3 bonded is trigonal planar

16. *Please see the answers on textbook Page 346.*

Angular: 3 regions of electron density, 2 bonded (one region has to be a double bond to satisfy the octet rule); bent: 4 regions of electron density, 2 bonded

17. Electron pair: Linear; Molecular: Linear

The green-blue-green bond angle appears to be 180°. This is consistent with a linear geometry. See Table 12.1 on Page 326 for electron-pair angles.

18. Electron pair: Tetrahedral; Molecular: Bent (Angular) *or* Electron pair: Trigonal planar; Molecular Angular (Bent)

It is difficult to distinguish whether the white-red-white bond angle is 120° or 109.5°. If you think it is 120°, it has a trigonal planar electron-pair geometry (three regions of electron density) and an angular molecular geometry (2 bonded). If you think it is 109.5°, it has a tetrahedral electron-pair geometry (four regions of electron density) and a bent molecular geometry (2 bonded). See Table 12.1 on Page 326 for electron-pair angles.

19. Electron pair: Tetrahedral; Molecular: Bent (Angular) *or* Electron pair: Trigonal planar; Molecular Angular (Bent)

It is difficult to distinguish whether the white-red-white bond angle is 120° or 109.5°. If you think it is 120°, it has a trigonal planar electron-pair geometry (three regions of electron density) and an angular molecular geometry (2 bonded). If you think it is 109.5°, it has a tetrahedral electron-pair geometry (four regions of electron density) and a bent molecular geometry (2 bonded). See Table 12.1 on Page 326 for electron-pair angles.

20. a and c

A polar molecule is one in which there is an asymmetrical distribution of charge. See Section 12.5, Polarity of Molecules, on Pages 334-336. In part (a), there are four electron pairs around the central atom, with two bonded, which yields a bent (109.5°) bond angle. Even though the two-dimensional Lewis diagram can be drawn with the symbols in a straight line, the molecule is not linear.

21. Charge is balanced in the linear geometry of $BeCl_2$. Charge is unbalanced in the bent geometry of SCl_2.

A polar molecule is one in which there is an asymmetrical distribution of charge. See Section 12.5, Polarity of Molecules, on Pages 334-336. In SCl_2, there are four electron pairs around the central atom, with two bonded, which yields a bent (109.5°) bond angle. Even though the two-dimensional Lewis diagram can be drawn with the symbols in a straight line, the molecule is not linear.

22. All the molecules are linear. BrF is the most polar, followed by ClF. The polarity of ICl is next greatest, and BrCl and IBr have about the same polarity. The first element in each formula is the positive end of the molecule.

A polar molecule is one in which there is an asymmetrical distribution of charge. See Section 12.5, Polarity of Molecules, on Pages 334-336. In all cases, the atoms at each end of the bond are different, so there is an electronegativity difference and the molecule is polar. To determine the extent of the polarity, see the electronegativity values in Figure 11.7 on Page 301. The greater the electronegativity difference, the more polar the bond.

23. *Please see the answers on textbook Page 346.*

In both C_2H_6O molecules, there are four electron pairs around the oxygen atom, with two pairs bonded, giving a bent structure. This results in an asymmetric distribution of charge and a polar molecule. If there are more carbons in the chain, as in C_4H_9OH, the molecule is less polar, but the bent geometry still results in some polarity.

24. e

A linear molecular geometry results from two electron pairs around the central atom, both bonded, as in BeF_2.

25. Structure f is trigonal planar; c has a trigonal pyramidal shape.

A trigonal planar molecular geometry results from three electron pairs around the central atom, all three bonded, as in BF_3. A trigonal pyramidal molecular geometry results from four electron pairs around the central atom with three bonded, as in PCl_3.

26. The carbon atom can form four stable covalent bonds to other carbon atoms or to atoms of other elements.

See Section 12.6, The Structure of Some Organic Compounds, on Pages 336-340. In particular, review The Bonding Capabilities of the Carbon Atom subsection on Page 336.

27. *Hydro-* in *hydrocarbon* refers to hydrogen. A hydrocarbon is a compound of hydrogen and carbon. The ending *-hydrate* in *carbohydrate* refers to water. A carbohydrate is a compound of carbon and the elements that make up water, hydrogen and oxygen. One cannot be an example of the other. Carbohydrates contain oxygen but hydrocarbons do not.

See Section 12.6, The Structure of Some Organic Compounds, on Pages 336-340. In particular, review the Hydrocarbons subsection on Pages 336-338.

28. A Lewis diagram shows the structure of a molecule—the relative positions of the atoms to one another—but it does not show the shape.

See the last paragraph in the Hydrocarbons subsection on Page 338.

29. A line formula is one written as a "mini-Lewis diagram" in which the structure of the molecule is suggested. They are especially useful for organic molecules because of the importance of their structures, particularly for isomers.

Line formulas are discussed in the full paragraph on Page 327.

30. An alcohol is an alkane with a hydroxyl group, —OH, substituted for a hydrogen. The properties of an alcohol are the properties of the hydroxyl group.

See the Alcohols and Ethers subsection of Section 12.6 on Pages 338-339.

31. Carboxylic acids contain a carboxyl group, —COOH. The geometry around the carbon is trigonal planar, and it is bent around the oxygen. *Please see the answer on textbook Page 346 for the Lewis diagram of the carboxyl group.*

See the Carboxylic Acids subsection of Section 12.6 on Pages 339-340.

33. True: a, d, e, g, h*, i, j. False: b, c, f.
*Statement h is true if central atoms are limited to four electron pairs, as they are in this book.

b) Molecular geometry is the direct effect of electron-pair geometry.
c) If the geometry of a molecule is linear, the central atom in the molecule must be surrounded by two regions of electron density, no matter whether those regions are single, double, or triple bonds.
f) A molecule is polar if it has an asymmetric distribution of charge.

34. *Please see the answer on textbook Page 346 for the three-dimensional ball-and-stick representation.*
Bond angles between carbon atoms in an alkane are tetrahedral, so the atoms cannot lie in a straight line.

See the last paragraph in the Hydrocarbons subsection on Page 338.

35. *Please see the answer on textbook Page 346.*

1) 6(4) + 6(1) = 30 valence electrons available
2) C atoms are central (H cannot be a central atom)
3) Arrange the 6 Cs in a ring in the shape of a hexagon, single bond Cs to one another, and single bond Hs to Cs; add one unshared pair on each C
4) Tentative diagram has 36 electrons, 6 too many; erase 2 lone pairs from neighboring Cs and add an additional bond, repeat for a total of 3 double bonds (leaving 3 single bonds)
5) Hs have 2 electrons, Cs have 8 electrons

36. *Please see the answer on textbook Page 346.*

$HCOO^-$: 1) 4 + 2(6) + 1 + 1 = 18 valence electrons available
2) C is the least electronegative atom and therefore the central atom (H cannot be central)
3) The line formula gives information about the structure; a H, an O, and another O are bonded to the C.
4) Doing this with all single bonds gives 20 valence electrons in the tentative structure. Erase a lone pair from the C bonded to an O and from the O, and add a double bond. The structure now has 18 valence electrons
5) The H has 2 electrons; the C and each O have 8 electrons

37. *Please see the answer on textbook Page 346.*

1) 4(4) + 10(1) + 6 = 32 valence electrons available
2) C is the least electronegative atom and therefore the central atom (H cannot be central)
3) To attach the same group of atoms on either side of the oxygen atom, half the carbons and half the hydrogens must be attached on each side, C_2H_5—O—C_2H_5, and add two lone pairs to the O

4) Doing this with all single bonds gives 32 valence electrons in the tentative structure, which matches the 32 valence electrons available
5) Hs have 2 electrons, Cs and O have 8 electrons

38. (a) Trigonal planar with 120° angles around both carbon atoms. (b) Linear. (c) Zigzag carbon chain; all bond angles tetrahedral.

$C_2H_2Br_2$: 1) 2(4) + 2(1) + 2(7) = 24 valence electrons available
2) C is the least electronegative atom and therefore the central atom (H cannot be central)
3) Single bond Cs to one another and single bond Hs and Brs to Cs; the *cis*- prefix indicates that the bromines are attached to the same carbon, but this does not have an effect on the molecular geometry; add 3 unshared pairs to each Br, add 1 unshared pair to each C
4) Tentative diagram has 26 valence electrons; erase a lone pair from each C and add an additional bond between them; the structure now has 24 valence electrons, matching the number available
5) Hs have 2 electrons, Cs and Brs have 8 electrons
Electron-pair geometry around each C: 3 regions of electron density is trigonal planar
Molecular geometry around each C: 3 regions, 3 bonded is trigonal planar

C_2H_2: 1) 2(4) + 2(1) = 10 valence electrons available
2) C is the least electronegative atom and therefore the central atom (H cannot be central)
3) Single bond Cs to one another and single bond Hs to Cs; add two unshared electron pairs to each C
4) Tentative diagram has 14 valence electrons, 4 too many; erase an unshared pair from each C and replace with a bonding pair, repeat; the structure now has 10 valence electrons, matching the number available
5) Each H has 2 electrons, each C has 8 electrons

$CH_3CH_2CH_2CH_3$: 1) (C_4H_{10}) 4(4) + 10(1) = 26 valence electrons available
2) C is the least electronegative atom and therefore the central atom (H cannot be central)
3) Single bond Cs to one another and single bond Hs to Cs (the *n*- prefix indicates that the Cs are attached in a continuous chain with no branching)
4) Tentative diagram has 26 valence electrons, matching the number available
5) Each H has 2 electrons, each C has 8 electrons

Chapter 13

The Ideal Gas Law and Its Applications

1. At constant pressure and temperature, the volume of a gas is directly proportional to amount (moles). The volume of a liquid or solid is the constant volume of the particles plus the negligible space between them. The volume of a gas is the volume of its container, which consists of the volume of the particles plus the relatively huge volume of space between them.

   See Section 13.1, Gases Revisited, on Page 348, for a comparison of the properties of a gas with those of a liquid or solid. See Section 13.2, The Volume-Amount (Avogadro's) Law, on Pages 349-350 for a discussion of the Volume-Amount Law.

2. *GIVEN:* 9.81 L; 23.5 mol; 23°C (296 K) *WANTED:* atm

   $PV = nRT \quad \frac{PV}{V} = \frac{nRT}{V} \quad P = \frac{nRT}{V}$

   EQUATION: $P = \frac{nRT}{V} = 23.5 \text{ mol} \times \frac{0.0821 \text{ L} \cdot \text{atm}}{\text{mol} \cdot \text{K}} \times \frac{296 \text{ K}}{9.81 \text{ L}} = 58.2 \text{ atm}$

   See Section 13.4, The Ideal Gas Equation: Determination of a Single Variable, on Pages 352-354.

3. *GIVEN:* 8.50×10^2 torr; 0.446 mol; 39°C (312 K) *WANTED:* Volume (assume L)

   $PV = nRT \quad \frac{PV}{P} = \frac{nRT}{P} \quad V = \frac{nRT}{P}$

   EQUATION: $V = \frac{nRT}{P} = 0.446 \text{ mol} \times \frac{62.4 \text{ L} \cdot \text{torr}}{\text{mol} \cdot \text{K}} \times \frac{312 \text{ K}}{8.50 \times 10^2 \text{ torr}} = 10.2 \text{ L}$

   See Section 13.4, The Ideal Gas Equation: Determination of a Single Variable, on Pages 352-354.

4. *GIVEN:* 5.24 L; 1.62 atm; 17°C (290 K) *WANTED:* mol

   $PV = nRT \quad \frac{PV}{RT} = \frac{nRT}{RT} \quad \frac{PV}{RT} = n$

   EQUATION: $n = \frac{PV}{RT} = 1.62 \text{ atm} \times \frac{\text{mol} \cdot \text{K}}{0.0821 \text{ L} \cdot \text{atm}} \times \frac{5.24 \text{ L}}{290 \text{ K}} = 0.357 \text{ mol}$

   See Section 13.4, The Ideal Gas Equation: Determination of a Single Variable, on Pages 352-354.

5. *GIVEN:* 0.258 mol; 1.00 L; 6.43 atm *WANTED:* $T_{°C}$

   $PV = nRT \quad \frac{PV}{nR} = \frac{nRT}{nR} \quad \frac{PV}{nR} = T$

   EQUATION: $T = \frac{PV}{nR} = 6.43 \text{ atm} \times \frac{\text{mol} \cdot \text{K}}{0.0821 \text{ L} \cdot \text{atm}} \times \frac{1.00 \text{ L}}{0.258 \text{ mol}} = 304 \text{ K} = 31°\text{C}$

   See Section 13.4, The Ideal Gas Equation: Determination of a Single Variable, on Pages 352-354. $T_{°C} = T_K - 273 = 304 - 273 = 31$

6. *GIVEN:* 6.64 L; 3.62×10^3 torr; 25°C (298 K) *WANTED:* mol

$$PV = nRT \quad \frac{PV}{RT} = \frac{nRT}{RT} \quad \frac{PV}{RT} = n$$

$$\textit{EQUATION: } n = \frac{PV}{RT} = 3.62 \times 10^3 \text{ torr} \times \frac{\text{mol} \cdot \text{K}}{62.4 \text{ L} \cdot \text{torr}} \times \frac{6.64 \text{ L}}{298 \text{ K}} = 1.29 \text{ mol}$$

See Section 13.4, The Ideal Gas Equation: Determination of a Single Variable, on Pages 352-354.

7. *GIVEN:* 1.10 mol; 553 torr; 29°C (302 K) *WANTED:* Volume (assume L)

$$PV = nRT \quad \frac{PV}{P} = \frac{nRT}{P} \quad V = \frac{nRT}{P}$$

$$\textit{EQUATION: } V = \frac{nRT}{P} = 1.10 \text{ mol} \times \frac{62.4 \text{ L} \cdot \text{torr}}{\text{mol} \cdot \text{K}} \times \frac{302 \text{ K}}{553 \text{ torr}} = 37.5 \text{ L}$$

See Section 13.4, The Ideal Gas Equation: Determination of a Single Variable, on Pages 352-354.

8. (a) *GIVEN:* NO, 14.01 + 16.00 = 30.01 g/mol; 273 K; 1 atm *WANTED:* D = m/V (assume g/L)

$$PV = nRT \quad PV = \frac{m}{MM}RT \quad PV(MM) = mRT \quad \frac{PV(MM)}{V} = \frac{mRT}{V} \quad P(MM) = \frac{m}{V}RT$$

$$\frac{P(MM)}{RT} = \frac{m}{V}\frac{RT}{RT} \quad \frac{P(MM)}{RT} = \frac{m}{V}$$

$$\textit{EQUATION: } \frac{m}{V} = \frac{(MM)P}{RT} = \frac{30.01 \text{ g}}{\text{mol}} \times \frac{\text{mol} \cdot \text{K}}{0.0821 \text{ L} \cdot \text{atm}} \times \frac{1 \text{ atm}}{273 \text{ K}} = 1.34 \text{ g/L}$$

(b) *GIVEN:* NO, 14.01 + 16.00 = 30.01 g/mol; –6°C (273 K); 719 torr
WANTED: D = m/V (assume g/L)

$$\frac{m}{V} = \frac{(MM)P}{RT} = \frac{30.01 \text{ g}}{\text{mol}} \times \frac{\text{mol} \cdot \text{K}}{62.4 \text{ L} \cdot \text{torr}} \times \frac{719 \text{ torr}}{267 \text{ K}} = 1.30 \text{ g/L}$$

See Section 13.5, Gas Density and Molar Volume, on Pages 354-357.

9. 136.201 g – 135.831 g = 0.370 g gas; 385.42 g – 135.831 g = 249.59 g H_2O

$$249.59 \text{ g} \times \frac{1 \text{ mL}}{1.00 \text{ g}} = 2.50 \times 10^2 \text{ mL} = 0.250 \text{ L}$$

GIVEN: 0.370 g; 0.250 L; 273 K; 1 atm *WANTED:* MM (g/mol)

$$PV = nRT \quad PV = \frac{m}{MM}RT \quad PV(MM) = mRT \quad \frac{PV(MM)}{PV} = \frac{mRT}{PV} \quad MM = \frac{mRT}{PV}$$

$$\textit{EQUATION: } MM = \frac{mRT}{PV} = \frac{0.370 \text{ g}}{0.250 \text{ L}} \times \frac{0.0821 \text{ L} \cdot \text{atm}}{\text{mol} \cdot \text{K}} \times \frac{273 \text{ K}}{1 \text{ atm}} = 33.2 \text{ g/mol}$$

See Section 13.5, Gas Density and Molar Volume, on Pages 354-357.

10. *GIVEN:* 1.61 g/L; 273 K; 1 atm *WANTED:* MM (g/mol)

$$PV = nRT \quad PV = \frac{m}{MM}RT \quad PV(MM) = mRT \quad \frac{PV(MM)}{PV} = \frac{mRT}{PV} \quad MM = \frac{mRT}{PV}$$

$$\textit{EQUATION: } MM = \frac{mRT}{PV} = \frac{1.61 \text{ g}}{\text{L}} \times \frac{0.0821 \text{ L} \cdot \text{atm}}{\text{mol} \cdot \text{K}} \times \frac{273 \text{ K}}{1 \text{ atm}} = 36.1 \text{ g/mol}$$

See Section 13.5, Gas Density and Molar Volume, on Pages 354-357.

11. (a) *GIVEN:* 0.686 g; 549 mL = 0.549 L *WANTED:* D (assume g/L)

EQUATION: $D \equiv \frac{m}{V} = \frac{0.686 \text{ g}}{0.549 \text{ L}} = 1.25 \text{ g/L}$

(b) *GIVEN:* 0.686 g; 549 mL = 0.549 L; 34°C (307 K); 0.665 atm *WANTED:* MM (g/mol)

$PV = nRT \quad PV = \frac{m}{MM}RT \quad PV(MM) = mRT \quad \frac{PV(MM)}{PV} = \frac{mRT}{PV} \quad MM = \frac{mRT}{PV}$

EQUATION: $MM = \frac{mRT}{PV} = \frac{0.686 \text{ g}}{0.549 \text{ L}} \times \frac{0.0821 \text{ L} \cdot \text{atm}}{\text{mol} \cdot \text{K}} \times \frac{307 \text{ K}}{0.665 \text{ atm}} = 47.4 \text{ g/mol}$

See Section 13.5, Gas Density and Molar Volume, on Pages 354-357.

12. *GIVEN:* 98°C (371 K); 1.08 atm; 2.84 g/L *WANTED:* MM (g/mol)

$PV = nRT \quad PV = \frac{m}{MM}RT \quad PV(MM) = mRT \quad \frac{PV(MM)}{PV} = \frac{mRT}{PV} \quad MM = \frac{mRT}{PV}$

$MM = \frac{mRT}{PV} = \frac{2.84 \text{ g}}{\text{L}} \times \frac{0.0821 \text{ L} \cdot \text{atm}}{\text{mol} \cdot \text{K}} \times \frac{371 \text{ K}}{1.08 \text{ atm}} = 80.1 \text{ g/mol}$

The gas must be SO_3, which has a molar mass of 80.07 g/mol.

SO_2: 32.07 + 2(16.00) = 64.07 g/mol SO_3: 32.07 + 3(16.00) = 80.07 g/mol
See Section 13.5, Gas Density and Molar Volume, on Pages 354-357.

13. *GIVEN:* 795 torr; 19°C (292 K) *WANTED:* MV (L/mol)

$PV = nRT \quad \frac{PV}{n} = \frac{nRT}{n} \quad \frac{PV}{n} = RT \quad \frac{PV}{Pn} = \frac{RT}{P} \quad \frac{V}{n} = \frac{RT}{P}$

EQUATION: $MV \equiv \frac{V}{n} = \frac{RT}{P} = \frac{62.4 \text{ L} \cdot \text{torr}}{\text{mol} \cdot \text{K}} \times \frac{292 \text{ K}}{795 \text{ torr}} = 22.9 \text{ L/mol}$

See Section 13.5, Gas Density and Molar Volume, on Pages 354-357.

14. *GIVEN:* 6.32 atm; –2°C (271 K) *WANTED:* MV (L/mol)

$PV = nRT \quad \frac{PV}{n} = \frac{nRT}{n} \quad \frac{PV}{n} = RT \quad \frac{PV}{Pn} = \frac{RT}{P} \quad \frac{V}{n} = \frac{RT}{P}$

EQUATION: $MV \equiv \frac{V}{n} = \frac{RT}{P} = \frac{0.0821 \text{ L} \cdot \text{atm}}{\text{mol} \cdot \text{K}} \times \frac{271 \text{ K}}{6.32 \text{ atm}} = 3.52 \text{ L/mol}$

See Section 13.5, Gas Density and Molar Volume, on Pages 354-357.

15. *GIVEN:* 71.8 g HgO *WANTED:* STP Volume of O_2 (assume L)

PER/PATH: g HgO $\xrightarrow{216.6 \text{ g HgO/mol HgO}}$ mol HgO $\xrightarrow{1 \text{ mol } O_2/2 \text{ mol HgO}}$ mol O_2 $\xrightarrow{22.4 \text{ L } O_2/\text{mol } O_2}$ L O_2

$71.8 \text{ g HgO} \times \frac{1 \text{ mol HgO}}{216.6 \text{ g HgO}} \times \frac{1 \text{ mol } O_2}{2 \text{ mol HgO}} \times \frac{22.4 \text{ L } O_2}{\text{mol } O_2} = 3.71 \text{ L } O_2$

See Section 13.6, Gas Stoichiometry at Standard Temperature and Pressure, on Pages 358-359.

16. *GIVEN:* 0.652 L CO_2 *WANTED:* g HCl

PER/PATH: $\text{L } CO_2 \xrightarrow{22.4 \text{ L } CO_2/\text{mol } CO_2} \text{mol } CO_2 \xrightarrow{2 \text{ mol HCl}/1 \text{ mol } CO_2} \text{mol HCl} \xrightarrow{36.46 \text{ g HCl/mol HCl}} \text{g HCl}$

$$0.652 \text{ L } CO_2 \times \frac{1 \text{ mol } CO_2}{22.4 \text{ L } CO_2} \times \frac{2 \text{ mol HCl}}{1 \text{ mol } CO_2} \times \frac{36.46 \text{ g HCl}}{\text{mol HCl}} = 2.12 \text{ g HCl}$$

See Section 13.6, Gas Stoichiometry at Standard Temperature and Pressure, on Pages 358–359.

Questions 17–20 may be solved by the molar volume method (Section 13.7) or by the ideal gas equation method (Section 13.8). In the answer section, the setups are given first for the molar volume method. Then the answers are given for the ideal gas equation method. Check your work according to the section you studied.

MOLAR VOLUME METHOD

17. *GIVEN:* 1.04 atm; 343°C *WANTED:* MV (L/mol)

$$PV = nRT \quad \frac{PV}{n} = \frac{nRT}{n} \quad \frac{PV}{n} = RT \quad \frac{PV}{Pn} = \frac{RT}{P} \quad \frac{V}{n} = \frac{RT}{P}$$

EQUATION: $MV = \frac{RT}{P} = \frac{0.0821 \text{ L} \cdot \text{atm}}{\text{mol} \cdot \text{K}} \times \frac{616 \text{ K}}{1.04 \text{ atm}} = 48.6 \text{ L/mol}$

GIVEN: 4.83 g $NaHCO_3$ *WANTED:* Volume CO_2 (assume L)

PER/PATH: $\text{g } NaHCO_3 \xrightarrow{84.01 \text{ g } NaHCO_3/\text{mol } NaHCO_3} \text{mol } NaHCO_3 \xrightarrow{1 \text{ mol } CO_2/1 \text{ mol } NaHCO_3} \text{mol } CO_2 \xrightarrow{48.6 \text{ L } CO_2/\text{mol } CO_2} \text{L } CO_2$

$$4.83 \text{ g } NaHCO_3 \times \frac{1 \text{ mol } NaHCO_3}{84.01 \text{ g } NaHCO_3} \times \frac{1 \text{ mol } CO_2}{1 \text{ mol } NaHCO_3} \times \frac{48.6 \text{ L } CO_2}{\text{mol } CO_2} = 2.79 \text{ L } CO_2$$

See Section 13.7, Gas Stoichiometry: Molar Volume Method, on Pages 360–361.

18. $CH_4 + 2\,O_2 \rightarrow CO_2 + 2\,H_2O$

GIVEN: 0.813 atm; 26°C (299 K) *WANTED:* MV (L/mol)

$$PV = nRT \quad \frac{PV}{n} = \frac{nRT}{n} \quad \frac{PV}{n} = RT \quad \frac{PV}{Pn} = \frac{RT}{P} \quad \frac{V}{n} = \frac{RT}{P}$$

$$MV = \frac{RT}{P} = \frac{0.0821 \text{ L} \cdot \text{atm}}{\text{mol} \cdot \text{K}} \times \frac{299 \text{ K}}{0.813 \text{ atm}} = 30.2 \text{ L/mol}$$

GIVEN: 19.2 L CH_4 *WANTED:* g CO_2

PER/PATH: $\text{L } CH_4 \xrightarrow{30.2 \text{ L } CH_4/\text{mol } CH_4} \text{mol } CH_4 \xrightarrow{1 \text{ mol } CO_2/1 \text{ mol } CH_4} \text{mol } CO_2 \xrightarrow{44.01 \text{ g } CO_2/\text{mol } CO_2} \text{g } CO_2$

$$19.2 \text{ L } CH_4 \times \frac{1 \text{ mol } CH_4}{30.2 \text{ L } CH_4} \times \frac{1 \text{ mol } CO_2}{1 \text{ mol } CH_4} \times \frac{44.01 \text{ g } CO_2}{\text{mol } CO_2} = 28.0 \text{ g } CO_2$$

See Section 13.7, Gas Stoichiometry: Molar Volume Method, on Pages 360–361.

19. *GIVEN:* 264°C (537 K); 1.09 atm *WANTED:* MV (L/mol)

$$PV = nRT \quad \frac{PV}{n} = \frac{nRT}{n} \quad \frac{PV}{n} = RT \quad \frac{PV}{Pn} = \frac{RT}{P} \quad \frac{V}{n} = \frac{RT}{P}$$

$$MV = \frac{RT}{P} = \frac{0.0821 \text{ L} \cdot \text{atm}}{\text{mol} \cdot \text{K}} \times \frac{537 \text{ K}}{1.09 \text{ atm}} = 40.4 \text{ L/mol}$$

GIVEN: 1 kg $CaCO_3 \cdot MgCO_3$ *WANTED:* L CO_2

PER/PATH: kg $CaCO_3 \cdot MgCO_3$ $\xrightarrow{184.41\ kg\ CaCO_3 \cdot MgCO_3/kmol\ CaCO_3 \cdot MgCO_3}$ kmol $CaCO_3 \cdot MgCO_3$ $\xrightarrow{2\ kmol\ CO_2/1\ kmol\ CaCO_3 \cdot MgCO_3}$ kmol CO_2 $\xrightarrow{40.4\ kL\ CO_2/kmol\ CO_2}$ kL CO_2

$$1\ kg\ CaCO_3 \cdot MgCO_3 \times \frac{1\ kmol\ CaCO_3 \cdot MgCO_3}{184.41\ kg\ CaCO_3 \cdot MgCO_3} \times \frac{2\ kmol\ CO_2}{1\ kmol\ CaCO_3 \cdot MgCO_3} \times \frac{40.4\ kL\ CO_2}{1\ kmol\ CO_2} = 0.438\ kL = 438\ L\ CO_2$$

See Section 13.7, Gas Stoichiometry: Molar Volume Method, on Pages 360-361.

20. a) *GIVEN:* 691 torr; 19°C (292 K) *WANTED:* MV (L/mol)

$$PV = nRT \quad \frac{PV}{n} = \frac{nRT}{n} \quad \frac{PV}{n} = RT \quad \frac{PV}{Pn} = \frac{RT}{P} \quad \frac{V}{n} = \frac{RT}{P}$$

$$MV = \frac{RT}{P} = \frac{62.4\ L \cdot torr}{mol \cdot K} \times \frac{292\ K}{691\ torr} = 26.4\ L/mol$$

GIVEN: 1.94 L NO_2 *WANTED:* g Sn

PER/PATH: L NO_2 $\xrightarrow{26.4\ L\ NO_2/mol\ NO_2}$ mol NO_2 $\xrightarrow{1\ mol\ Sn/4\ mol\ NO_2}$ mol Sn $\xrightarrow{118.7\ g\ Sn/mol\ Sn}$ g Sn

$$1.94\ L\ NO_2 \times \frac{1\ mol\ NO_2}{26.4\ L\ NO_2} \times \frac{1\ mol\ Sn}{4\ mol\ NO_2} \times \frac{118.7\ g\ Sn}{mol\ Sn} = 2.18\ g\ Sn$$

b) *GIVEN:* 2.18 g Sn; 4.77 g solder *WANTED:* % Sn

EQUATION: $\% \ Sn = \frac{g\ Sn}{g\ solder} \times 100 = \frac{2.18\ g}{4.77\ g} \times 100 = 45.7\%\ Sn$

a) See Section 13.7, Gas Stoichiometry: Molar Volume Method, on Pages 360-361.
b) The definition of percent is discussed on Page 180. See Equation 7.2.

IDEAL GAS EQUATION METHOD

17. *GIVEN:* 4.83 g $NaHCO_3$ *WANTED:* mol CO_2

PER/PATH: g $NaHCO_3$ $\xrightarrow{84.01\ g\ NaHCO_3/mol\ NaHCO_3}$ mol $NaHCO_3$ $\xrightarrow{1\ mol\ CO_2/1\ mol\ NaHCO_3}$ mol CO_2

$$4.83\ g\ NaHCO_3 \times \frac{1\ mol\ NaHCO_3}{84.01\ g\ NaHCO_3} \times \frac{1\ mol\ CO_2}{1\ mol\ NaHCO_3} = 0.0575\ mol\ CO_2$$

GIVEN: 0.0575 mol CO_2; 1.04 atm; 343°C (616 K) *WANTED:* Volume CO_2 (assume L)

$$PV = nRT \quad \frac{PV}{P} = \frac{nRT}{P} \quad V = \frac{nRT}{P}$$

EQUATION: $V = \frac{nRT}{P} = 0.0575\ mol\ CO_2 \times \frac{0.0821\ L \cdot atm}{mol \cdot K} \times \frac{616\ K}{1.04\ atm} = 2.80\ L\ CO_2$

See Section 13.8, Gas Stoichiometry: Ideal Gas Equation Method, on Pages 362-364.

18. $CH_4 + 2\,O_2 \rightarrow CO_2 + 2\,H_2O$

GIVEN: 19.2 L; 0.813 atm; 26°C (299 K) *WANTED:* mol CH_4

$$PV = nRT \qquad \frac{PV}{RT} = \frac{nRT}{RT} \qquad \frac{PV}{RT} = n$$

EQUATION: $n = \frac{PV}{RT} = 0.813 \text{ atm} \times \frac{19.2 \text{ L}}{299 \text{ K}} \times \frac{\text{mol} \cdot \text{K}}{0.0821 \text{ L} \cdot \text{atm}} = 0.636 \text{ mol } CH_4$

GIVEN: 0.636 mol CH_4 *WANTED:* g CO_2

PER/PATH: mol CH_4 $\xrightarrow{1 \text{ mol } CO_2/1 \text{ mol } CH_4}$ mol CO_2 $\xrightarrow{44.01 \text{ g } CO_2/\text{mol } CO_2}$ g CO_2

$$0.636 \text{ mol } CH_4 \times \frac{1 \text{ mol } CO_2}{1 \text{ mol } CH_4} \times \frac{44.01 \text{ g } CO_2}{\text{mol } CO_2} = 28.0 \text{ g } CO_2$$

See Section 13.8, Gas Stoichiometry: Ideal Gas Equation Method, on Pages 362-364.

19. *GIVEN:* 1 kg $CaCO_3 \cdot MgCO_3$ *WANTED:* mol CO_2

PER/PATH: kg $CaCO_3 \cdot MgCO_3$ $\xrightarrow{184.41 \text{ kg } CaCO_3 \cdot MgCO_3/\text{mol } CaCO_3 \cdot MgCO_3}$ kmol $CaCO_3 \cdot MgCO_3$ $\xrightarrow{2 \text{ kmol } CO_2/1 \text{ kmol } CaCO_3 \cdot MgCO_3}$ kmol CO_2

$$1 \text{ kg } CaCO_3 \cdot MgCO_3 \times \frac{1 \text{ kmol } CaCO_3 \cdot MgCO_3}{184.41 \text{ kg } CaCO_3 \cdot MgCO_3} \times \frac{2 \text{ kmol } CO_2}{1 \text{ kmol } CaCO_3 \cdot MgCO_3} = 0.0108 \text{ kmol } CO_2 = 10.8 \text{ mol } CO_2$$

GIVEN: 10.8 mol; 264°C (537 K); 1.09 atm *WANTED:* L CO_2

$$PV = nRT \qquad \frac{PV}{P} = \frac{nRT}{P} \qquad V = \frac{nRT}{P}$$

EQUATION: $V = \frac{nRT}{P} = 10.8 \text{ mol} \times \frac{0.0821 \text{ L} \cdot \text{atm}}{\text{mol} \cdot \text{K}} \times \frac{537 \text{ K}}{1.09 \text{ atm}} = 437 \text{ L } CO_2$

See Section 13.8, Gas Stoichiometry: Ideal Gas Equation Method, on Pages 362-364.

20. a) *GIVEN:* 1.94 L; 691 torr; 19°C (292 K) *WANTED:* mol NO_2

$$PV = nRT \qquad \frac{PV}{RT} = \frac{nRT}{RT} \qquad \frac{PV}{RT} = n$$

EQUATION: $n = \frac{PV}{RT} = 691 \text{ torr} \times \frac{1.94 \text{ L}}{292 \text{ K}} \times \frac{\text{mol} \cdot \text{K}}{62.4 \text{ L} \cdot \text{torr}} = 0.0736 \text{ mol } NO_2$

GIVEN: 0.0736 mol NO_2 *WANTED:* g Sn

PER/PATH: mol NO_2 $\xrightarrow{1 \text{ mol Sn}/4 \text{ mol } NO_2}$ mol Sn $\xrightarrow{118.7 \text{ g Sn/mol Sn}}$ g Sn

$$0.0736 \text{ mol } NO_2 \times \frac{1 \text{ mol Sn}}{4 \text{ mol } NO_2} \times \frac{118.7 \text{ g Sn}}{\text{mol Sn}} = 2.18 \text{ g Sn}$$

b) *GIVEN:* 2.18 g Sn; 4.77 g solder *WANTED:* % Sn

EQUATION: $\% \text{ Sn} = \frac{\text{g Sn}}{\text{g solder}} \times 100 = \frac{2.18 \text{ g}}{4.77 \text{ g}} \times 100 = 45.7\% \text{ Sn}$

See Section 13.8, Gas Stoichiometry: Ideal Gas Equation Method, on Pages 362-364.
b) The definition of percent is discussed on Page 180. See Equation 7.2.

21. a) *GIVEN:* 207 L O_2 *WANTED:* L NO_2

PER/PATH: L O_2 $\xrightarrow{2 \text{ L } NO_2/1 \text{ L } O_2}$ L NO_2

$$207 \text{ L } O_2 \times \frac{2 \text{ L } NO_2}{1 \text{ L } O_2} = 414 \text{ L } NO_2$$

b)

	Volume	Temperature	Pressure
Initial Value (1)	207 L	18 + 273 = 291 K	0.877 atm
Final Value (2)	V_2	84 + 273 = 357 K	1.16 atm

$$V_2 = V_1 \times \frac{T_2}{T_1} \times \frac{P_1}{P_2} = 207\ L\ O_2 \times \frac{357\ K}{291\ K} \times \frac{0.877\ atm}{1.16\ atm} \times \frac{2\ L\ NO_2}{1\ L\ O_2} = 384\ L\ NO_2$$

See Section 13.9, Volume-Volume Stoichiometry, on Pages 364-367.

22.

	Volume	Temperature	Pressure
Initial Value (1)	4.29×10^3 ft^3	392 + 273 = 665 K	1.31 atm
Final Value (2)	V_2	36 + 273 = 309 K	0.904 atm

$$V_2 = V_1 \times \frac{T_2}{T_1} \times \frac{P_1}{P_2} = 4.29 \times 10^3\ ft^3\ SO_2 \times \frac{309\ K}{665\ K} \times \frac{1.31\ atm}{0.904\ atm} = 2.89 \times 10^3\ ft^3\ SO_2$$

GIVEN: 2.89×10^3 ft^3 SO_2 *WANTED:* ft^3 O_2

PER/PATH: ft^3 SO_2 $\xrightarrow{1\ ft^3\ O_2/2\ ft^3\ SO_2}$ ft^3 O_2

$$2.89 \times 10^3\ ft^3\ SO_2 \times \frac{1\ ft^3\ O_2}{2\ ft^3\ SO_2} = 1.45 \times 10^3\ ft^3\ O_2$$

If both steps are combined into a single setup:

$$4.29 \times 10^3\ ft^3\ SO_2 \times \frac{309\ K}{665\ K} \times \frac{1.31\ atm}{0.904\ atm} \times \frac{1\ ft^3\ O_2}{2\ ft^3\ SO_2} = 1.44 \times 10^3\ ft^3\ O_2$$

See Section 13.9, Volume-Volume Stoichiometry, on Pages 364-367.

23.

	Volume	Temperature	Pressure
Initial Value (1)	895 L	1700 + 273 = 1973 K	846 torr
Final Value (2)	V_2	29 + 273 = 302 K	182 torr

$$V_2 = V_1 \times \frac{T_2}{T_1} \times \frac{P_1}{P_2} = 895\ L\ CO \times \frac{302\ K}{1973\ K} \times \frac{846\ torr}{182\ torr} = 637\ L\ CO$$

GIVEN: 637 L CO *WANTED:* L O_2

PER/PATH: L CO $\xrightarrow{1\ L\ O_2/2\ L\ CO}$ L O_2

$$637\ L\ CO \times \frac{1\ L\ O_2}{2\ L\ CO} = 3.2 \times 10^2\ L\ O_2$$

If both steps are combined into a single setup:

$$895\ L\ CO \times \frac{846\ torr}{182\ torr} \times \frac{302\ K}{1973\ K} \times \frac{1\ L\ O_2}{2\ L\ CO} = 3.2 \times 10^2\ L\ O_2$$

(1700°C is assumed to be a two-significant-figure number.)

See Section 13.9, Volume-Volume Stoichiometry, on Pages 364-367.

25. True: a, c. False: b, d.

b) The number of particles in 5.00 L of NH_3 is the same as the number of particles in 5.00 L of CO if both volumes are measured at the same temperature and pressure.

d) To change liters of a gas to moles, PV = nRT, so $n = V \times \frac{P}{RT}$, and thus you multiply by P/RT.

26. *GIVEN:* 6.74 g C_2H_4; 41°C; 733 torr *WANTED:* V (assume L)

$$PV = nRT \quad PV = \frac{m}{MM}RT \quad PV(MM) = mRT \quad \frac{PV(MM)}{P(MM)} = \frac{mRT}{P(MM)} \quad V = \frac{mRT}{P(MM)}$$

EQUATION: $V = \frac{mRT}{(MM)P} = 6.74 \text{ g} \times \frac{62.4 \text{ L} \cdot \text{torr}}{\text{mol} \cdot \text{K}} \times \frac{314 \text{ K}}{733 \text{ torr}} \times \frac{1 \text{ mol}}{28.05 \text{ g}} = 6.42 \text{ L}$

C_2H_4: 2(12.01) + 4(1.008) = 28.05 g/mol

See Section 13.3, The Ideal Gas Law, on Pages 351-352. In particular, note Equation 13.6 on Page 352. Also see Section 13.4, The Ideal Gas Equation: Determination of a Single Variable, on Pages 352-354.

27. *GIVEN:* 0.972 atm; 14°C (287 K) *WANTED:* D = m/V (assume g/L)

$$PV = nRT \quad PV = \frac{m}{MM}RT \quad PV(MM) = mRT \quad \frac{PV(MM)}{V} = \frac{mRT}{V} \quad P(MM) = \frac{m}{V}RT$$

$$\frac{P(MM)}{RT} = \frac{m}{V}\frac{RT}{RT} \quad \frac{P(MM)}{RT} = \frac{m}{V}$$

EQUATION: $\frac{m}{V} = \frac{(MM)P}{RT} = \frac{34.09 \text{ g}}{\text{mol}} \times \frac{0.972 \text{ atm}}{287 \text{ K}} \times \frac{\text{mol} \cdot \text{K}}{0.0821 \text{ L} \cdot \text{atm}} = 1.41 \text{ g/L}$

H_2S: 2(1.008) + 32.07 = 34.09 g/mol

See Section 13.5, Gas Density and Molar Volume, on Pages 354-357.

28. In algebra, if $x \propto y$, then $x = ky$, where k is a proportionality constant. At a given temperature and pressure, solving Equation 13.6 for density, m/V, yields

$$D = \frac{m}{V} = \frac{P(MM)}{RT} = \frac{P}{RT} \times MM = k \times MM \quad D = k \times MM$$

In this equation, P/RT has the role of a proportionality constant. Hence, molar mass and density are directly proportional to each other at a given temperature and pressure.

See Section 13.5, Gas Density and Molar Volume, on Pages 354-357.

29. Butane, C_4H_{10}, has a higher molar mass, 58.12 g/mol, than propane, C_3H_8, 44.09 g/mol. At a given temperature and pressure, density is proportional to molar mass. Therefore B, the higher-density gas, must be butane, Its density is $1.37 \text{ g/L} \times \frac{58.12 \text{ g/mol}}{44.09 \text{ g/mol}} = 1.81 \text{ g/L}$.

See Section 13.5, Gas Density and Molar Volume, on Pages 354-357.

Chapter 14

Combined Gas Law Applications

1. Molar volume is the volume of one mole of gas at a specified temperature and pressure. Its units are liters per mole, L/mol.

   See Section 14.2, Molar Volume, on Pages 375-377. Molar volume is defined in Equation 14.1 on Page 375.

2. *GIVEN:* 27.2 L O_2 *WANTED:* mol O_2

   PER/PATH: $L\ O_2 \xrightarrow{22.4\ L\ O_2/mol\ O_2} mol\ O_2$

   $$27.2\ L \times \frac{1\ mol}{22.4\ L} = 1.21\ mol$$

   See Section 14.2, Molar Volume, on Pages 375-377.

3. *GIVEN:* 795 torr; 19°C (292 K) *WANTED:* MV (L/mol)

	Molar Volume	Temperature	Pressure
Initial Value (1)	22.4 L/mol	0°C (273 K)	760 torr
Final Value (2)	V_2	19 + 273 = 292 K	795 torr

$$MV_2 = MV_1 \times \frac{P_1}{P_2} \times \frac{T_2}{T_1} = \frac{22.4\ L}{mol} \times \frac{760\ torr}{795\ torr} \times \frac{292\ K}{273\ K} = 22.9\ L/mol$$

See Section 14.2, Molar Volume, on Pages 375-377.

4. *GIVEN:* 6.32 atm; –2°C (271 K) *WANTED:* MV (L/mol)

	Molar Volume	Temperature	Pressure
Initial Value (1)	22.4 L/mol	0°C (273 K)	1 atm
Final Value (2)	V_2	–2 + 273 = 271 K	6.32 atm

$$MV_2 = MV_1 \times \frac{P_1}{P_2} \times \frac{T_2}{T_1} = \frac{22.4\ L}{mol} \times \frac{1\ atm}{6.32\ atm} \times \frac{271\ K}{273\ K} = 3.52\ L/mol$$

See Section 14.2, Molar Volume, on Pages 375-377.

5. *GIVEN:* 4.47 L Cl_2; 3.36 g Cl_2/L Cl_2 *WANTED:* mass Cl_2 (assume g)

   PER/PATH: $L\ Cl_2 \xrightarrow{3.36\ g\ Cl_2/L\ Cl_2} g\ Cl_2$

   $$4.47\ L\ Cl_2 \times \frac{3.36\ g}{L} = 15.0\ g\ Cl_2$$

   See Section 14.3, Three Ratios: Gas Density, Molar Volume, and Molar Mass, on Pages 377-379. In particular, see the Gas Density subsection on Page 378.

6. *GIVEN:* 1.10 g/L; 807 g *WANTED:* volume (assume L)

PER/PATH: g $\xrightarrow{1.10\ \text{g/L}}$ L

$$807\ \text{g} \times \frac{1\ \text{L}}{1.10\ \text{g}} = 734\ \text{L}$$

See Section 14.3, Three Ratios: Gas Density, Molar Volume, and Molar Mass, on Pages 377-379. In particular, see the Gas Density subsection on Page 378.

7. (a) *GIVEN:* 30.01 g NO/mol NO; 22.4 L NO/mol NO *WANTED:* D (assume g/L)

$$\textit{EQUATION: } D \equiv \frac{m}{V} = \frac{30.01\ \text{g}}{\text{mol}} \times \frac{1\ \text{mol}}{22.4\ \text{L}} = 1.34\ \text{g/L}$$

(b) *GIVEN:* 719 torr; –6°C (267 K) *WANTED:* MV (L/mol)

	Molar Volume	Temperature	Pressure
Initial Value (1)	22.4 L/mol	0°C (273 K)	760 torr
Final Value (2)	V_2	–6 + 273 = 267 K	719 torr

$$MV_2 = MV_1 \times \frac{P_1}{P_2} \times \frac{T_2}{T_1} = \frac{22.4\ \text{L}}{\text{mol}} \times \frac{760\ \text{torr}}{719\ \text{torr}} \times \frac{267\ \text{K}}{273\ \text{K}} = 23.2\ \text{L/mol}$$

GIVEN: 30.01 g/mol; 23.3 L/mol *WANTED:* D (assume g/L)

$$\textit{EQUATION: } D \equiv \frac{m}{V} = \frac{30.01\ \text{g}}{\text{mol}} \times \frac{1\ \text{mol}}{23.2\ \text{L}} = 1.30\ \text{g/L}$$

See Section 14.4, Three-Ratio Problems, on Pages 380-383. In particular, see the Three-Ratio Problems at Standard Temperature and Pressure (STP) subsection on Pages 380-381 for part (a), and see the Three-Ratio Problems at a Given Temperature and Pressure subsection on Pages 381-383 for part (b).

8. *GIVEN:* 1.42 g/L; 30.07 g C_2H_6/mol C_2H_6 *WANTED:* MV (L/mol)

$$\textit{EQUATION: } MV \equiv \frac{V}{n} = \frac{1\ \text{L}}{1.42\ \text{g}} \times \frac{30.07\ \text{g}\ C_2H_6}{\text{mol}\ C_2H_6} = 21.2\ \text{L/mol}$$

See Section 14.4, Three-Ratio Problems, on Pages 380-383.

9. *GIVEN:* 1.61 g/L; 22.4 L/mol *WANTED:* MM (g/mol)

$$\textit{EQUATION: } MM \equiv \frac{\text{g}}{\text{mol}} = \frac{1.61\ \text{g}}{\text{L}} \times \frac{22.4\ \text{L}}{\text{mol}} = 36.1\ \text{g/mol}$$

See Section 14.4, Three-Ratio Problems, on Pages 380-383.

10. *GIVEN:* 27°C (300 K); 797 torr *WANTED:* MV (L/mol)

	Molar Volume	Temperature	Pressure
Initial Value (1)	22.4 L/mol	0°C (273 K)	760 torr
Final Value (2)	V_2	27 + 273 = 300 K	797 torr

$$MV_2 = MV_1 \times \frac{P_1}{P_2} \times \frac{T_2}{T_1} = \frac{22.4\ \text{L}}{\text{mol}} \times \frac{760\ \text{torr}}{797\ \text{torr}} \times \frac{300\ \text{K}}{273\ \text{K}} = 23.5\ \text{L/mol}$$

GIVEN: 17.03 g NH_3/mol NH_3; 23.5 L/mol *WANTED:* D (assume g/L)

$$\textit{EQUATION: } D \equiv \frac{m}{V} = \frac{17.03\ \text{g}\ NH_3}{\text{mol}\ NH_3} \times \frac{1\ \text{mol}}{23.5\ \text{L}} = 0.725\ \text{g/L}$$

See Section 14.4, Three-Ratio Problems, on Pages 380-383. In particular, see the Three-Ratio Problems at a Given Temperature and Pressure subsection on Pages 381-383.

11. (a) *GIVEN:* 0.686 g; 549 mL = 0.549 L *WANTED:* D (assume g/L)

EQUATION: $D \equiv \frac{m}{V} = \frac{0.686\ g}{0.549\ L} = 1.25\ g/L$

(b) *GIVEN:* 34°C (307 K); 0.655 atm *WANTED:* MV (L/mol)

	Molar Volume	Temperature	Pressure
Initial Value (1)	22.4 L/mol	0°C (273 K)	1 atm
Final Value (2)	V_2	34 + 273 = 307 K	0.655 atm

$$MV_2 = MV_1 \times \frac{P_1}{P_2} \times \frac{T_2}{T_1} = \frac{22.4\ L}{mol} \times \frac{1\ atm}{0.655\ atm} \times \frac{307\ K}{273\ K} = 38.5\ L/mol$$

GIVEN: 0.686 g; 549 mL = 0.549 L; 38.5 L/mol *WANTED:* MM (g/mol)

EQUATION: $MM \equiv \frac{g}{mol} = \frac{0.686\ g}{0.549\ L} \times \frac{38.5\ L}{mol} = 48.1\ g/mol$

See Section 14.4, Three-Ratio Problems, on Pages 380–383. In particular, see the Three-Ratio Problems at a Given Temperature and Pressure subsection on Pages 381–383.

12. *GIVEN:* 32°C (305 K); 0.402 atm *WANTED:* MV (L/mol)

	Molar Volume	Temperature	Pressure
Initial Value (1)	22.4 L/mol	0°C (273 K)	1 atm
Final Value (2)	V_2	32 + 273 = 305 K	0.402 atm

$$MV_2 = MV_1 \times \frac{P_1}{P_2} \times \frac{T_2}{T_1} = \frac{22.4\ L}{mol} \times \frac{1\ atm}{0.402\ atm} \times \frac{305\ K}{273\ K} = 62.3\ L/mol$$

GIVEN: 46.01 g NO_2/mol NO_2; 62.3 L/mol *WANTED:* D (assume g/L)

EQUATION: $D \equiv \frac{m}{V} = \frac{46.01\ g\ NO_2}{mol\ NO_2} \times \frac{1\ mol}{62.3\ L} = 0.739\ g/L$

See Section 14.4, Three-Ratio Problems, on Pages 380–383. In particular, see the Three-Ratio Problems at a Given Temperature and Pressure subsection on Pages 381–383.

13. *GIVEN:* 98°C (371 K); 1.08 atm *WANTED:* MV (L/mol)

	Molar Volume	Temperature	Pressure
Initial Value (1)	22.4 L/mol	0°C (273 K)	1 atm
Final Value (2)	V_2	98 + 273 = 371 K	1.08 atm

$$MV_2 = MV_1 \times \frac{P_1}{P_2} \times \frac{T_2}{T_1} = \frac{22.4\ L}{mol} \times \frac{1\ atm}{1.08\ atm} \times \frac{371\ K}{273\ K} = 28.2\ L/mol$$

GIVEN: 2.84 g/L; 28.2 L/mol *WANTED:* MM (g/mol)

EQUATION: $MM \equiv \frac{g}{mol} = \frac{2.84\ g}{L} \times \frac{28.2\ L}{mol} = 80.1\ g/mol$

The gas must be SO_3, which has a molar mass of 80.07 g/mol.

SO_2: 32.07 + 2(16.00) = 64.07 g/mol SO_3: 32.07 + 3(16.00) = 80.07 g/mol

See Section 14.4, Three-Ratio Problems, on Pages 380–383. In particular, see the Three-Ratio Problems at a Given Temperature and Pressure subsection on Pages 381–383.

14. *GIVEN:* 71.8 g HgO *WANTED:* STP Volume of O_2 (assume L)

PER/PATH: g HgO $\xrightarrow{216.6\text{ g HgO/mol HgO}}$ mol HgO $\xrightarrow{1\text{ mol }O_2/2\text{ mol HgO}}$ mol O_2 $\xrightarrow{22.4\text{ L }O_2/\text{mol }O_2}$ L O_2

$$71.8\text{ g HgO} \times \frac{1\text{ mol HgO}}{216.6\text{ g HgO}} \times \frac{1\text{ mol }O_2}{2\text{ mol HgO}} \times \frac{22.4\text{ L }O_2}{\text{mol }O_2} = 3.71\text{ L }O_2$$

See Section 14.5, Gas Stoichiometry at Standard Temperature and Pressure, on Pages 383-385.

15. *GIVEN:* 0.652 L CO_2 *WANTED:* g HCl

PER/PATH: L CO_2 $\xrightarrow{22.4\text{ L }CO_2/\text{mol }CO_2}$ mol CO_2 $\xrightarrow{2\text{ mol HCl}/1\text{ mol }CO_2}$ mol HCl $\xrightarrow{36.46\text{ g HCl/mol HCl}}$ g HCl

$$0.652\text{ L }CO_2 \times \frac{1\text{ mol }CO_2}{22.4\text{ L }CO_2} \times \frac{2\text{ mol HCl}}{1\text{ mol }CO_2} \times \frac{36.46\text{ g HCl}}{\text{mol HCl}} = 2.12\text{ g HCl}$$

See Section 14.5, Gas Stoichiometry at Standard Temperature and Pressure, on Pages 383-385.

Questions 16–19 may be solved by the molar volume method (Section 14.6) or by the combined gas equation method (Section 14.7). In the answer section, the setups are given first for the molar volume method. Then the answers are given for the combined gas equation method. Check your work according to the section you studied.

MOLAR VOLUME METHOD

16. *GIVEN:* 1.04 atm; 343°C (616 K) *WANTED:* MV (L/mol)

	Molar Volume	Temperature	Pressure
Initial Value (1)	22.4 L/mol	0°C (273 K)	1 atm
Final Value (2)	V_2	343 + 273 = 616 K	1.04 atm

$$MV_2 = MV_1 \times \frac{P_1}{P_2} \times \frac{T_2}{T_1} = \frac{22.4\text{ L}}{\text{mol}} \times \frac{1\text{ atm}}{1.04\text{ atm}} \times \frac{616\text{ K}}{273\text{ K}} = 48.6\text{ L/mol}$$

GIVEN: 4.83 g $NaHCO_3$ *WANTED:* Volume CO_2 (assume L)

PER/PATH: g $NaHCO_3$ $\xrightarrow{84.01\text{ g }NaHCO_3/\text{mol }NaHCO_3}$ mol $NaHCO_3$ $\xrightarrow{1\text{ mol }CO_2/1\text{ mol }NaHCO_3}$ mol CO_2 $\xrightarrow{48.6\text{ L }CO_2/\text{mol }CO_2}$ L CO_2

$$4.83\text{ g }NaHCO_3 \times \frac{1\text{ mol }NaHCO_3}{84.01\text{ g }NaHCO_3} \times \frac{1\text{ mol }CO_2}{1\text{ mol }NaHCO_3} \times \frac{48.6\text{ L }CO_2}{\text{mol }CO_2} = 2.79\text{ L }CO_2$$

See Section 14.6, Gas Stoichiometry: Molar Volume Method, on Pages 385-387.

17. $CH_4 + 2\ O_2 \rightarrow CO_2 + 2\ H_2O$

GIVEN: 0.813 atm; 26°C (299 K) *WANTED:* MV (L/mol)

	Molar Volume	Temperature	Pressure
Initial Value (1)	22.4 L/mol	0°C (273 K)	1 atm
Final Value (2)	V_2	26 + 273 = 299 K	0.813 atm

$$MV_2 = MV_1 \times \frac{P_1}{P_2} \times \frac{T_2}{T_1} = \frac{22.4\text{ L}}{\text{mol}} \times \frac{1\text{ atm}}{0.813\text{ atm}} \times \frac{299\text{ K}}{273\text{ K}} = 30.2\text{ L/mol}$$

GIVEN: 19.2 L CH_4 *WANTED:* g CO_2

PER/PATH: L CH_4 $\xrightarrow{30.2\ L\ CH_4/mol\ CH_4}$ mol CH_4 $\xrightarrow{1\ mol\ CO_2/1\ mol\ CH_4}$ mol CO_2 $\xrightarrow{44.01\ g\ CO_2/mol\ CO_2}$ g CO_2

$$19.2\ L\ CH_4 \times \frac{1\ mol\ CH_4}{30.2\ L\ CH_4} \times \frac{1\ mol\ CO_2}{1\ mol\ CH_4} \times \frac{44.01\ g\ CO_2}{mol\ CO_2} = 28.0\ g\ CO_2$$

See Section 14.6, Gas Stoichiometry: Molar Volume Method, on Pages 385-387.

18. *GIVEN:* 264°C (537 K); 1.09 atm *WANTED:* MV (L/mol)

	Molar Volume	Temperature	Pressure
Initial Value (1)	22.4 L/mol	0°C (273 K)	1 atm
Final Value (2)	V_2	264 + 273 = 537 K	1.09 atm

$$MV_2 = MV_1 \times \frac{P_1}{P_2} \times \frac{T_2}{T_1} = \frac{22.4\ L}{mol} \times \frac{1\ atm}{1.09\ atm} \times \frac{537\ K}{273\ K} = 40.4\ L/mol$$

GIVEN: 1 kg $CaCO_3 \cdot MgCO_3$ *WANTED:* L CO_2

PER/PATH: kg $CaCO_3 \cdot MgCO_3$ $\xrightarrow{184.41\ kg\ CaCO_3 \cdot MgCO_3/kmol\ CaCO_3 \cdot MgCO_3}$ kmol $CaCO_3 \cdot MgCO_3$ $\xrightarrow{2\ kmol\ CO_2/1\ kmol\ CaCO_3 \cdot MgCO_3}$ kmol CO_2 $\xrightarrow{40.4\ kL\ CO_2/kmol\ CO_2}$ kL CO_2

$$1\ kg\ CaCO_3 \cdot MgCO_3 \times \frac{1\ kmol\ CaCO_3 \cdot MgCO_3}{184.41\ kg\ CaCO_3 \cdot MgCO_3} \times \frac{2\ kmol\ CO_2}{1\ kmol\ CaCO_3 \cdot MgCO_3} \times \frac{40.4\ kL\ CO_2}{1\ kmol\ CO_2} = 0.438\ kL = 438\ L\ CO_2$$

See Section 14.6, Gas Stoichiometry: Molar Volume Method, on Pages 385-387.

19. a) *GIVEN:* 691 torr; 19°C (292 K) *WANTED:* MV (L/mol)

	Molar Volume	Temperature	Pressure
Initial Value (1)	22.4 L/mol	0°C (273 K)	760 torr
Final Value (2)	V_2	19 + 273 = 292 K	691 torr

$$MV_2 = MV_1 \times \frac{P_1}{P_2} \times \frac{T_2}{T_1} = \frac{22.4\ L}{mol} \times \frac{760\ torr}{691\ torr} \times \frac{292\ K}{273\ K} = 26.4\ L/mol$$

GIVEN: 1.94 L NO_2 *WANTED:* g Sn

PER/PATH: L NO_2 $\xrightarrow{26.4\ L\ NO_2/mol\ NO_2}$ mol NO_2 $\xrightarrow{1\ mol\ Sn/4\ mol\ NO_2}$ mol Sn $\xrightarrow{118.7\ g\ Sn/mol\ Sn}$ g Sn

$$1.94\ L\ NO_2 \times \frac{1\ mol\ NO_2}{26.4\ L\ NO_2} \times \frac{1\ mol\ Sn}{4\ mol\ NO_2} \times \frac{118.7\ g\ Sn}{mol\ Sn} = 2.18\ g\ Sn$$

b) *GIVEN:* 2.18 g Sn; 4.77 g solder *WANTED:* % Sn

EQUATION: $\%\ Sn = \frac{g\ Sn}{g\ solder} \times 100 = \frac{2.18\ g}{4.77\ g} \times 100 = 45.7\%\ Sn$

a) See Section 14.6, Gas Stoichiometry: Molar Volume Method, on Pages 385-387.
b) The definition of percent is discussed on Page 180. See Equation 7.2.

COMBINED GAS EQUATION METHOD

16. *GIVEN:* 4.83 g $NaHCO_3$ *WANTED:* STP volume of CO_2 (assume L)

PER/PATH: g $NaHCO_3$ $\xrightarrow{84.01\text{ g NaHCO}_3\text{/mol NaHCO}_3}$ mol $NaHCO_3$ $\xrightarrow{1\text{ mol CO}_2\text{/1 mol NaHCO}_3}$ mol CO_2 $\xrightarrow{22.4\text{ L CO}_2\text{/mol CO}_2}$ L CO_2

$$4.83\text{ g NaHCO}_3 \times \frac{1\text{ mol NaHCO}_3}{84.01\text{ g NaHCO}_3} \times \frac{1\text{ mol CO}_2}{1\text{ mol NaHCO}_3} \times \frac{22.4\text{ L CO}_2}{\text{mol CO}_2} = 1.29\text{ L CO}_2$$

	Volume	Temperature	Pressure
Initial Value (1)	1.29 L	0 + 273 = 273 K	1 atm
Final Value (2)	V_2	343 + 273 = 616 K	1.04 atm

$$V_2 = V_1 \times \frac{P_1}{P_2} \times \frac{T_2}{T_1} = 1.29\text{ L CO}_2 \times \frac{1\text{ atm}}{1.04\text{ atm}} \times \frac{616\text{ K}}{273\text{ K}} = 2.80\text{ L CO}_2$$

If both steps are combined into a single setup:

$$4.83\text{ g NaHCO}_3 \times \frac{1\text{ mol NaHCO}_3}{84.01\text{ g NaHCO}_3} \times \frac{1\text{ mol CO}_2}{1\text{ mol NaHCO}_3} \times \frac{22.4\text{ L CO}_2}{\text{mol CO}_2} \times \frac{1\text{ atm}}{1.04\text{ atm}} \times \frac{616\text{ K}}{273\text{ K}} = 2.79\text{ L CO}_2$$

See Section 14.7, Gas Stoichiometry: Combined Gas Equation Method, on Pages 387-389.

17. $CH_4 + 2\,O_2 \rightarrow CO_2 + 2\,H_2O$

	Volume	Temperature	Pressure
Initial Value (1)	19.2 L	26 + 273 = 299 K	0.813 atm
Final Value (2)	V_2	0 + 273 = 273 K	1 atm

$$V_2 = V_1 \times \frac{P_1}{P_2} \times \frac{T_2}{T_1} = 19.2\text{ L CH}_4 \times \frac{0.813\text{ atm}}{1\text{ atm}} \times \frac{273\text{ K}}{299\text{ K}} = 14.3\text{ L CH}_4$$

GIVEN: 14.3 L CH_4 *WANTED:* g CO_2

PER/PATH: L CH_4 $\xrightarrow{22.4\text{ L CH}_4\text{/mol CH}_4}$ mol CH_4 $\xrightarrow{1\text{ mol CO}_2\text{/1 mol CH}_4}$ mol CO_2 $\xrightarrow{44.01\text{ g CO}_2\text{/mol CO}_2}$ g CO_2

$$14.3\text{ L CH}_4 \times \frac{1\text{ mol CH}_4}{22.4\text{ L CH}_4} \times \frac{1\text{ mol CO}_2}{1\text{ mol CH}_4} \times \frac{44.01\text{ g CO}_2}{\text{mol CO}_2} = 28.1\text{ g CO}_2$$

If both steps are combined into a single setup:

$$19.2\text{ L CH}_4 \times \frac{0.813\text{ atm}}{1\text{ atm}} \times \frac{273\text{ K}}{299\text{ K}} \times \frac{1\text{ mol CH}_4}{22.4\text{ L CH}_4} \times \frac{1\text{ mol CO}_2}{1\text{ mol CH}_4} \times \frac{44.01\text{ g CO}_2}{\text{mol CO}_2} = 28.0\text{ g CO}_2$$

See Section 14.7, Gas Stoichiometry: Combined Gas Equation Method, on Pages 387-389.

18. *GIVEN:* 1 kg $CaCO_3 \cdot MgCO_3$ *WANTED:* kL CO_2

PER/PATH: kg $CaCO_3 \cdot MgCO_3$ $\xrightarrow{184.41\text{ kg CaCO}_3\cdot\text{MgCO}_3\text{/mol CaCO}_3\cdot\text{MgCO}_3}$ kmol $CaCO_3 \cdot MgCO_3$ $\xrightarrow{2\text{ kmol CO}_2\text{/1 kmol CaCO}_3\cdot\text{MgCO}_3}$ kmol CO_2 $\xrightarrow{22.4\text{ kL CO}_2\text{/kmol CO}_2}$ kL CO_2

$$1\text{ kg CaCO}_3\cdot\text{MgCO}_3 \times \frac{1\text{ kmol CaCO}_3\cdot\text{MgCO}_3}{184.41\text{ kg CaCO}_3\cdot\text{MgCO}_3} \times \frac{2\text{ kmol CO}_2}{1\text{ kmol CaCO}_3\cdot\text{MgCO}_3} \times \frac{22.4\text{ kL CO}_2}{1\text{ kmol CO}_2} = 0.243\text{ kL CO}_2$$

	Volume	Temperature	Pressure
Initial Value (1)	0.243 kL	0 + 273 = 273 K	1 atm
Final Value (2)	V_2	264 + 273 = 537 K	1.09 atm

$$V_2 = V_1 \times \frac{P_1}{P_2} \times \frac{T_2}{T_1} = 0.243 \text{ kL } CO_2 \times \frac{1 \text{ atm}}{1.09 \text{ atm}} \times \frac{537 \text{ K}}{273 \text{ K}} = 0.439 \text{ kL} = 439 \text{ L } CO_2$$

If both steps are combined into a single setup:

$$1 \text{ kg } CaCO_3 \cdot MgCO_3 \times \frac{1 \text{ kmol } CaCO_3 \cdot MgCO_3}{184.41 \text{ kg } CaCO_3 \cdot MgCO_3} \times \frac{2 \text{ kmol } CO_2}{1 \text{ kmol } CaCO_3 \cdot MgCO_3} \times \frac{22.4 \text{ kL } CO_2}{1 \text{ kmol } CO_2} \times \frac{1 \text{ atm}}{1.09 \text{ atm}} \times \frac{537 \text{ K}}{273 \text{ K}} = 0.438 \text{ kL} = 438 \text{ L } CO_2$$

See Section 14.7, *Gas Stoichiometry: Combined Gas Equation Method*, on Pages 387-389.

19. a)

	Volume	Temperature	Pressure
Initial Value (1)	1.94 L	19 + 273 = 292 K	691 torr
Final Value (2)	V_2	0 + 273 = 273 K	760 torr

$$V_2 = V_1 \times \frac{P_1}{P_2} \times \frac{T_2}{T_1} = 1.94 \text{ L } NO_2 \times \frac{691 \text{ torr}}{760 \text{ torr}} \times \frac{273 \text{ K}}{292 \text{ K}} = 1.65 \text{ L } NO_2$$

GIVEN: 1.65 L NO_2 *WANTED:* g Sn

PER/PATH: L NO_2 $\xrightarrow{22.4 \text{ L } NO_2/\text{mol } NO_2}$ mol NO_2 $\xrightarrow{1 \text{ mol Sn}/4 \text{ mol } NO_2}$ mol Sn $\xrightarrow{118.7 \text{ g Sn/mol Sn}}$ g Sn

$$1.65 \text{ L } NO_2 \times \frac{1 \text{ mol } NO_2}{22.4 \text{ L } NO_2} \times \frac{1 \text{ mol Sn}}{4 \text{ mol } NO_2} \times \frac{118.7 \text{ g Sn}}{\text{mol Sn}} = 2.19 \text{ g Sn}$$

If both steps are combined into a single setup:

$$1.94 \text{ L } NO_2 \times \frac{691 \text{ torr}}{760 \text{ torr}} \times \frac{273 \text{ K}}{292 \text{ K}} \times \frac{1 \text{ mol } NO_2}{22.4 \text{ L } NO_2} \times \frac{1 \text{ mol Sn}}{4 \text{ mol } NO_2} \times \frac{118.7 \text{ g Sn}}{\text{mol Sn}} = 2.18 \text{ g Sn}$$

b) *GIVEN:* 2.18 g Sn; 4.77 g solder *WANTED:* % Sn

EQUATION: $\% \text{ Sn} = \frac{\text{g Sn}}{\text{g solder}} \times 100 = \frac{2.18 \text{ g}}{4.77 \text{ g}} \times 100 = 45.7\% \text{ Sn}$

a) See Section 14.7, *Gas Stoichiometry: Combined Gas Equation Method*, on Pages 387-389.
b) The definition of percent is discussed on Page 180. See Equation 7.2.

20. The Volume–Amount Law states that equal volumes of all gases at the same temperature and pressure contain the same number of molecules. Gas particles are widely separated from each other. The volume they occupy is determined by the volume of the container that holds them, not by their particle size. Therefore, it is *possible* for the same number of different-sized particles to occupy the same volume. Experimental evidence confirms this possibility. Particles in liquids and solids, however, touch each other. The volume occupied by a given number of liquid or solid particles depends primarily on their particle size and the nature of the attractions among the particles. Solid volume also depends on how they are arranged or packed. The room in which you are reading this answer can easily hold a dozen golf balls, a dozen baseballs, or a dozen basketballs in a "gaseous" state. But there is no way that a box just big enough for a dozen golf balls can hold a dozen baseballs or basketballs, nor will a one-dozen baseball box accommodate twelve basketballs.

See Section 14.8, *Volume-Volume Gas Stoichiometry*, on Pages 389-393.

21. a) *GIVEN:* 207 L O_2 *WANTED:* L NO_2

PER/PATH: $L\ O_2 \xrightarrow{2\ L\ NO_2/1\ L\ O_2} L\ NO_2$

$$207\ L\ O_2 \times \frac{2\ L\ NO_2}{1\ L\ O_2} = 414\ L\ NO_2$$

b)

	Volume	Temperature	Pressure
Initial Value (1)	207 L	18 + 273 = 291 K	0.877 atm
Final Value (2)	V_2	84 + 273 = 357 K	1.16 atm

$$V_2 = V_1 \times \frac{T_2}{T_1} \times \frac{P_1}{P_2} = 207\ L\ O_2 \times \frac{357\ K}{291\ K} \times \frac{0.877\ atm}{1.16\ atm} \times \frac{2\ L\ NO_2}{1\ L\ O_2} = 384\ L\ NO_2$$

See Section 14.8, Volume-Volume Gas Stoichiometry, on Pages 389-393.

22.

	Volume	Temperature	Pressure
Initial Value (1)	4.29×10^3 ft^3	392 + 273 = 665 K	1.31 atm
Final Value (2)	V_2	36 + 273 = 309 K	0.904 atm

$$V_2 = V_1 \times \frac{T_2}{T_1} \times \frac{P_1}{P_2} = 4.29 \times 10^3\ ft^3\ SO_2 \times \frac{309\ K}{665\ K} \times \frac{1.31\ atm}{0.904\ atm} = 2.89 \times 10^3\ ft^3\ SO_2$$

GIVEN: 2.89×10^3 ft^3 SO_2 *WANTED:* ft^3 O_2

PER/PATH: $ft^3\ SO_2 \xrightarrow{1\ ft^3\ O_2/2\ ft^3\ SO_2} ft^3\ O_2$

$$2.89 \times 10^3\ ft^3\ SO_2 \times \frac{1\ ft^3\ O_2}{2\ ft^3\ SO_2} = 1.45 \times 10^3\ ft^3\ O_2$$

If both steps are combined into a single setup:

$$4.29 \times 10^3\ ft^3\ SO_2 \times \frac{309\ K}{665\ K} \times \frac{1.31\ atm}{0.904\ atm} \times \frac{1\ ft^3\ O_2}{2\ ft^3\ SO_2} = 1.44 \times 10^3\ ft^3\ O_2$$

See Section 14.8, Volume-Volume Gas Stoichiometry, on Pages 389-393.

23.

	Volume	Temperature	Pressure
Initial Value (1)	895 L	1700 + 273 = 1973 K	846 torr
Final Value (2)	V_2	29 + 273 = 302 K	182 torr

$$V_2 = V_1 \times \frac{T_2}{T_1} \times \frac{P_1}{P_2} = 895\ L\ CO \times \frac{302\ K}{1973\ K} \times \frac{846\ torr}{182\ torr} = 637\ L\ CO$$

GIVEN: 637 L CO *WANTED:* L O_2

PER/PATH: $L\ CO \xrightarrow{1\ L\ O_2/2\ L\ CO} L\ O_2$

$$637\ L\ CO \times \frac{1\ L\ O_2}{2\ L\ CO} = 3.2 \times 10^2\ L\ O_2$$

If both steps are combined into a single setup:

$$895\ L\ CO \times \frac{846\ torr}{182\ torr} \times \frac{302\ K}{1973\ K} \times \frac{1\ L\ O_2}{2\ L\ CO} = 3.2 \times 10^2\ L\ O_2$$

(1700°C is assumed to be a two-significant-figure number.)

See Section 14.8, Volume-Volume Gas Stoichiometry, on Pages 389-393.

25. True: a, c, d. False: b.

b) The number of particles in 5.00 L of NH_3 is the same as the number of particles in 5.00 L of CO if both volumes are measured at the same temperature and pressure.

26. *GIVEN:* 41°C (314 K); 733 torr *WANTED:* MV (L/mol)

	Molar Volume	Temperature	Pressure
Initial Value (1)	22.4 L/mol	0°C (273 K)	760 torr
Final Value (2)	V_2	41 + 273 = 314 K	733 torr

$$MV_2 = MV_1 \times \frac{P_1}{P_2} \times \frac{T_2}{T_1} = \frac{22.4 \text{ L}}{\text{mol}} \times \frac{760 \text{ torr}}{733 \text{ torr}} \times \frac{314 \text{ K}}{273 \text{ K}} = 26.7 \text{ L/mol}$$

GIVEN: 6.74 g; 28.05 g/mol; 26.7 L/mol *WANTED:* volume (assume L)

PER/PATH: $\text{g} \xrightarrow{28.05 \text{ g/mol}} \text{mol} \xrightarrow{26.7 \text{ L/mol}} \text{L}$

$$6.74 \text{ g} \times \frac{1 \text{ mol}}{28.05 \text{ g}} \times \frac{26.7 \text{ L}}{\text{mol}} = 6.42 \text{ L}$$

See Section 14.4, Three-Ratio Problems, on Pages 380-383. In particular, see the Three-Ratio Problems at a Given Temperature and Pressure subsection on Pages 381-383.

27. *GIVEN:* 0.972 atm; 14°C (287 K) *WANTED:* MV (L/mol)

	Molar Volume	Temperature	Pressure
Initial Value (1)	22.4 L/mol	0°C (273 K)	1 atm
Final Value (2)	V_2	14 + 273 = 287 K	0.972 atm

$$MV_2 = MV_1 \times \frac{P_1}{P_2} \times \frac{T_2}{T_1} = \frac{22.4 \text{ L}}{\text{mol}} \times \frac{1 \text{ atm}}{0.972 \text{ atm}} \times \frac{287 \text{ K}}{273 \text{ K}} = 24.2 \text{ L/mol}$$

GIVEN: 34.09 g H_2S/mol H_2S; 24.2 L/mol *WANTED:* D (assume g/L)

$$\textit{EQUATION:}\ D \equiv \frac{m}{V} = \frac{34.09 \text{ g } H_2S}{\text{mol } H_2S} \times \frac{1 \text{ mol}}{24.2 \text{ L}} = 1.41 \text{ g}$$

See Section 14.4, Three-Ratio Problems, on Pages 380-383. In particular, see the Three-Ratio Problems at a Given Temperature and Pressure subsection on Pages 381-383.

28. Molar mass = molar volume × density; MM = k × D. At a given temperature and pressure, molar volume (MV) is constant. In this equation, the molar volume has the role of a proportionality constant. Hence, molar mass and density are directly proportional to each other at a given temperature and pressure.

See Section 14.3, Three Ratios: Gas Density, Molar Volume, and Molar Mass on Pages 377-379. In particular, see the Relationships among the Ratios subsection on page 379.

29. Butane, C_4H_{10}, has a higher molar mass, 58.12 g/mol, than propane, C_3H_8, 44.09 g/mol. At a given temperature and pressure, density is proportional to molar mass. Therefore B, the higher-density gas, must be butane, Its density is $1.37 \text{ g/L} \times \frac{58.12 \text{ g/mol}}{44.09 \text{ g/mol}} = 1.81 \text{ g/L}$.

See Section 14.3, Three Ratios: Gas Density, Molar Volume, and Molar Mass on Pages 377-379. In particular, see the Relationships among the Ratios subsection on page 379.

Chapter 15

Gases, Liquids, and Solids

1. The total pressure exerted by a mixture of gases is the sum of the partial pressures of the gases in the mixture. See the first paragraph in Section 15.1 for an explanation of how Dalton's Law fits the ideal gas model.

   See Section 15.1, Dalton's Law of Partial Pressures, on Pages 400-402. Equation 15.1 (Page 401) gives Dalton's Law of Partial Pressures as an equation.

2. $P_{total} = p_{methane} + p_{ethane} + p_{propane} = 0.319 \text{ atm} + 0.605 \text{ atm} + 0.456 \text{ atm} = 1.380 \text{ atm}$

   See Section 15.1, Dalton's Law of Partial Pressures, on Pages 400-402. Equation 15.1 (Page 401) gives Dalton's Law of Partial Pressures as an equation.

3. $P_{total} = p_{oxygen} + p_{water}$
   $p_{oxygen} = P_{total} - p_{water} = 755 \text{ torr} - 25 \text{ torr} = 7.30 \times 10^2 \text{ torr}$

   See Section 15.1, Dalton's Law of Partial Pressures, on Pages 400-402. Equation 15.1 (Page 401) gives Dalton's Law of Partial Pressures as an equation.

4. Gas particles are very widely spaced; liquid particles are "touchingly close." Density is mass per unit volume. The same number of liquid particles will occupy a smaller volume than they occupy as a gas.

   Figure 2.5 on Page 20 shows the three states of matter as illustrated by water. Also review items 3 and 4 on the list of the main features of the ideal gas model on Page 94. In Section 15.2, Properties of Liquids, item 3 on the list of comparison of properties of liquids with properties of gases (Page 402) is also important in answering this question.

5. There is no space between water molecules; they cannot be forced closer together. Molecules in air are widely separated; they can be pushed closer together.

   Figure 2.5 on Page 20 shows the three states of matter as illustrated by water. In Section 15.2, Properties of Liquids, item 1 on the list of comparison of properties of liquids with properties of gases (Page 402) is also important in answering this question.

6. The stronger the intermolecular attractions, the slower the evaporation rate, other things being equal. Condensation rate therefore equals evaporation rate at a lower vapor concentration, which means lower partial pressure at equilibrium.

   See Section 15.2, Properties of Liquids, on Pages 402-404 In particular, the first paragraph of the section (Page 402) discusses the intermolecular attractions in a liquid, and the Vapor Pressure subsection (Pages 402-403) explains vapor pressure.

7. Liquids with strong intermolecular attractions have high viscosity, an internal resistance to flow.

   See Section 15.2, Properties of Liquids, on Pages 402-404. In particular, the first paragraph of the section (Page 402) discusses the intermolecular attractions in a liquid, and the Viscosity subsection (Page 403) explains viscosity.

8. Motor oil is more viscous than water. This predicts that intermolecular attractions are stronger in motor oil, as strong attractions lead to internal resistance to flow, which is viscosity.

   See Section 15.2, Properties of Liquids, on Pages 402-404. In particular, the first paragraph of the section (Page 402) discusses the intermolecular attractions in a liquid, and the Viscosity subsection (Page 403) explains viscosity.

9. An isolated drop is normally spherical because of surface tension. When sitting on a plate, a drop tends to be flattened by gravity. The honey drop remains closer to spherical than the water drop, indicating stronger surface tension and stronger intermolecular attractions.

   See Section 15.2, Properties of Liquids, on Pages 402-404. In particular, the first paragraph of the section (Page 402) discusses the intermolecular attractions in a liquid, and the Surface Tension subsection (Page 403) explains surface tension. Figure 15.3 on Page 403 discusses spherical liquid drops, and Figure 15.4 on Page 404 illustrates surface tension.

10. Soap reduces intermolecular attractions, thus lowering the surface tension of water. This makes the soapy water able to penetrate fabrics and clean them throughout.

    See Section 15.2, Properties of Liquids, on Pages 402-404. In particular, the first paragraph of the section (Page 402) discusses the intermolecular attractions in a liquid, and the Surface Tension subsection (Page 403) explains surface tension. Figure 15.4 on Page 404 illustrates surface tension.

11. NO, N_2O, NO_2

    In order for all substances to be in the solid state initially, they must all be at a temperature below the lowest melting point of the three. Think of a number line that goes from smaller to larger, such as ..., -3, -2, -1, 0, 1, 2, 3, ... -164°C is the smallest of the melting points, so when the temperature rises to this point, NO will melt. The next higher melting point is -90.8°C; and the highest melting point is -11.2°C.

12. Of the three compounds, only N_2O is a liquid at –90°C. Therefore, only N_2O possesses the property of viscosity. If a solid is considered more "viscous" than a liquid, NO_2 is the most viscous.

    Note that NO is a gas at -90°C (boiling point -152°C), N_2O has just melted, and NO_2 is a solid. See Section 15.2, Properties of Liquids, on Pages 402-404. In particular, see the Viscosity subsection on Page 403.

13. A compound with a large nonpolar section and a —OH or an —NH_2 group at the end would exhibit hydrogen bonding, but the induced dipole forces resulting from the large nonpolar section would dominate the intermolecular attractive forces. An example is $CH_3CH_2CH_2CH_2CH_2CH_2CH_2OH$.

The -OH or $-NH_2$ group is required for hydrogen bonding. See Figure 15.7 on Page 406 for examples of molecules that have this group of atoms. The relative strength of the types of intermolecular forces is discussed in the next-to-last paragraph on Page 406. In general, see Section 15.3, Types of Intermolecular Forces, on Pages 404-408. Figure 15.9 on Page 407 provides an important summary of the section.

14. HBr and NF_3, dipole; C_2H_2, induced dipoles; C_2H_5OH, hydrogen bonds.

HBr and NF_3 are polar molecules, so they have dipole forces and induced dipoles, with the dipole forces being the strongest. C_2H_2 is a nonpolar molecule, so it has induced dipoles. C_2H_5OH has hydrogen bonds, dipole forces, and induced dipole forces, with hydrogen bonds being the strongest. See Section 15.3, Types of Intermolecular Forces, on Pages 404-408.

15. Ionic compounds have ionic bonds as the interparticle forces. Polar molecular compounds have dipole forces acting between the particles. Ionic bonds are much stronger than any intermolecular forces, including dipole forces. The ionic compound would have the higher melting point because more energy is needed to overcome the stronger forces holding the solid together.

In general, bonds (ionic, covalent, metallic) are much stronger than intermolecular forces. See Section 11.2, Ionic Bonds, on Pages 296-298, to review ionic compounds. See Section 15.3, Types of Intermolecular Forces, on Pages 404-408, to review intermolecular forces.

16. CCl_4, because it is larger, as suggested by its higher molecular mass.

CH_4 and CCl_4 are nonpolar molecules, so they both have induced dipoles as the primary intermolecular force. CCl_4 is larger than CH_4, so the induced dipole force in CCl_4 is greater than in CH_4. The greater the intermolecular attractions, the higher the boiling point. See Section 15.3, Types of Intermolecular Forces, on Pages 404-408.

17. H_2S, because it is slightly more polar.

H_2S and PH_3 are polar molecules, so they both have dipole forces as the primary intermolecular force. They have similar molecular masses, 34 g/mol for both H_2S and PH_3, so the contribution of induced dipoles is similar. The H-S bond is slightly more polar than the H-P bond, and the bent structure of H_2S leads to a more asymmetric distribution of the charge than the trigonal pyramdial structure of PH_3, making H_2S more polar. See Section 15.3, Types of Intermolecular Forces, on Pages 404-408.

18. For hydrogen bonding to occur, a hydrogen atom must be bonded to another element with a high enough electronegativity to shift the bonding electron pair away from the hydrogen atom. Fluorine, oxygen, and nitrogen are the only elements that satisfy this requirement.

See Section 15.3, Types of Intermolecular Forces, on Pages 404-408. In particular, see Figure 15.7 and item 3 in the list of kinds of intermolecular forces on Page 406

19. a, induced dipole forces. b, dipole forces and induced dipole forces.

See Section 15.3, Types of Intermolecular Forces, on Pages 404-408.

20. a, c, and d, dipole forces and induced dipole forces. b and e, induced dipole forces.

    a, c, and d are polar molecules. b and e are nonpolar molecules. See Section 15.3, Types of Intermolecular Forces, on Pages 404-408.

21. C_6H_{14}, a larger molecule with higher molecular mass than C_3H_8, will have stronger intermolecular attractions and therefore higher melting and boiling points.

    C_3H_8 and C_6H_{14} are nonpolar molecules, so they both have induced dipoles as the primary intermolecular force. C_6H_{14} is larger than C_3H_8, so the induced dipole force in C_6H_{14} is greater than in C_3H_8. The greater the intermolecular attractions, the higher the melting and boiling points. See Section 15.3, Types of Intermolecular Forces, on Pages 404-408.

22. CO_2 molecules are smaller than SO_2 molecules, and CO_2 molecules are linear and nonpolar while SO_2 molecules are bent and polar. Both differences predict weaker intermolecular attractions and therefore higher vapor pressure for CO_2.

    See Section 15.3, Types of Intermolecular Forces, on Pages 404-408.

23. The term *dynamic* refers to the "active" character of the equilibrium. The rate of change from liquid to vapor equals the rate of change from vapor to liquid. Although the concentrations remain constant, there is continual change between the two states at the particulate level.

    See Section 15.4, Liquid-Vapor Equilibrium, on Pages 408-412. The dynamic equilibrium concept is discussed in the last paragraph on Page 409.

24. The greater the concentration of particles in the vapor state, the greater the rate of condensation. Temperature affects both evaporation and condensation rates. The partial pressure at equilibrium for a liquid–vapor system is temperature-dependent.

    See Section 15.4, Liquid-Vapor Equilibrium, on Pages 408-412. In particular, see Figure 15.11 on Page 410.

25. Boiling point is the temperature at which the vapor pressure of a liquid is equal to the pressure above its surface. Your vapor-pressure-versus-temperature curve should be similar to one from Figure 15.12. The normal boiling point is found at a vapor pressure of 760 torr.

    See The Effect of Temperature subsection and Figure 15.12 on Page 411. Also see Section 15.5, The Boiling Process, on Page 412.

26. A gas can be condensed by increasing the pressure. Higher pressure forces the particles closer to one another, where intermolecular attractive forces become significant, thus condensing a gas into a liquid.

    See Section 15.5, The Boiling Process, on Page 412.

27. High-boiling liquids have strong intermolecular attractive forces and therefore require high energy to escape from the liquid to form a gas. Evaporation rate is therefore slow, quickly equaled by condensation rate at low vapor concentration, or vapor pressure.

    Vapor pressure is described in a paragraph that starts on the bottom of Page 402. Section 15.4, Liquid-Vapor Equilibrium, describes the evaporation process. Also see Section 15.5, The Boiling Process, on Page 412.

28. More energy is required to vaporize X, so it would have the higher boiling point and lower vapor pressure.

Molar heat of vaporization is described in a paragraph on Page 403. For the definition of normal boiling point, see Section 15.5, The Boiling Process, on Page 412. Vapor pressure is described in a paragraph that starts on the bottom of Page 402.

29. Amorphous solids have no long-range ordering on the particulate level. Crystalline solids have particles arranged in a repeating pattern. Polycrystalline solids have small orderly crystals arranged in a random fashion. Because of the disorderly arrangement of particles in an amorphous solid, there is a range of strengths of intermolecular attractive forces throughout the solid, and thus no definite physical properties. Crystalline solids have distinct physical properties.

See Section 15.7, The Solid State, on Page 414.

30. A, molecular; B, metallic.

Solid A is either molecular or metallic based on its low melting point and insolubility in water, but it nonconductivity eliminates a metallic classification, making it molecular. Solid B is metallic because it is an excellent conductor. The other data verify this. See Section 15.8, Types of Crystalline Solids, on Pages 415-417. Table 15.3, General Properties of Crystals, on Page 417, summarizes the properties of crystalline solids.

31. *GIVEN:* 29.3 kJ; 6.04 g *WANTED:* ΔH_{vap} (assume kJ)

EQUATION: $\Delta H_{vap} \equiv \frac{q}{m} = \frac{29.3 \text{ kJ}}{6.04 \text{ g}} = 4.85 \text{ kJ/g}$

See Section 15.9, Energy and Change of State, on Pages 417-420.

32. *GIVEN:* 16 g Cu; $\Delta H_{vap} = 4.81$ kJ/g Cu (Table 15.4, Page 418) *WANTED:* energy (assume kJ)

EQUATION: $q = m \times \Delta H_{vap} = 16 \text{ g Cu} \times \frac{4.81 \text{ kJ}}{\text{g Cu}} = 77 \text{ kJ}$

See Section 15.9, Energy and Change of State, on Pages 417-420.

33. *GIVEN:* 18.3 kJ; 0.371 kJ/g hexane *WANTED:* mass hexane (assume g)

EQUATION: $m = \frac{q}{\Delta H_{vap}} = q \times \frac{1}{\Delta H_{vap}} = 18.3 \text{ kJ} \times \frac{1 \text{ g hexane}}{0.371 \text{ kJ}} = 49.3 \text{ g hexane}$

See Section 15.9, Energy and Change of State, on Pages 417-420.

34. *GIVEN:* 744 g CCl_2F_2; 35 kJ/mol CCl_2F_2 *WANTED:* energy (assume kJ)

PER/PATH: $\text{g } CCl_2F_2 \xrightarrow{120.91 \text{ g } CCl_2F_2/\text{mol } CCl_2F_2} \text{mol } CCl_2F_2 \xrightarrow{35 \text{ kJ/mol } CCl_2F_2} \text{kJ}$

$$744 \text{ g } CCl_2F_2 \times \frac{1 \text{ mol } CCl_2F_2}{120.91 \text{ g } CCl_2F_2} \times \frac{35 \text{ kJ}}{\text{mol } CCl_2F_2} = 2.2 \times 10^2 \text{ kJ}$$

See Section 15.9, Energy and Change of State, on Pages 417-420.

35. *GIVEN:* 35.4 g Au; 64 J/g (Table 15.4, Page 418) *WANTED:* energy (assume kJ)

EQUATION: $q = m \times \Delta H_{fus} = 35.4 \text{ g Au} \times \frac{64 \text{ J}}{\text{g Au}} = 2.3 \times 10^3 \text{ J} = 2.3 \text{ kJ}$

See Section 15.9, Energy and Change of State, on Pages 417-420.

36. *GIVEN:* 7.08 kJ; 46.9 g *WANTED:* ΔH_{fus} (assume J/g)

EQUATION: $\Delta H_{fus} \equiv \frac{q}{m} = \frac{7.08 \text{ kJ}}{46.9 \text{ g}} = 0.151 \text{ kJ/g} = 151 \text{ J/g}$

See Section 15.9, Energy and Change of State, on Pages 417-420.

37. *GIVEN:* 11.3 kJ; 105 J/g Ag (Table 15.4, Page 418) *WANTED:* g Ag

EQUATION: $m = \frac{q}{\Delta H_{fus}} = q \times \frac{1}{\Delta H_{fus}} = 11.3 \text{ kJ} \times \frac{1000 \text{ J}}{\text{kJ}} \times \frac{1 \text{ g Ag}}{105 \text{ J}} = 108 \text{ g Ag}$

See Section 15.9, Energy and Change of State, on Pages 417-420.

38. Energy change is proportional to ΔT. If all other factors are the same, less energy will be required—less time in the microwave—for the smaller ΔT. Starting with hot water gives the smaller ΔT.

$q = m \times c \times \Delta T$, and if mass is constant in two samples, $m \times c$ = constant, so q = (constant) $\times \Delta T$. A smaller ΔT exists for taking hot water to boiling than for taking cold water to boiling. Thus a smaller q is needed to heat the hot water. If q is proportional to time, a shorter time is needed to heat the hot water. See Section 15.10, Energy and Change of Temperature: Specific Heat, on Page 420-422.

39. *GIVEN:* 204 g; $T_i = 22.8°C$; $T_f = 64.9°C$; $c = 0.16 \text{ J/g} \cdot °C$ (from Table 15.5, Page 421)
WANTED : q (assume kJ)

EQUATION: $q = m \times c \times \Delta T = 204 \text{ g} \times \frac{0.16 \text{ J}}{\text{g} \cdot °C} \times (64.9 - 22.8)°C = 1.4 \times 10^3 \text{ J} = 1.4 \text{ kJ}$

See Section 15.10, Energy and Change of Temperature: Specific Heat, on Page 420-422.

40. *GIVEN:* 2.55 kg; $T_i = 384°C$; $T_f = 25°C$; $c = 0.444 \text{ J/g} \cdot °C$ (from Table 15.5, Page 421)
WANTED: q (assume kJ)

EQUATION: $q = m \times c \times \Delta T = 2.55 \text{ kg} \times \frac{1000 \text{ g}}{\text{kg}} \times \frac{0.444 \text{ J}}{\text{g} \cdot °C} \times (25 - 384)°C =$

$-4.06 \times 10^5 \text{ J} = -406 \text{ kJ}$

See Section 15.10, Energy and Change of Temperature: Specific Heat, on Page 420-422.

41. *GIVEN:* 545 g; $T_i = 25.0°C$; 3.14 kJ; $c = 0.46 \text{ J/g} \cdot °C$ (from Table 15.5, Page 421)
WANTED: T_f (assume °C)

$q = m \times c \times \Delta T \quad \frac{q}{m \times c} = \frac{m \times c \times \Delta T}{m \times c} \quad \frac{q}{m \times c} = \Delta T \quad \Delta T = q \times \frac{1}{m} \times \frac{1}{c}$

EQUATION: $\Delta T = q \times \frac{1}{m} \times \frac{1}{c} = 3.14 \text{ kJ} \times \frac{1000 \text{ J}}{\text{kJ}} \times \frac{1}{545 \text{ g}} \times \frac{\text{g} \cdot °C}{0.46 \text{ J}} = 13°C$

$\Delta T = T_f - T_i \quad T_f = T_i + \Delta T = 25.0 + 13 = 38°C$

See Section 15.10, Energy and Change of Temperature: Specific Heat, on Page 420-422.

42. Vertically, temperature; horizontally, energy

See Section 15.11, Change in Temperature Plus Change in State, on Pages 422-425. In particular, see Figure 15.24 on Page 423.

43. E definitely; maybe D and F

The liquid section of the curve ranges from N to O. See Section 15.11, Change in Temperature Plus Change in State, on Pages 422-425. In particular, see Figure 15.24 on Page 423.

44. G

The gas-to-liquid transition occurs from O to P. See Section 15.11, Change in Temperature Plus Change in State, on Pages 422-425. In particular, see Figure 15.24 on Page 423.

45. The solid substance melts completely at constant temperature K

The energy represented by N - M is the energy needed to melt the sample. See Section 15.11, Change in Temperature Plus Change in State, on Pages 422-425. In particular, see Figure 15.24 on Page 423.

46. O – N

The freezing point is temperature K. The boiling point is temperature J. The substance is in the liquid state between energies N and O. See Section 15.11, Change in Temperature Plus Change in State, on Pages 422-425. In particular, see Figure 15.24 on Page 423.

47. Step 1: Heat liquid

GIVEN: 632 g H_2O; $T_i = 28°C$; $T_f = 100°C$; $c = 4.18\ J/g \cdot °C$ (from Table 15.5, Page 421)
WANTED: q (assume kJ)

EQUATION: $q = m \times c \times \Delta T = 632\ g \times \frac{4.18\ J}{g \cdot °C} \times (100 - 28)°C = 1.90 \times 10^5\ J = 1.90 \times 10^2\ kJ$

Step 2: Boil

GIVEN: 632 g H_2O; $\Delta H_{vap} = 2.26\ kJ/g$ (from Table 15.4, Page 418) *WANTED:* q (assume kJ)

EQUATION: $q = 632\ g \times \frac{2.26\ kJ}{g} = 1.43 \times 10^3\ kJ$

Step 3: Heat steam

GIVEN: 632 g H_2O; $T_i = 100°C$; $T_f = 168°C$; $c = 2.0\ J/g \cdot °C$ (from Table 15.5, Page 421)
WANTED: q (assume kJ)

EQUATION: $q = 632\ g \times \frac{2.0\ J}{g \cdot °C} \times (168 - 100)°C = 8.6 \times 10^4\ J = 86\ kJ$

Step 4: Sum the heat flows

$q_{total} = 1.90 \times 10^2\ kJ + 1.43 \times 10^3\ kJ + 86\ kJ = 1.71 \times 10^3\ kJ$

See Section 15.11, Change in Temperature Plus Change in State, on Pages 422-425.

48. Step 1: Cool liquid

GIVEN: 30.0 g Au; T_i = 1085°C; T_f = 1063°C; c = 0.148 J/ g • °C (from Table 15.5, Page 421)
WANTED: q (assume kJ)

EQUATION: $q = 30.0\ g \times \frac{0.148\ J}{g \cdot °C} \times (1063 - 1085)°C = -97.7\ J = -0.0977\ kJ$

Step 2: Freeze

GIVEN: 30.0 g Au; ΔH_{fus} = 64 J/g (from Table 15.4, Page 418) *WANTED:* q (assume kJ)

EQUATION: $q = 30.0\ g \times -\frac{64\ J}{g} = -1.9 \times 10^3\ J = -1.9\ kJ$

Step 3: Cool solid

GIVEN: 30.0 g Au; T_i = 1063°C; T_f = 18°C; c = 0.13 J/ g • °C (from Table 15.5, Page 421)
WANTED: q (assume kJ)

EQUATION: $q = 30.0\ g \times \frac{0.13\ J}{g \cdot °C} \times (18 - 1063)°C = -4.1 \times 10^3\ J = -4.1\ kJ$

Step 4: Sum the heat flows

$q_{total} = -0.0977\ kJ + (-1.9\ kJ) + (-4.1\ kJ) = -6.1\ kJ$

See Section 15.11, Change in Temperature Plus Change in State, on Pages 422-425.

49. Step 1: Cool liquid

GIVEN: 25.1 kg Fe; T_i = 1645°C; T_f = 1535°C; c = 0.452 J/ g • °C (from Table 15.5, Page 421)
WANTED: q (assume kJ)

EQUATION: $q = 25.1\ kg \times \frac{0.452\ kJ}{kg \cdot °C} \times (1535 - 1645)°C = -1.25 \times 10^3\ kJ$

Step 2: Freeze

GIVEN: 25.1 kg Fe; ΔH_{fus} = 267 J/g (from Table 15.4, Page 418) *WANTED:* q (assume kJ)

EQUATION: $q = 25.1\ kg \times -\frac{267\ kJ}{kg} = -6.70 \times 10^3\ kJ$

Step 3: Cool solid

GIVEN: 25.1 kg Fe; T_i = 1535°C; T_f = 33°C; c = 0.444 J/ g • °C (from Table 15.5, Page 421)
WANTED: q (assume kJ)

EQUATION: $q = 25.1\ kg \times \frac{0.444\ kJ}{kg \cdot °C} \times (33 - 1535)°C = -1.67 \times 10^4\ kJ$

Step 4: Sum the heat flows

$q_{total} = -1.25 \times 10^3\ kJ + (-6.70 \times 10^3\ kJ) + (-1.67 \times 10^4\ kJ) = -2.47 \times 10^4\ kJ$

See Section 15.11, Change in Temperature Plus Change in State, on Pages 422-425.

51. True: a, c, f, h, i, r. False: b, d, e, g, j, k, l, m, n, o, p, q.

b) Substances with weak intermolecular attractions generally have high vapor pressures.
d) A substance with a relatively high surface tension usually has a high boiling point.
e) All other things being equal, hydrogen bonds are stronger than induced dipole or dipole forces.
g) Other things being equal, nonpolar molecules have weaker intermolecular attractions than polar molecules.
j) The heat of vaporization is equal to the heat of condensation, but with opposite sign.
k) The boiling point of a liquid depends on the pressure above it.
l) If you break (shatter) an amorphous solid, it will break in curved lines, but if you break a crystal, it will break in straight lines.
m) Ionic crystals are usually soluble in water.

n) Molecular crystals are usually more soluble in organic solvents than in water.
o) The numerical value of heat of vaporization is the same as the numerical value of heat of condensation.
p) The units of heat of fusion are J/g.
q) The temperature of water is constant while it is freezing.

52. Both molecules have induced dipole and dipole forces. CH_3OH has hydrogen bonding and CH_3F does not. The molecules are about the same size. It is reasonable to predict stronger intermolecular forces in CH_3OH and therefore a higher boiling point, and CH_3F will have the higher equilibrium vapor pressure.

Section 15.3, Types of Intermolecular Forces, on Pages 404-408, explains intermolecular forces. The Vapor Pressure subsection of Section 15.2 (Pages 402-403) explains vapor pressure. The Boiling Point subsection of Section 15.2 (Page 403) explains boiling point.

53. Reducing volume increases vapor concentration, which causes an increase in the rate of condensation. Evaporation rate, which depends only on temperature, is not affected.

See Section 15.4, Liquid-Vapor Equilibrium, on Pages 408-412. Most importantly, see Figure 15.11 on Page 410.

54. (a) Evaporation at constant rate begins immediately when liquid is introduced. At that time condensation rate is zero. Net rate of increase in vapor concentration is at a maximum, so rate of vapor pressure increase is a maximum at start. Later condensation rate is more than zero but less than evaporation rate. Net rate of increase in vapor concentration is less than initially, so rate of vapor pressure increase is less than initially. At equilibrium, evaporation and condensation rates are equal. Vapor concentration and therefore vapor pressure remain constant.

See Section 15.4, Liquid-Vapor Equilibrium, on Pages 408-412.

55. All of the liquid evaporated before the vapor concentration was high enough to yield a condensation rate equal to the evaporation rate. At lower-than-equilibrium vapor concentration the vapor pressure is lower than the equilibrium vapor pressure. More liquid must be introduced to the flask until some excess remains and the pressure stabilizes in order to measure equilibrium vapor pressure.

See Section 15.4, Liquid-Vapor Equilibrium, on Pages 408-412.

56. Heat gained by water = Heat lost by tin

Heat gained by water $= q = m \times c \times \Delta T = 72.0 \text{ g} \times \dfrac{4.18 \text{ J}}{\text{g} \cdot {}^\circ\text{C}} \times (25.5 - 19.2){}^\circ\text{C}$

Heat lost by tin $= q = m \times c \times \Delta T = -[141 \text{ g} \times c \times (25.5 - 89.0){}^\circ\text{C}]$

$72.0 \text{ g} \times \dfrac{4.18 \text{ J}}{\text{g} \cdot {}^\circ\text{C}} \times (25.5 - 19.2){}^\circ\text{C} = -[141 \text{ g} \times c \times (25.5 - 89.0){}^\circ\text{C}]$ $\quad c = 0.21 \text{ J/g} \cdot {}^\circ\text{C}$

See Section 15.10, Energy and Change of Temperature: Specific Heat, on Page 420-422.

57. Dissolve the compounds and check for electrical conductivity. The ionic potassium sulfate solute will conduct, whereas the molecular sugar solute will not.

See Section 15.8, Types of Crystalline Solids, on Pages 415-417. Table 15.3, General Properties of Crystals, on Page 417, summarizes the properties of crystalline solids.

58. As temperature drops, the equilibrium vapor pressure drops below the atmospheric vapor pressure. The air becomes first saturated, then supersaturated, and condensation (dew) begins to form.

See Section 15.4, Liquid-Vapor Equilibrium, on Pages 408-412.

Chapter 16

Solutions

1. The properties of a solution will differ from those of the components. They will also be variable, depending on the solution concentration.

   See Section 2.4, Pure Substances and Mixtures, on Pages 24-26, to review their similarities and differences. Also see Section 16.1, The Characteristics of a Solution, on Page 434.

2. Ions are present in a solution of an ionic compound; molecules are present in a solution of a molecular compound unless the compound ionizes when it dissolves, as acids do. Then both ions and molecules are present.

   See Section 11.2, Ionic Bonds (Pages 296-298), and Section 11.3, Covalent Bonds (Pages 299-300), to review ionic and molecular compounds. Also see Section 16.1, The Characteristics of a Solution, on Page 434.

3. Generally, the substance present in the smallest amount is the solute. However, if gases or solids are dissolved in a liquid, the liquid is usually called the solvent.

   See the Solute and Solvent paragraph of Section 16.2 on Pages 434-435.

4. If solute A is very soluble, its 10 grams per 100 grams of solvent concentration may be quite *dilute* compared with the possible concentration. If solute B is only slightly soluble, its 5 grams per 100 grams of solvent may be close to its maximum solubility, and therefore *concentrated.*

   See the Concentrated and Dilute paragraph of Section 16.2 on Page 435.

5. Drop a small amount of solute into the solution. If the solution is unsaturated, the solute will dissolve; if saturated, it will simply settle to the bottom; if supersaturated, it will promote additional crystallization.

   See the Solubility, Saturated, and Unsaturated and the Supersaturated Solutions paragraphs of Section 16.2 on Page 435.

6. Any quantity units of solute divided by any quantity units of solvent may be used to express solubility.

   See the Solubility, Saturated, and Unsaturated paragraph of Section 16.2 on Page 435.

7. The solute is dispersed uniformly throughout the solution, so sugar is soluble in water. *Miscible* is a term usually reserved for solutions of liquids in liquids.

   See the Solubility, Saturated, and Unsaturated and the Miscible and Immiscible paragraphs of Section 16.2 on Page 435.

8. A solute particle that attracts polar water molecules to it is referred to as *hydrated.*

   The term *hydrated* in defined in Section 16.3, The Formation of a Solution, Pages 435-438. See Figure 16.2 on Page 436 for an illustration of hydrated ions.

9. Attractive forces between solute and solvent particles promote dissolving. For an ionic solid solute in water, the positively charged ions are attracted to the negatively charged regions of the polar water molecules, and the negatively charged ions are attracted to the positively charged regions of the water molecules.

   See Section 16.3, The Formation of a Solution, on Pages 435-438. See Figure 16.2 on Page 436 for an illustration of the forces that act to promote dissolving.

10. On the macroscopic level, the process appears to be static; nothing seems to be happening. On the particulate level, solute particles are continually changing between the solid and dissolved states, dissolving and crystallizing. This constant change makes the process dynamic.

    See Section 16.3, The Formation of a Solution, on Pages 435-438. See Figure 16.3 on Page 437 for a description of the formation of a dynamic equilibrium.

11. Solute ions pass from solute to solution faster than from solution to solute in an unsaturated solution.

    See Section 16.3, The Formation of a Solution, on Pages 435-438. See Figure 16.3 on Page 437 for a description of the formation of a dynamic equilibrium.

12. Stirring decreases the rate of crystallization and thus decreases the time needed for formation of an unsaturated solution. Any form of agitation, stirring or shaking, will be effective.

    See Section 16.3, The Formation of a Solution, on Pages 435-438. In particular, see the list of factors that affect the time required to dissolve a given amount of solute on Page 437.

13. Increase the amount of surface area of the solid by dividing it into smaller pieces, agitate the solution, or increase the temperature.

    See Section 16.3, The Formation of a Solution, on Pages 435-438. In particular, see the list of factors that affect the time required to dissolve a given amount of solute on Page 437.

14. (a) dimethyl ether and (d) hydrogen fluoride. Dimethyl ether is polar, as is water, and hydrogen fluoride will match the hydrogen bonding found in water. Hexane and tetrachloroethane are nonpolar and more apt to be soluble in nonpolar cyclohexane.

    Substances with similar intermolecular forces will usually dissolve in one another. See Section 16.4, Factors That Determine Solubility, on Pages 438-439. You may also find it helpful to review Section 15.3, Types of Intermolecular Forces, on Page 404-408.

15. Water is polar and has hydrogen bonding. The same is true of ethanol, so it would probably not work. A nonpolar solvent is more likely to dissolve what a polar solvent does not. Cyclohexane is more promising.

    Substances with similar intermolecular forces will usually dissolve in one another. See Section 16.4, Factors That Determine Solubility, on Pages 438-439. You may also find it helpful to review Section 15.3, Types of Intermolecular Forces, on Page 404-408.

16. The statement is true only if the air contains carbon dioxide. The solubility of a gas in a liquid depends on the partial pressure of that gas over the liquid. It is independent of the partial pressures of other gases and of the total pressure.

See Section 16.4, Factors That Determine Solubility, on Pages 438-439. In particular, see the Partial Pressure of Solute Gas over Liquid Solution subsection on Pages 438-439.

17. *GIVEN:* 135 g solution; 18.5 g solute *WANTED:* % by mass

EQUATION: $\% \text{ by mass} = \dfrac{\text{g solute}}{\text{g solution}} \times 100 = \dfrac{18.5 \text{ g salt}}{135 \text{ g soln}} \times 100 = 13.7\% \text{ salt}$

See Section 16.5, Solution Concentration: Percentage by Mass, on Pages 439-441.

18. *GIVEN:* 65.0 g solution; 13.0% solution *WANTED:* g solute

PER/PATH: $\text{g solution} \xrightarrow{13.0 \text{ g solute/100 g solution}} \text{g solute}$

$65.0 \text{ g soln} \times \dfrac{13.0 \text{ g solute}}{100 \text{ g soln}} = 8.45 \text{ g solute}$

See Section 16.5, Solution Concentration: Percentage by Mass, on Pages 439-441.

19. *GIVEN:* 6.00×10^2 mL solution; 23.5 g Na_2SO_4 *WANTED:* M

PER/PATH: $\text{g } Na_2SO_4 \xrightarrow{142.05 \text{ g } Na_2SO_4/\text{mol } Na_2SO_4} \text{mol } Na_2SO_4$

$23.5 \text{ g } Na_2SO_4 \times \dfrac{1 \text{ mol } Na_2SO_4}{142.05 \text{ g } Na_2SO_4} = 0.165 \text{ mol } Na_2SO_4$

EQUATION: $M \equiv \dfrac{\text{mol}}{\text{L}} = \dfrac{0.165 \text{ mol } Na_2SO_4}{6.00 \times 10^2 \text{ mL}} \times \dfrac{1000 \text{ mL}}{\text{L}} = 0.275 \text{ M } Na_2SO_4$

See Section 16.6, Solution Concentration: Molarity, on Pages 441-444.

20. *GIVEN:* 1.2×10^2 g $Na_2S_2O_3 \cdot 5\,H_2O$; 1.25×10^3 mL solution *WANTED:* M

PER/PATH: $\text{g } Na_2S_2O_3 \cdot 5\,H_2O \xrightarrow{248.20 \text{ g } Na_2S_2O_3 \cdot 5\,H_2O/\text{mol } Na_2S_2O_3 \cdot 5\,H_2O} \text{g } Na_2S_2O_3 \cdot 5\,H_2O$

$1.2 \times 10^2 \text{ g } Na_2S_2O_3 \cdot 5\,H_2O \times \dfrac{1 \text{ mol } Na_2S_2O_3}{248.20 \text{ g } Na_2S_2O_3 \cdot 5\,H_2O} = 0.48 \text{ mol } Na_2S_2O_3$

EQUATION: $M \equiv \dfrac{\text{mol}}{\text{L}} = \dfrac{0.48 \text{ mol } Na_2S_2O_3}{1.25 \times 10^3 \text{ mL}} \times \dfrac{1000 \text{ mL}}{\text{L}} = 0.38 \text{ M } Na_2S_2O_3$

See Section 16.6, Solution Concentration: Molarity, on Pages 441-444.

21. *GIVEN:* 4.00×10^2 mL solution; 0.800 M Na_2CO_3 *WANTED:* g Na_2CO_3

PER/PATH: $\text{mL} \xrightarrow{1000 \text{ mL/L}} \text{L} \xrightarrow{0.800 \text{ mol } Na_2CO_3/\text{L}} \text{mol } Na_2CO_3 \xrightarrow{105.99 \text{ g } Na_2CO_3/\text{mol } Na_2CO_3} \text{g } Na_2CO_3$

$4.00 \times 10^2 \text{ mL} \times \dfrac{1 \text{ L}}{1000 \text{ mL}} \times \dfrac{0.800 \text{ mol } Na_2CO_3}{\text{L}} \times \dfrac{105.99 \text{ g } Na_2CO_3}{\text{mol } Na_2CO_3} = 33.9 \text{ g } Na_2CO_3$

See Section 16.6, Solution Concentration: Molarity, on Pages 441-444.

22. *GIVEN:* 7.50×10^2 mL solution; 0.600 M $HC_2H_3O_2$ *WANTED:* g $HC_2H_3O_2$

PER/PATH: $\text{mL} \xrightarrow{1000\ \text{mL/L}} \text{L} \xrightarrow{0.600\ \text{mol}\ HC_2H_3O_2/\text{mol}\ HC_2H_3O_2} \text{mol}\ HC_2H_3O_2 \xrightarrow{60.05\ \text{g}\ HC_2H_3O_2/\text{mol}\ HC_2H_3O_2} \text{g}\ HC_2H_3O_2$

$$7.50 \times 10^2\ \text{mL} \times \frac{1\ \text{L}}{1000\text{mL}} \times \frac{0.600\ \text{mol}\ HC_2H_3O_2}{\text{L}} \times \frac{60.05\ \text{g}\ HC_2H_3O_2}{\text{mol}\ HC_2H_3O_2} = 27.0\ \text{g}\ HC_2H_3O_2$$

See Section 16.6, Solution Concentration: Molarity, on Pages 441-444.

23. *GIVEN:* 0.0150 mol H^+; 0.850 M HCl *WANTED:* mL

PER/PATH: $\text{mol H+} \xrightarrow{1\ \text{mol}\ H^+/1\ \text{mol HCl}} \text{mol HCl} \xrightarrow{0.850\ \text{mol HCl/L}} \text{L} \xrightarrow{1000\ \text{mL/L}} \text{mL}$

$$0.0150\ \text{mol}\ H^+ \times \frac{1\ \text{mol HCl}}{1\ \text{mol}\ H^+} \times \frac{1\ \text{L}}{0.850\ \text{mol HCl}} \times \frac{1000\ \text{mL}}{\text{L}} = 17.6\ \text{mL}$$

See Section 16.6, Solution Concentration: Molarity, on Pages 441-444.

24. *GIVEN:* 15 M NH_3; 75.0 g NH_3 *WANTED:* volume solution (assume L)

PER/PATH: $\text{g}\ NH_3 \xrightarrow{17.03\ \text{g}\ NH_3/\text{mol}\ NH_3} \text{mol}\ NH_3 \xrightarrow{15\ \text{mol}\ NH_3/\text{L}} \text{L}$

$$75.0\ \text{g}\ NH_3 \times \frac{1\ \text{mol}\ NH_3}{17.03\ \text{g}\ NH_3} \times \frac{1\ \text{L}}{15\ \text{mol}\ NH_3} = 0.29\ \text{L}$$

See Section 16.6, Solution Concentration: Molarity, on Pages 441-444.

25. *GIVEN:* 65.0 mL; 2.20 M NaOH *WANTED:* mol NaOH

PER/PATH: $\text{mL} \xrightarrow{1000\ \text{mL/L}} \text{L} \xrightarrow{2.20\ \text{mol NaOH/L}} \text{mol NaOH}$

$$65.0\ \text{mL} \times \frac{1\ \text{L}}{1000\ \text{mL}} \times \frac{2.20\ \text{mol NaOH}}{\text{L}} = 0.143\ \text{mol NaOH}$$

See Section 16.6, Solution Concentration: Molarity, on Pages 441-444.

26. *GIVEN:* 29.3 mL; 0.482 M H_2SO_4 *WANTED:* mol H_2SO_4

PER/PATH: $\text{mL} \xrightarrow{1000\ \text{mL/L}} \text{L} \xrightarrow{0.482\ \text{mol}\ H_2SO_4/\text{L}} \text{mol}\ H_2SO_4$

$$29.3\ \text{mL} \times \frac{1\ \text{L}}{1000\ \text{mL}} \times \frac{0.482\ \text{mol}\ H_2SO_4}{\text{L}} = 0.0141\ \text{mol}\ H_2SO_4$$

See Section 16.6, Solution Concentration: Molarity, on Pages 441-444.

27. *GIVEN:* 1.09 g solution/mL solution; 2.4 mol solute/L solution; 81.7 g solute/mol solute
WANTED: % concentration

PER/PATH: $81.7\ \text{g solute/mol solute} \xrightarrow{2.4\ \text{mol solute/L solution}} \text{g solute/L solution} \xrightarrow{1000\ \text{mL solution/L solution}} \text{g solute/mL solution} \xrightarrow{1.09\ \text{g solution/mL solution}} \text{g solute/g solution} = \%\ \text{concentration}$

$$\frac{81.7\ \text{g solute}}{\text{mol solute}} \times \frac{2.4\ \text{mol solute}}{\text{L soln}} \times \frac{1\ \text{L soln}}{1000\ \text{mL soln}} \times \frac{1\ \text{mL soln}}{1.09\ \text{g soln}} = 0.18\ \frac{\text{g solute}}{\text{g soln}}$$

$$0.18\ \frac{\text{g solute}}{\text{g soln}} \times \frac{100}{100} = \frac{18\ \text{g solute}}{100\ \text{g soln}} = 18\%\ \text{solute}$$

See Section 16.6, Solution Concentration: Molarity, on Pages 441-444.

28. *GIVEN:* 20.0 g $C_{12}H_{22}O_{11}$ *WANTED:* mol $C_{12}H_{22}O_{11}$

PER/PATH: g $C_{12}H_{22}O_{11}$ $\xrightarrow{342.30\ g\ C_{12}H_{22}O_{11}/mol\ C_{12}H_{22}O_{11}}$ mol $C_{12}H_{22}O_{11}$

$$20.0\ g\ C_{12}H_{22}O_{11} \times \frac{1\ mol\ C_{12}H_{22}O_{11}}{342.30\ g\ C_{12}H_{22}O_{11}} = 0.0584\ mol\ C_{12}H_{22}O_{11}$$

GIVEN: 1.00×10^2 mL H_2O *WANTED:* kg H_2O

PER/PATH: mL H_2O $\xrightarrow{1\ g\ H_2O/1\ mL\ H_2O}$ g H_2O $\xrightarrow{1000\ g\ H_2O/kg\ H_2O}$ kg H_2O

$$1.00 \times 10^2\ mL\ H_2O \times \frac{1\ g\ H_2O}{1\ mL\ H_2O} \times \frac{1\ kg\ H_2O}{1000\ g\ H_2O} = 0.100\ kg\ H_2O$$

EQUATION: $m \equiv \frac{mol\ C_{12}H_{22}O_{11}}{kg\ H_2O} = \frac{0.0584\ mol\ C_{12}H_{22}O_{11}}{0.100\ kg\ H_2O} = 0.584\ m$

See Section 16.7, Solution Concentration: Molality, on Pages 444-446.

29. *GIVEN:* 4.00 m $CO(NH_2)_2$; 80.0 mL H_2O *WANTED:* g $CO(NH_2)_2$

PER/PATH: mL H_2O $\xrightarrow{1\ g\ H_2O/1\ mL\ H_2O}$ g H_2O $\xrightarrow{1\ kg\ H_2O/1000\ g\ H_2O}$ kg H_2O $\xrightarrow{4.00\ mol\ CO(NH_2)_2/kg\ H_2O}$ mol $CO(NH_2)_2$ $\xrightarrow{60.06\ g\ CO(NH_2)_2/mol\ CO(NH_2)_2}$ g $CO(NH_2)_2$

$$80.0\ mL\ H_2O \times \frac{1\ g\ H_2O}{mL\ H_2O} \times \frac{1\ kg\ H_2O}{1000\ g\ H_2O} \times \frac{4.00\ mol\ CO(NH_2)_2}{kg\ H_2O} \times \frac{60.06\ g\ CO(NH_2)_2}{mol\ CO(NH_2)_2} = 19.2\ g\ CO(NH_2)_2$$

See Section 16.7, Solution Concentration: Molality, on Pages 444-446.

30. *GIVEN:* 90.9 g $HC_2H_3O_2$; 1.40 m $HC_2H_3O_2$ *WANTED:* mL H_2O

PER/PATH: g $HC_2H_3O_2$ $\xrightarrow{60.05\ g\ HC_2H_3O_2/mol\ HC_2H_3O_2}$ mol $HC_2H_3O_2$ $\xrightarrow{1.40\ mol\ HC_2H_3O_2/kg\ H_2O}$ kg H_2O $\xrightarrow{1000\ g\ H_2O/kg\ H_2O}$ g H_2O $\xrightarrow{1\ g\ H_2O/1\ mL\ H_2O}$ mL H_2O

$$90.9\ g\ HC_2H_3O_2 \times \frac{1\ mol\ HC_2H_3O_2}{60.05\ g\ HC_2H_3O_2} \times \frac{1\ kg\ H_2O}{1.40\ mol\ HC_2H_3O_2} \times \frac{1000\ g\ H_2O}{kg\ H_2O} \times \frac{1\ mL\ H_2O}{1\ g\ H_2O} = 1.08 \times 10^3\ mL\ H_2O$$

See Section 16.7, Solution Concentration: Molality, on Pages 444-446.

31. The number of equivalents per mole depends on a specific reaction, not just the number of moles of H^+ or OH^- in a compound.

See Section 16.8, Solution Concentration: Normality, on Pages 446-451.

32. 1 eq/mol HF; 2 eq/mol $H_2C_2O_4$

HF has 1 eq/mol because there is only one H^+ available to react. Two H^+ react in the $H_2C_2O_4$ reaction, so there are 2 eq/mol. See Section 16.8, Solution Concentration: Normality, on Pages 446-451.

33. 2 eq/mol $Zn(OH)_2$; 1 eq/mol RbOH

The maximum number of equivalents in a hydroxide-ion base is the number of hydroxide ions in the formula unit. See Section 16.8, Solution Concentration: Normality, on Pages 446-451.

34. 20.01 g HF/eq; 90.04 g $H_2C_2O_4$/2 eq = 45.02 g $H_2C_2O_4$/eq

HF: 1.008 + 19.00 = 20.01 g/mol; $H_2C_2O_4$: 2(1.008) + 2(12.01) + 4(16.00) = 90.04 g/mol. See Section 16.8, Solution Concentration: Normality, on Pages 446-451.

35. 99.41 g $Zn(OH)_2$/2 eq = 49.71 g $Zn(OH)_2$/eq; 102.48 g RbOH/eq

$Zn(OH)_2$: 65.39 + 2(16.00) + 2(1.008) = 99.41 g/mol; RbOH: 85.47 + 16.00 + 1.008 = 102.48 g/mol. See Section 16.8, Solution Concentration: Normality, on Pages 446-451.

36. *GIVEN:* 17.2 g $HC_2H_3O_2$ *WANTED:* eq $HC_2H_3O_2$

PER/PATH: $\text{g } HC_2H_3O_2 \xrightarrow{60.05 \text{ g } HC_2H_3O_2/\text{eq } HC_2H_3O_2} \text{eq } HC_2H_3O_2$

$$17.2 \text{ g } HC_2H_3O_2 \times \frac{1 \text{ eq } HC_2H_3O_2}{60.05 \text{ g } HC_2H_3O_2} = 0.286 \text{ eq } HC_2H_3O_2$$

GIVEN: 0.286 eq $HC_2H_3O_2$; 3.00×10^2 mL *WANTED:* N $HC_2H_3O_2$

EQUATION: $N \equiv \frac{\text{eq}}{\text{L}} = \frac{0.286 \text{ eq } HC_2H_3O_2}{3.00 \times 10^2 \text{ mL}} \times \frac{1000 \text{ mL}}{\text{L}} = 0.953 \text{ N } HC_2H_3O_2$

See Section 16.8, Solution Concentration: Normality, on Pages 446-451.

37. *GIVEN:* 9.79 g $NaHCO_3$ *WANTED:* eq $NaHCO_3$

PER/PATH: $\text{g } NaHCO_3 \xrightarrow{84.01 \text{ g } NaHCO_3/\text{eq } NaHCO_3} \text{eq } NaHCO_3$

$$9.79 \text{ g } NaHCO_3 \times \frac{1 \text{ eq } NaHCO_3}{84.01 \text{ g } NaHCO_3} = 0.117 \text{ eq } NaHCO_3$$

GIVEN: 0.117 eq $NaHCO_3$; 5.00×10^2 mL *WANTED:* N NaHCO3

EQUATION: $N \equiv \frac{\text{eq}}{\text{L}} = \frac{0.117 \text{ eq } NaHCO_3}{5.00 \times 10^2 \text{ mL}} \times \frac{1000 \text{ mL}}{\text{L}} = 0.234 \text{ N } NaHCO_3$

See Section 16.8, Solution Concentration: Normality, on Pages 446-451.

38. *GIVEN:* 6.00×10^2 mL; 2.00 N KOH *WANTED:* g KOH

PER/PATH: $\text{mL} \xrightarrow{1000 \text{ mL/L}} \text{L} \xrightarrow{2.00 \text{ eq KOH/L}} \text{eq KOH} \xrightarrow{56.11 \text{ g KOH/eq KOH}} \text{g KOH}$

$$6.00 \times 10^2 \text{ mL} \times \frac{1 \text{ L}}{1000 \text{ mL}} \times \frac{2.00 \text{ eq KOH}}{\text{L}} \times \frac{56.11 \text{ g KOH}}{\text{eq KOH}} = 67.3 \text{ g KOH}$$

See Section 16.8, Solution Concentration: Normality, on Pages 446-451.

39. (a) 0.423 N HCl; (b) 0.846 N H_2SO_4

HCl: 1 eq/mol; H_2SO_4: 2 eq/mol, so $\frac{0.423 \text{ mol } H_2SO_4}{L} \times \frac{2 \text{ eq } H_2SO_4}{\text{mol } H_2SO_4} = \frac{0.846 \text{ eq } H_2SO_4}{L}$

See Section 16.8, Solution Concentration: Normality, on Pages 446-451.

40. *GIVEN:* 2.25 L *WANTED:* eq H_2SO_4

PER/PATH: $L \xrightarrow{0.871 \text{ eq } H_2SO_4/L} \text{eq } H_2SO_4$

$$2.25 \text{ L} \times \frac{0.871 \text{ eq } H_2SO_4}{L} = 1.96 \text{ eq } H_2SO_4$$

See Section 16.8, Solution Concentration: Normality, on Pages 446-451.

41. *GIVEN:* 0.0385 eq *WANTED:* volume (assume L)

PER/PATH: $\text{eq} \xrightarrow{0.371 \text{ eq HCl/L}} L$

$$0.0385 \text{ eq} \times \frac{1 \text{ L}}{0.371 \text{ eq HCl}} = 0.104 \text{ L}$$

See Section 16.8, Solution Concentration: Normality, on Pages 446-451.

42. $V_c = 1.00 \times 10^2$ mL $M_c = 12$ M

$V_d = 2.00$ L M_d = wanted

$$V_c \times M_c = V_d \times M_d \qquad \frac{V_c \times M_c}{V_d} = \frac{V_d \times M_d}{V_d} \qquad \frac{V_c \times M_c}{V_d} = M_d$$

$$M_d = \frac{V_c \times M_c}{V_d} = \frac{1.00 \times 10^2 \text{ mL} \times 12 \text{ M}}{2.00 \text{ L}} \times \frac{1 \text{ L}}{1000 \text{ mL}} = 0.60 \text{ M HCl}$$

See Section 16.10, Dilution of Concentrated Solutions, on Pages 452-453.

43. V_c = wanted $M_c = 15$ M

$V_d = 5.00 \times 10^2$ mL $M_d = 6.0$ M

$$V_c \times M_c = V_d \times M_d \qquad \frac{V_c \times M_c}{M_c} = \frac{V_d \times M_d}{M_c} \qquad V_c = \frac{V_d \times M_d}{M_c}$$

$$V_c = \frac{V_d \times M_d}{M_c} = \frac{5.00 \times 10^2 \text{ mL} \times 6.0 \text{ M}}{15 \text{ M}} = 2.0 \times 10^2 \text{ mL } NH_3$$

See Section 16.10, Dilution of Concentrated Solutions, on Pages 452-453.

44. V_c = wanted $M_c = 12$ M

$V_d = 2.0$ L "M_d" = 0.50 N

$$V_c \times M_c = V_d \times M_d \qquad \frac{V_c \times M_c}{M_c} = \frac{V_d \times M_d}{M_c} \qquad V_c = \frac{V_d \times M_d}{M_c}$$

$$V_c = \frac{V_d \times M_d}{M_c} \doteq \frac{2.0 \text{ L} \times 0.50 \text{ eq HCl/L}}{12 \text{ mol HCl/L}} \times \frac{1 \text{ mol HCl}}{\text{eq HCl}} \times \frac{1000 \text{ mL}}{L} = 83 \text{ mL HCl}$$

See Section 16.10, Dilution of Concentrated Solutions, on Pages 452-453.

45. $V_c = 25.0$ mL $\quad M_c = 15$ M

$V_d = 4.00 \times 10^2$ mL $\quad$ "M_d" = wanted (N)

$$V_c \times M_c = V_d \times M_d \qquad \frac{V_c \times M_c}{V_d} = \frac{V_d \times M_d}{V_d} \qquad \frac{V_c \times M_c}{V_d} = M_d\ (N_d)$$

$$N_d = \frac{V_c \times M_c}{V_d} = \frac{25.0 \text{ mL} \times 15 \text{ mol HNO}_3\text{/L}}{4.00 \times 10^2 \text{ mL}} \times \frac{1 \text{ eq HNO}_3}{1 \text{ mol HNO}_3} = 0.94 \text{ N HNO}_3$$

See Section 16.10, Dilution of Concentrated Solutions, on Pages 452-453.

46. $AgNO_3 + NaCl \rightarrow AgCl + NaNO_3$

GIVEN: 50.0 mL $\quad$ *WANTED:* g AgCl

PER/PATH: $\text{mL} \xrightarrow{1000 \text{ mL/L}} \text{L} \xrightarrow{0.855 \text{ mol AgNO}_3\text{/L}} \text{mol AgNO}_3 \xrightarrow{1 \text{ mol AgCl/1 mol AgNO}_3} \text{mol AgCl} \xrightarrow{143.4 \text{ g AgCl/mol AgCl}} \text{g AgCl}$

$$50.0 \text{ mL} \times \frac{1 \text{ L}}{1000 \text{ mL}} \times \frac{0.855 \text{ mol AgNO}_3}{\text{L}} \times \frac{1 \text{ mol AgCl}}{1 \text{ mol AgNO}_3} \times \frac{143.4 \text{ g AgCl}}{\text{mol AgCl}} = 6.13 \text{ g AgCl}$$

See Section 16.11, Solution Stoichiometry, on Pages 453-456.

47. $Ba(NO_3)_2 + 2\ NaF \rightarrow BaF_2 + 2\ NaNO_3$

GIVEN: 40.0 mL $\quad$ *WANTED:* mass BaF_2 (assume g)

PER/PATH: $\text{mL} \xrightarrow{1000 \text{ mL/L}} \text{L} \xrightarrow{0.436 \text{ mol NaF/L}} \text{mol NaF} \xrightarrow{1 \text{ mol BaF}_2\text{/2 mol NaF}} \text{mol BaF}_2 \xrightarrow{175.3 \text{ g BaF}_2\text{/mol BaF}_2} \text{g BaF}_2$

$$40.0 \text{ mL} \times \frac{1 \text{ L}}{1000 \text{ mL}} \times \frac{0.436 \text{ mol NaF}}{\text{L}} \times \frac{1 \text{ mol BaF}_2}{2 \text{ mol NaF}} \times \frac{175.3 \text{ g BaF}_2}{\text{mol BaF}_2} = 1.53 \text{ g BaF}_2$$

See Section 16.11, Solution Stoichiometry, on Pages 453-456.

48. *GIVEN:* 50.0 mL $\quad$ *WANTED:* volume Cl_2 (assume L)

PER/PATH: $\text{mL} \xrightarrow{1000 \text{ mL/L}} \text{L} \xrightarrow{1.20 \text{ mol HCl/L}} \text{mol HCl} \xrightarrow{1 \text{ mol Cl}_2\text{/4 mol HCl}} \text{mol Cl}_2 \xrightarrow{22.4 \text{ L Cl}_2\text{/mol Cl}_2} \text{L Cl}_2$

$$50.0 \text{ mL} \times \frac{1 \text{ L}}{1000 \text{ mL}} \times \frac{1.20 \text{ mol HCl}}{\text{L}} \times \frac{1 \text{ mol Cl}_2}{4 \text{ mol HCl}} \times \frac{22.4 \text{ L Cl}_2}{\text{mol Cl}_2} = 0.336 \text{ L Cl}_2$$

See Section 16.11, Solution Stoichiometry, on Pages 453-456.

49. $Na_2CO_3 + 2\ HCl \rightarrow 2\ NaCl + H_2O + CO_2$

GIVEN: 1.24 g Na_2CO_3 $\quad$ *WANTED:* mL HCl

PER/PATH: $\text{g Na}_2\text{CO}_3 \xrightarrow{105.99 \text{ g Na}_2\text{CO}_3\text{/mol Na}_2\text{CO}_3} \text{mol Na}_2\text{CO}_3 \xrightarrow{2 \text{ mol HCl/1 mol Na}_2\text{CO}_3} \text{mol HCl} \xrightarrow{0.715 \text{ mol HCl/L}} \text{L} \xrightarrow{1000 \text{ mL/L}} \text{mL}$

$$1.24 \text{ g Na}_2\text{CO}_3 \times \frac{1 \text{ mol Na}_2\text{CO}_3}{105.99 \text{ g Na}_2\text{CO}_3} \times \frac{2 \text{ mol HCl}}{1 \text{ mol Na}_2\text{CO}_3} \times \frac{1 \text{ L}}{0.715 \text{ mol HCl}} \times \frac{1000 \text{ mL}}{\text{L}} = 32.7 \text{ mL}$$

See Section 16.11, Solution Stoichiometry, on Pages 453-456.

50. *GIVEN:* 1.359 g $KH(IO_3)_2$ *WANTED:* mol KOH

PER/PATH: g $KH(IO_3)_2$ $\xrightarrow{389.9\text{ g }KH(IO_3)_2/\text{mol }KH(IO_3)_2}$ mol $KH(IO_3)_2$ $\xrightarrow{1\text{ mol KOH}/1\text{ mol }KH(IO_3)_2}$ mol KOH

$$1.359\text{ g }KH(IO_3)_2 \times \frac{1\text{ mol }KH(IO_3)_2}{389.9\text{ g }KH(IO_3)_2} \times \frac{1\text{ mol KOH}}{1\text{ mol }KH(IO_3)_2} = 0.003486\text{ mol KOH}$$

GIVEN: 0.003486 mol KOH; 32.14 mL *WANTED:* M KOH

EQUATION: $M \equiv \frac{\text{mol}}{\text{L}} = \frac{0.003486\text{ mol KOH}}{32.14\text{ mL}} \times \frac{1000\text{ mL}}{\text{L}} = 0.1085\text{ M KOH}$

See Section 16.12, Titration Using Molarity, on Pages 456-459.

51. $H_2C_2O_4 + 2\ NaOH \rightarrow 2\ H_2O + Na_2C_2O_4$

GIVEN: 3.290 g $H_2C_2O_4 \cdot 2\ H_2O$ *WANTED:* mol NaOH

PER/PATH: g $H_2C_2O_4 \cdot 2\ H_2O$ $\xrightarrow{126.07\text{ g }H_2C_2O_4 \cdot 2\ H_2O/\text{mol }H_2C_2O_4 \cdot 2\ H_2O}$ mol $H_2C_2O_4 \cdot 2\ H_2O$ $\xrightarrow{2\text{ mol NaOH}/1\text{ mol }H_2C_2O_4}$ mol NaOH

$$3.290\text{ g }H_2C_2O_4 \cdot 2\ H_2O \times \frac{1\text{ mol }H_2C_2O_4 \cdot 2\ H_2O}{126.07\text{ g }H_2C_2O_4 \cdot 2\ H_2O} \times \frac{2\text{ mol NaOH}}{1\text{ mol }H_2C_2O_4} = 0.05219\text{ mol NaOH}$$

GIVEN: 0.05219 mol NaOH; 30.10 mL *WANTED:* M NaOH

EQUATION: $M \equiv \frac{\text{mol}}{\text{L}} = \frac{0.05219\text{ mol NaOH}}{30.10\text{ mL}} \times \frac{1000\text{ mL}}{\text{L}} = 1.734\text{ M NaOH}$

See Section 16.12, Titration Using Molarity, on Pages 456-459.

52. *GIVEN:* 15.0 mL *WANTED:* mol $H_2C_2O_4$

PER/PATH: mL $\xrightarrow{1000\text{ mL/L}}$ L $\xrightarrow{0.100\text{ mol NaOH/L}}$ mol NaOH $\xrightarrow{1\text{ mol }H_2C_2O_4/2\text{ mol NaOH}}$ mol $H_2C_2O_4$

$$15.0\text{ mL} \times \frac{1\text{ L}}{1000\text{ mL}} \times \frac{0.100\text{ mol NaOH}}{\text{L}} \times \frac{1\text{ mol }H_2C_2O_4}{2\text{ mol NaOH}} = 0.000750\text{ mol }H_2C_2O_4$$

GIVEN: 0.000750 mol $H_2C_2O_4$; 25.0 mL *WANTED:* M $H_2C_2O_4$

EQUATION: $M \equiv \frac{\text{mol}}{\text{L}} = \frac{0.000750\text{ mol }H_2C_2O_4}{25.0\text{ mL}} \times \frac{1000\text{ mL}}{\text{L}} = 0.0300\text{ M }H_2C_2O_4$

See Section 16.12, Titration Using Molarity, on Pages 456-459.

53. At 1 eq/mol, 0.1085 M KOH = 0.1085 N KOH

The base in Question 50 is KOH, and you determined its concentration to be 0.1085 M. One mole of KOH reacts with one mole of acid, so there is 1 eq/mol. See Section 16.8, Solution Concentration: Normality, on Pages 446-451.

54. *GIVEN:* 3.290 g $H_2C_2O_4 \cdot 2\ H_2O$ *WANTED:* eq NaOH

PER/PATH: g $H_2C_2O_4 \cdot 2\ H_2O$ $\xrightarrow{126.07\text{ g }H_2C_2O_4 \cdot 2\ H_2O/\text{mol }H_2C_2O_4 \cdot 2\ H_2O}$ mol $H_2C_2O_4 \cdot 2\ H_2O$ $\xrightarrow{2\text{ mol NaOH}/1\text{ mol }H_2C_2O_4}$ mol NaOH $\xrightarrow{1\text{ eq NaOH}/1\text{ mol NaOH}}$ eq NaOH

$$3.290\text{ g }H_2C_2O_4 \cdot 2\ H_2O \times \frac{1\text{ mol }H_2C_2O_4 \cdot 2\ H_2O}{126.07\text{ g }H_2C_2O_4 \cdot 2\ H_2O} \times \frac{2\text{ mol NaOH}}{1\text{ mol }H_2C_2O_4} \times \frac{1\text{ eq NaOH}}{\text{mol NaOH}} = 0.05219\text{ eq NaOH}$$

GIVEN: 0.05219 eq NaOH; 30.10 mL *WANTED:* N NaOH

EQUATION: $N \equiv \frac{eq}{L} = \frac{0.05219 \text{ eq NaOH}}{30.10 \text{ mL}} \times \frac{1000 \text{ mL}}{L} = 1.734 \text{ N NaOH}$

See Section 16.13, Titration Using Normality, on Pages 459-461.

55. $V_1 = 15.0$ mL $N_1 = 0.882$ N
$V_2 = 12.8$ mL N_2 = wanted

$V_1 \times N_1 = V_2 \times N_2 \quad \frac{V_1 \times N_1}{V_2} = \frac{V_2 \times N_2}{V_2} \quad \frac{V_1 \times N_1}{V_2} = N_2$

$N_2 = \frac{V_1 \times N_1}{V_2} = \frac{15.0 \text{ mL} \times 0.882 \text{ N}}{12.8 \text{ mL}} = 1.03 \text{ N acid}$

See Section 16.13, Titration Using Normality, on Pages 459-461.

56. $V_1 = 28.4$ mL $N_1 = 0.424$ N
$V_2 = 25.0$ mL N_2 = wanted

$V_1 \times N_1 = V_2 \times N_2 \quad \frac{V_1 \times N_1}{V_2} = \frac{V_2 \times N_2}{V_2} \quad \frac{V_1 \times N_1}{V_2} = N_2$

$N_2 = \frac{V_1 \times N_1}{V_2} = \frac{28.4 \text{ mL} \times 0.424 \text{ N}}{25.0 \text{ mL}} = 0.482 \text{ N } Ba(OH)_2$

See Section 16.13, Titration Using Normality, on Pages 459-461.

57. $V_1 = 32.6$ mL $N_1 = 0.208$ N
$V_2 = 20.0$ mL N_2 = wanted

$V_1 \times N_1 = V_2 \times N_2 \quad \frac{V_1 \times N_1}{V_2} = \frac{V_2 \times N_2}{V_2} \quad \frac{V_1 \times N_1}{V_2} = N_2$

$N_2 = \frac{32.6 \text{ mL} \times 0.208 \text{ N}}{20.0 \text{ mL}} = 0.339 \text{ N } H_3PO_4$

See Section 16.13, Titration Using Normality, on Pages 459-461.

58. *GIVEN:* 15.6 mL *WANTED:* eq NaOH (= eq acid)

PER/PATH: mL $\xrightarrow{1000 \text{ mL/L}}$ L $\xrightarrow{0.562 \text{ eq/L}}$ eq

$15.6 \text{ mL} \times \frac{1 \text{ L}}{1000 \text{ mL}} \times \frac{0.562 \text{ eq}}{L} = 0.00877 \text{ eq}$

GIVEN: 0.00877 eq; 0.631 g *WANTED:* equivalent mass (g/eq)

EQUATION: Equivalent mass $\equiv \frac{g}{eq} = \frac{0.631 \text{ g}}{0.00877 \text{ eq}} = 71.9 \text{ g/eq}$

Using Equation 16.10, VN = eq, $\frac{0.631 \text{ g}}{15.6 \text{ mL} \times 0.562 \text{ eq/L}} \times \frac{1000 \text{ mL}}{L} = 72.0 \text{ g/eq}$

Equivalent mass is defined on Page 448. Also see Section 16.13, Titration Using Normality, on Pages 459-461.

59. A colligative property of a solution is independent of the identity of the solute. Specific gravity, with opposite effects from different solutes, is not a colligative property.

See Section 16.14, Colligative Properties of Solutions, on Pages 461-464.

60. Find solution molality, mol solute/kg solvent:

GIVEN: 50.0 g $C_6H_{12}O_6$ *WANTED:* mol $C_6H_{12}O_6$

PER/PATH: g $C_6H_{12}O_6$ $\xrightarrow{180.16\text{ g }C_6H_{12}O_6/\text{mol }C_6H_{12}O_6}$ mol $C_6H_{12}O_6$

$$50.0\text{ g }C_6H_{12}O_6 \times \frac{1\text{ mol }C_6H_{12}O_6}{180.16\text{ g }C_6H_{12}O_6} = 0.278\text{ mol }C_6H_{12}O_6$$

GIVEN: 0.278 mol $C_6H_{12}O_6$; 1.00×10^2 g H_2O *WANTED:* m

$$\textit{EQUATION:}\ m \equiv \frac{\text{mol solute}}{\text{kg solvent}} = \frac{\text{mol }C_6H_{12}O_6}{\text{kg }H_2O} = \frac{0.278\text{ mol }C_6H_{12}O_6}{1.00 \times 10^2\text{ g }H_2O} \times \frac{1000\text{ g }H_2O}{\text{kg }H_2O} = 2.78\text{ m}$$

Find freezing point depression/boiling point elevation:

$$\Delta T_b = K_b \times m = \frac{0.52°C}{m} \times 2.78\text{ m} = 1.4°C \qquad T_b = 100.0 + 1.4 = 101.4°C$$

$$\Delta T_f = K_f \times m = \frac{1.86°C}{m} \times 2.78\text{ m} = 5.17°C \qquad T_f = 0.00 - 5.17 = -5.17°C$$

See Section 16.14, Colligative Properties of Solutions, on Pages 461-464.

61. Find solution molality, mol solute/kg solvent:

GIVEN: 4.34 g $C_6H_4Cl_2$ *WANTED:* mol $C_6H_4Cl_2$

PER/PATH: g $C_6H_4Cl_2$ $\xrightarrow{146.99\text{ g }C_6H_4Cl_2/\text{mol }C_6H_4Cl_2}$ mol $C_6H_4Cl_2$

$$4.34\text{ g }C_6H_4Cl_2 \times \frac{1\text{ mol }C_6H_4Cl_2}{146.99\text{ g }C_6H_4Cl_2} = 0.0295\text{ mol }C_6H_4Cl_2$$

GIVEN: 0.0295 mol $C_6H_4Cl_2$; 65.0 g $C_{10}H_8$ *WANTED:* m

$$\textit{EQUATION:}\ m \equiv \frac{\text{mol solute}}{\text{kg solvent}} = \frac{\text{mol }C_6H_4Cl_2}{\text{kg }C_{10}H_8} = \frac{0.0295\text{ mol }C_6H_4Cl_2}{65.0\text{ g }C_{10}H_8} \times \frac{1000\text{ g }C_{10}H_8}{\text{kg }C_{10}H_8} = 0.454\text{ m}$$

Find freezing point depression:

$$\Delta T_f = K_f \times m = \frac{6.9°C}{m} \times 0.454\text{ m} = 3.13°C \qquad T_f = 80.2 - 3.13 = 77.1°C$$

See Section 16.14, Colligative Properties of Solutions, on Pages 461-464.

62. $\Delta T_b = 100.84 - 100.00 = 0.84°C$

$$\Delta T_b = K_b \times m \qquad m = \frac{\Delta T_b}{K_b}$$

$$m = \frac{\Delta T_b}{K_b} = \Delta T_b \times \frac{1}{K_b} = 0.84°C \times \frac{1\text{ m}}{0.52°C} = 1.62\text{ m}$$

"An aqueous solution" means that the solvent is water. You therefore use K_b for water. See Section 16.14, Colligative Properties of Solutions, on Pages 461-464.

63. Step 1: Calculate molality

$\Delta T_f = K_f \times m \quad m = \dfrac{\Delta T_f}{K_f}$

$m = \dfrac{\Delta T_f}{K_f} = \Delta T_f \times \dfrac{1}{K_f} = 1.18°C \times \dfrac{1\ m}{1.86°C} = 0.634\ m$

Step 2: Find moles of solute

GIVEN: 3.80×10^2 g H_2O *WANTED:* mol solute

PER/PATH: $g\ H_2O \xrightarrow{1000\ g\ H_2O/kg\ H_2O} kg\ H_2O \xrightarrow{0.634\ mol\ solute/kg\ H_2O}$ mol solute

$3.80 \times 10^2\ g\ H_2O \times \dfrac{1\ kg\ H_2O}{1000\ g\ H_2O} \times \dfrac{0.634\ mol\ solute}{kg\ H_2O} = 0.241\ mol\ solute$

Step 3: MM ≡ g/mol

EQUATION: $MM \equiv \dfrac{g}{mol} = \dfrac{26.0\ g}{0.241\ mol} = 108\ g/mol$

See Section 16.14, Colligative Properties of Solutions, on Pages 461-464. In particular, see the Procedure: How to Calculate the Molar Mass of a Solution from Freezing-Point Depression or Boiling-Point Elevation Data box on Page 463.

64. Step 1: Calculate molality

$\Delta T_f = K_f \times m \quad m = \dfrac{\Delta T_f}{K_f}$

$m = \dfrac{\Delta T_f}{K_f} = \Delta T_f \times \dfrac{1}{K_f} = (80.2 - 71.3)°C \times \dfrac{1\ m}{6.9°C} = 1.3\ m$

Step 2: Find moles of solute

GIVEN: 80.0 g $C_{10}H_8$ *WANTED:* mol solute

PER/PATH: $g\ C_{10}H_8 \xrightarrow{1000\ g\ C_{10}H_8/kg\ C_{10}H_8} kg\ C_{10}H_8 \xrightarrow{1.3\ mol\ solute/kg\ C_{10}H_8}$ mol solute

$80.0\ g\ C_{10}H_8 \times \dfrac{1\ kg\ C_{10}H_8}{1000\ g\ C_{10}H_8} \times \dfrac{1.3\ mol\ solute}{kg\ C_{10}H_8} = 0.10\ mol\ solute$

Step 3: MM ≡ g/mol

EQUATION: $MM \equiv \dfrac{g}{mol} = \dfrac{12.0\ g}{0.10\ mol} = 1.2 \times 10^2\ g/mol$

See Section 16.14, Colligative Properties of Solutions, on Pages 461-464. In particular, see the Procedure: How to Calculate the Molar Mass of a Solution from Freezing-Point Depression or Boiling-Point Elevation Data box on Page 463.

65. Find solution molality, mol solute/kg solvent:

GIVEN: 1.40 g NH_2CONH_2 *WANTED:* NH_2CONH_2

PER/PATH: $g\ NH_2CONH_2 \xrightarrow{60.06\ g\ NH_2CONH_2/mol\ NH_2CONH_2} mol\ NH_2CONH_2$

$1.40\ g\ NH_2CONH_2 \times \dfrac{1\ mol\ NH_2CONH_2}{60.06\ g\ NH_2CONH_2} = 0.0233\ mol\ NH_2CONH_2$

GIVEN: 0.0233 mol NH_2CONH_2; 16.3 g solvent *WANTED:* m

EQUATION: $m \equiv \dfrac{mol\ solute}{kg\ solvent} = \dfrac{mol\ NH_2CONH_2}{kg\ solvent} = \dfrac{0.0233\ mol\ NH_2CONH_2}{16.3\ g\ solvent} \times \dfrac{1000\ g\ solvent}{kg\ solvent} = 1.43\ m\ NH_2CONH_2$

Find K_b:

$\Delta T_b = K_b \times m \qquad \frac{\Delta T_b}{m} = \frac{K_b \times m}{m} \qquad \frac{\Delta T_b}{m} = K_b$

$K_b = \frac{\Delta T_b}{m} = \frac{3.92°C}{1.43\ m} = 2.74°C/m$

See Section 16.14, Colligative Properties of Solutions, on Pages 461-464.

67. True: a, d, h, j, m. False: b, c, e, f, g, i, k, l.

b) The term *concentrated* compares solutions of the same solute and solvent. It cannot be used to compare solutions of solute A and solute B.
c) Under carefully controlled conditions, a solution can have a concentration greater than its solubility at a given temperature. Such a solution is said to be supersaturated.
e) Stirring a solution prevents a concentration buildup at the solute surface, maximizing the net dissolving rate.
f) The crystallization rate is equal to the dissolving rate when equilibrium is established.
g) The solubility of most solids increases with rising temperature, but there are notable exceptions.
i) An ionic solute is more likely to dissolve in a polar solvent than in a nonpolar solvent.
k) The number of equivalents of all species in a reaction is the same. No matter the mole ratio of acid to base, there are an equal number of equivalents of acid as there are of base in the reaction.
l) The concentration of a primary standard is found by weighing something that can be weighed accurately.

68. The boiling point rises.

The solute that is dissolved in the radiator water reduces the freezing temperature well below the normal freezing point of pure water. It also raises the boiling point above the normal boiling point. See Section 16.14, Colligative Properties of Solutions, on Pages 461-464.

69. Distillation is one method. It is used to separate petroleum into its components, which include natural gas, gasoline, lubricating oil, asphalt, and many other products.

Miscible liquids are those that dissolve in each other in all proportions. See the Miscible and Immiscible paragraph in Section 16.2 on Page 435. Distillation is discussed in Section 2.5, Separation of Mixtures, on Pages 26-28. In particular, see Figure 2.12 on Page 27.

70. Attractions between solute particles and attractions between solvent particles.

The dissolving process is explained in Section 16.3, The Formation of a Solution, on Page 435-438.

71. Finely powdered sugar has more surface area than granular sugar and therefore dissolves more quickly. Both will sweeten coffee equally.

See the list of factors that affect the time required to dissolve a given amount of solute on Page 437.

72. *GIVEN:* 1.00×10^2 mL benzene (B) *WANTED:* g C_4H_8O

PER/PATH: mL B $\xrightarrow{0.879 \text{ g B/mL B}}$ g B $\xrightarrow{1000 \text{ g B/kg B}}$ kg B $\xrightarrow{0.254 \text{ mol } C_4H_8O\text{/kg B}}$ mol C_4H_8O $\xrightarrow{72.10 \text{ g } C_4H_8O\text{/mol } C_4H_8O}$ g C_4H_8O

$$1.00 \times 10^2 \text{ mL B} \times \frac{0.879 \text{ g B}}{\text{mL B}} \times \frac{1 \text{ kg B}}{1000 \text{ g B}} \times \frac{0.254 \text{ mol } C_4H_8O}{\text{kg B}} \times \frac{72.10 \text{ g } C_4H_8O}{\text{mol } C_4H_8O} = 1.61 \text{ g } C_4H_8O$$

See Section 16.7, Solution Concentration: Molality, on Pages 444-446.

73. $2\ Fe(NO_3)_3 \rightarrow 2\ Fe(OH)_3 \rightarrow Fe_2O_3$

GIVEN: 35.0 mL *WANTED:* g Fe_2O_3

PER/PATH: mL $\xrightarrow{1000 \text{ mL/L}}$ L $\xrightarrow{0.516 \text{ mol } Fe(NO_3)_3\text{/L}}$ mol $Fe(NO_3)_3$ $\xrightarrow{1 \text{ mol } Fe_2O_3\text{/2 mol } Fe(NO_3)_3}$ mol Fe_2O_3 $\xrightarrow{159.70 \text{ g } Fe_2O_3\text{/mol } Fe_2O_3}$ g Fe_2O_3

$$35.0 \text{ mL} \times \frac{1 \text{ L}}{1000 \text{ mL}} \times \frac{0.516 \text{ mol } Fe(NO_3)_3}{\text{L}} \times \frac{1 \text{ mol } Fe_2O_3}{2 \text{ mol } Fe(NO_3)_3} \times \frac{159.70 \text{ g } Fe_2O_3}{\text{mol } Fe_2O_3} = 1.44 \text{ g } Fe_2O_3$$

See Section 16.6, Solution Concentration: Molarity, on Pages 441-444.

74. (This limiting reactant problem is solved by the methods described in Sections 9.6 and 9.8.)

$2\ NaOH + CuSO_4 \rightarrow Cu(OH)_2 + Na_2SO_4$

GIVEN: 25.0 mL *WANTED:* g $Cu(OH)_2$

PER/PATH: mL $\xrightarrow{1000 \text{ mL/L}}$ L $\xrightarrow{0.350 \text{ mol NaOH/L}}$ mol NaOH $\xrightarrow{1 \text{ mol } Cu(OH)_2\text{/2 mol NaOH}}$ mol $Cu(OH)_2$ $\xrightarrow{97.57 \text{ g } Cu(OH)_2\text{/mol } Cu(OH)_2}$ g $Cu(OH)_2$

$$25.0 \text{ mL} \times \frac{1 \text{ L}}{1000 \text{ mL}} \times \frac{0.350 \text{ mol NaOH}}{\text{L}} \times \frac{1 \text{ mol } Cu(OH)_2}{2 \text{ mol NaOH}} \times \frac{97.57 \text{ g } Cu(OH)_2}{\text{mol } Cu(OH)_2} = 0.427 \text{ g } Cu(OH)_2$$

GIVEN: 45.0 mL *WANTED:* g $Cu(OH)_2$

PER/PATH: mL $\xrightarrow{1000 \text{ mL/L}}$ L $\xrightarrow{0.125 \text{ mol } CuSO_4\text{/L}}$ mol $CuSO_4$ $\xrightarrow{1 \text{ mol } Cu(OH)_2\text{/1 mol } CuSO_4}$ mol $Cu(OH)_2$ $\xrightarrow{97.57 \text{ g } Cu(OH)_2\text{/mol } Cu(OH)_2}$ g $Cu(OH)_2$

$$45.0 \text{ mL} \times \frac{1 \text{ L}}{1000 \text{ mL}} \times \frac{0.125 \text{ mol } CuSO_4}{\text{L}} \times \frac{1 \text{ mol } Cu(OH)_2}{1 \text{ mol } CuSO_4} \times \frac{97.57 \text{ g } Cu(OH)_2}{\text{mol } Cu(OH)_2} = 0.549 \text{ g } Cu(OH)_2$$

0.427 g of $Cu(OH)_2$ will precipitate

See Section 9.6, Limiting Reactants: The Problem, on Pages 241-243, and either Section 9.7, Limiting Reactants: Comparison-of-Moles Method, on Pages 243-246, or Section 9.8, Limiting Reactants: Smaller-Amount Method, on Pages 247-249. Also see Section 16.6, Solution Concentration: Molarity, on Pages 441-444.

75. *GIVEN:* 16.80 mL *WANTED:* mol Cl^-

PER/PATH: mL $\xrightarrow{1000\text{ mL/L}}$ L $\xrightarrow{0.629\text{ mol AgNO}_3\text{/L}}$ mol $AgNO_3$ $\xrightarrow{1\text{ mol Cl}^-/1\text{ mol AgNO}_3}$ mol Cl^-

$$16.80\text{ mL} \times \frac{1\text{ L}}{1000\text{ mL}} \times \frac{0.629\text{ mol AgNO}_3}{\text{L}} \times \frac{1\text{ mol Cl}^-}{1\text{ mol AgNO}_3} = 0.0106\text{ mol Cl}^-$$

GIVEN: 0.0106 mol Cl^-; 25.00 mL *WANTED:* M Cl^-

EQUATION: $M \equiv \frac{\text{mol}}{\text{L}} = \frac{0.0106\text{ mol Cl}^-}{25.00\text{ mL}} \times \frac{1000\text{ mL}}{\text{L}} = 0.424\text{ M Cl}^-$

See Section 16.12, Titration Using Molarity, on Pages 456-459.

76. $Na_2CO_3 + 2\,HCl \rightarrow 2\,NaCl + H_2O + CO_2$

GIVEN: 41.24 mL *WANTED:* mg Na_2CO_3

PER/PATH: mL $\xrightarrow{1000\text{ mL/L}}$ L $\xrightarrow{0.244\text{ mol HCl/L}}$ mol HCl $\xrightarrow{1\text{ mol Na}_2\text{CO}_3\text{/2 mol HCl}}$ mol Na_2CO_3 $\xrightarrow{105.99\text{ g Na}_2\text{CO}_3\text{/mol Na}_2\text{CO}_3}$ g Na_2CO_3 $\xrightarrow{1000\text{ mg Na}_2\text{CO}_3\text{/g Na}_2\text{CO}_3}$ mg Na_2CO_3

$$41.24\text{ mL} \times \frac{1\text{ L}}{1000\text{ mL}} \times \frac{0.244\text{ mol HCl}}{\text{L}} \times \frac{1\text{ mol Na}_2\text{CO}_3}{2\text{ mol HCl}} \times \frac{105.99\text{ g Na}_2\text{CO}_3}{\text{mol Na}_2\text{CO}_3} \times \frac{1000\text{ mg Na}_2\text{CO}_3}{\text{g Na}_2\text{CO}_3} = 533\text{ mg Na}_2\text{CO}_3$$

EQUATION: $\%\ Na_2CO_3 = \frac{\text{mg Na}_2\text{CO}_3}{\text{mg sample}} \times 100 = \frac{533\text{ mg Na}_2\text{CO}_3}{694\text{ mg sample}} \times 100 = 76.8\%\ Na_2CO_3$

See Section 16.12, Titration Using Molarity, on Pages 456-459. See Page 180 for the definition of percent, if necessary.

77. a) $2\,KI + Pb(NO_3)_2 \rightarrow PbI_2 + 2\,KNO_3$

GIVEN: 60.0 mL *WANTED:* g PbI_2

PER/PATH: mL $\xrightarrow{1000\text{ mL/L}}$ L $\xrightarrow{0.322\text{ mol KI/L}}$ mol KI $\xrightarrow{1\text{ mol PbI}_2\text{/2 mol KI}}$ mol PbI_2 $\xrightarrow{461.0\text{ g PbI}_2\text{/mol PbI}_2}$ g PbI_2

$$60.0\text{ mL} \times \frac{1\text{ L}}{1000\text{ mL}} \times \frac{0.322\text{ mol KI}}{\text{L}} \times \frac{1\text{ mol PbI}_2}{2\text{ mol KI}} \times \frac{461.0\text{ g PbI}_2}{\text{mol PbI}_2} = 4.45\text{ g PbI}_2$$

GIVEN: 20.0 mL *WANTED:* g PbI_2

PER/PATH: mL $\xrightarrow{1000\text{ mL/L}}$ L $\xrightarrow{0.530\text{ mol Pb(NO}_3)_2\text{/L}}$ mol $Pb(NO_3)_2$ $\xrightarrow{1\text{ mol PbI}_2\text{/1 mol Pb(NO}_3)_2}$ mol PbI_2 $\xrightarrow{461.0\text{ g PbI}_2\text{/mol PbI}_2}$ g PbI_2

$$20.0\text{ mL} \times \frac{1\text{ L}}{1000\text{ mL}} \times \frac{0.530\text{ mol Pb(NO}_3)_2}{\text{L}} \times \frac{1\text{ mol PbI}_2}{1\text{ mol Pb(NO}_3)_2} \times \frac{461.0\text{ g PbI}_2}{\text{mol PbI}_2} = 4.89\text{ g PbI}_2$$

4.45 g PbI_2 will precipitate

b)	$2\ KI$ +	$Pb(NO_3)_2$ →	PbI_2 +	$2\ KNO_3$
Volume at start, mL	60.0	20.0		
Volume at start, L	0.0600	0.0200		
Molarity, mol/L	0.322	0.530		
Moles at start	0.0193	0.0106		
Moles used (–), produced (+)	–0.0193	–0.00965	+0.00965	+0.0193
Moles at end	0	0.0010	0.00965	0.0193

Total volume = 0.0600 L + 0.0200 L = 0.0800 L

EQUATION: $M \equiv \frac{\text{mol}}{\text{L}} = \frac{0.0193 \text{ mol } KNO_3}{0.0800 \text{ L}} \times \frac{1 \text{ mol } K^+}{1 \text{ mol } KNO_3} = 0.241 \text{ M } K^+$

c) *EQUATION:* $M \equiv \frac{\text{mol}}{\text{L}} = \frac{0.0010 \text{ mol } Pb(NO_3)_2}{0.0800 \text{ L}} \times \frac{1 \text{ mol } Pb^{2+}}{1 \text{ mol } Pb(NO_3)_2} = 0.013 \text{ M } Pb^{2+}$

See Section 9.6, Limiting Reactants: The Problem, on Pages 241-243, and either Section 9.7, Limiting Reactants: Comparison-of-Moles Method, on Pages 243-246, or Section 9.8, Limiting Reactants: Smaller-Amount Method, on Pages 247-249. Also see Section 16.6, Solution Concentration: Molarity, on Pages 441-444.

Chapter 17

Net Ionic Equations

1. When A dissolves in water, it forms ions. When B dissolves in water, it does not form ions.

   A strong electrolyte is a substance whose solution is a good conductor. The ability of a solution to conduct electricity is regarded as positive evidence that ions are present in the solution. A nonelectrolyte is a nonconductor; the concentration of ions present in a solution of a nonconductor is too small to conduct. See Section 17.1, Electrolytes and Solution Conductivity, on Pages 474-475.

2. When electricity passes through a wire, electrons "flow" through the wire. When electricity passes through a solution, electrons are "carried" by charged ions that move freely. If a liquid can carry an electrical current, it must be a solution containing ions.

   The ability of a solution to conduct electricity is regarded as positive evidence that ions are present in the solution. See Section 17.1, Electrolytes and Solution Conductivity, on Pages 474-475. In particular, see Figure 17.3 on Page 475.

3. $Mg(NO_3)_2(s) \xrightarrow{H_2O} Mg^{2+}(aq) + 2\ NO_3^-(aq)$; Major species: $Mg^{2+}(aq)$, $NO_3^-(aq)$
   $FeCl_3(s) \xrightarrow{H_2O} Fe^{3+}(aq) + 3\ Cl^-(aq)$; Major species: $Fe^{3+}(aq)$, $Cl^-(aq)$

   See Section 17.2, Solutions of Ionic Compounds, on Pages 475-477.

4. $Ca(OH)_2(s) \xrightarrow{H_2O} Ca^{2+}(aq) + 2\ OH^-(aq)$; Major species: $Ca^{2+}(aq)$, $OH^-(aq)$
   $Li_2SO_3(s) \xrightarrow{H_2O} 2\ Li^+(aq) + SO_3^{2-}(aq)$; Major species: $Li^+(aq)$, $SO_3^{2-}(aq)$

   See Section 17.2, Solutions of Ionic Compounds, on Pages 475-477.

5. $HCHO_2(aq)$ is a weak acid, so the molecule is the major species
   $HC_4H_4O_6(aq)$ is a weak acid, so the molecule is the major species

   See Section 17.2, Solutions of Ionic Compounds, on Pages 475-477. Also see Section 17.3, Strong and Weak Acids, on Pages 477-479.

6. $HClO_3(aq) \xrightarrow{H_2O} H^+(aq) + ClO_3^-(aq)$; $HClO_3(aq)$ is a strong acid;
   Major species: $H^+(aq) + ClO_3^-(aq)$
   $H_3C_2H_5O_7(aq)$ is a weak acid, so the molecule is the major species

   See Section 17.2, Solutions of Ionic Compounds, on Pages 475-477. Also see Section 17.3, Strong and Weak Acids, on Pages 477-479.

7. $CaCl_2(aq) + K_2CO_3(aq) \rightarrow CaCO_3(s) + 2\ KCl(aq)$
$Ca^{2+}(aq) + 2\ Cl^-(aq) + 2\ K^+(aq) + CO_3^{2-}(aq) \rightarrow CaCO_3(s) + 2\ K^+(aq) + 2\ Cl^-(aq)$
$Ca^{2+}(aq) + CO_3^{2-}(aq) \rightarrow CaCO_3(s)$

See Section 17.4, Net Ionic Equations: What They Are and How to Write Them, on Pages 480-482.

8. $Pb(s) + 2\ AgNO_3(aq) \rightarrow 2\ Ag(s) + Pb(NO_3)_2(aq)$
$Pb(s) + 2\ Ag^+(aq) + 2\ NO_3^-(aq) \rightarrow 2\ Ag(s) + Pb^{2+}(aq) + 2\ NO_3^-(aq)$
$Pb(s) + 2\ Ag+(aq) \rightarrow Pb^{2+}(aq) + 2\ Ag(s)$

See Section 17.4, Net Ionic Equations: What They Are and How to Write Them, on Pages 480-482.

9. $NaCHO_2(aq) + HCl(aq) \rightarrow HCHO_2(aq) + NaCl(aq)$
$Na^+(aq) + CHO_2^-(aq) + H^+(aq) + Cl^-(aq) \rightarrow HCHO_2(aq) + Na^+(aq) + Cl^-(aq)$
$CHO_2^-(aq) + H^+(aq) \rightarrow HCHO_2(aq)$

See Section 17.4, Net Ionic Equations: What They Are and How to Write Them, on Pages 480-482.

10. $Zn(s) + 2\ AgNO_3(aq) \rightarrow Zn(NO_3)_2(aq) + 2\ Ag(s)$
$Zn(s) + 2\ Ag^+(aq) + 2\ NO_3^-(aq) \rightarrow Zn^{2+}(aq) + 2\ NO_3^-(aq) + 2\ Ag(s)$
$Zn(s) + 2\ Ag^+(aq) \rightarrow Zn^{2+}(aq) + 2\ Ag(s)$

See Section 17.4, Net Ionic Equations: What They Are and How to Write Them, on Pages 480-482. Also see Section 17.5, Single-Replacement Oxidation-Reduction (Redox) Reactions, on Pages 483-486.

11. $Pb(s) + Ca(NO_3)_2(aq) \rightarrow NR$

Ca is above Pb on the activity series (Table 17.2, Page 485), so Pb will not replace Ca^{2+} in the ionic compound. See Section 17.5, Single-Replacement Oxidation-Reduction (Redox) Reactions, on Pages 483-486.

12. $3\ Mg(s) + Al_2(SO_4)_3(aq) \rightarrow 3\ MgSO_4(aq) + 2\ Al(s)$
$3\ Mg(s) + 2\ Al^{3+}(aq) + 3\ SO_4^{2-}(aq) \rightarrow 3\ Mg^{2+}(aq) + 3\ SO_4^{2-}(aq) + 2\ Al(s)$
$3\ Mg(s) + 2\ Al^{3+}(aq) \rightarrow 3\ Mg^{2+}(aq) + 2\ Al(s)$

See Section 17.4, Net Ionic Equations: What They Are and How to Write Them, on Pages 480-482. Also see Section 17.5, Single-Replacement Oxidation-Reduction (Redox) Reactions, on Pages 483-486.

13. $BaCl_2(aq) + Na_2CO_3(aq) \rightarrow BaCO_3(s) + 2\ NaCl(aq)$
$Ba^{2+}(aq) + 2\ Cl^-(aq) + 2\ Na^+(aq) + CO_3^{2-}(aq) \rightarrow BaCO_3(s) + 2\ Na^+(aq) + 2\ Cl^-(aq)$
$Ba^{2+}(aq) + CO_3^{2-}(aq) \rightarrow BaCO_3(s)$

See Section 17.4, Net Ionic Equations: What They Are and How to Write Them, on Pages 480-482. Also see Section 17.6, Ion Combinations That Form Precipitates, on Pages 486-491.

14. $CoSO_4(aq) + 2\ NaOH(aq) \rightarrow Co(OH)_2(s) + Na_2SO_4(aq)$
$Co^{2+}(aq) + SO_4^{2-}(aq) + 2\ Na^+(aq) + 2\ OH^-(aq) \rightarrow Co(OH)_2(s) + 2\ Na^+(aq) + SO_4^{2-}(aq)$
$Co^{2+}(aq) + 2\ OH^-(aq) \rightarrow Co(OH)_2(s)$

See Section 17.4, Net Ionic Equations: What They Are and How to Write Them, on Pages 480-482. Also see Section 17.6, Ion Combinations That Form Precipitates, on Pages 486-491.

15. $FeCl_2(aq) + (NH_4)_2S(aq) \rightarrow FeS(s) + 2\ NH_4Cl(aq)$
$Fe^{2+}(aq) + 2\ Cl^-(aq) + 2\ NH_4^+(aq) + S^{2-}(aq) \rightarrow FeS(s) + 2\ NH_4^+(aq) + 2\ Cl^-(aq)$
$Fe^{2+}(aq) + S^{2-}(aq) \rightarrow FeS(s)$

See Section 17.4, Net Ionic Equations: What They Are and How to Write Them, on Pages 480-482. Also see Section 17.6, Ion Combinations That Form Precipitates, on Pages 486-491.

16. $NiCl_2(aq) + CuSO_4(aq) \rightarrow NiSO_4(aq) + CuCl_2(aq)$
$Ni^{2+}(aq) + 2\ Cl^-(aq) + Cu^{2+}(aq) + SO_4^{2-}(aq) \rightarrow Ni^{2+}(aq) + SO_4^{2-}(aq) + Cu^{2+}(aq) + 2\ Cl^-(aq)$
$NiCl_2(aq) + CuSO_4(aq) \rightarrow NR$

See Section 17.4, Net Ionic Equations: What They Are and How to Write Them, on Pages 480-482. Also see Section 17.6, Ion Combinations That Form Precipitates, on Pages 486-491.

17. $Mg^{2+}(aq) + 2\ F^-(aq) \rightarrow MgF_2(s)$; $3\ Zn^{2+}(aq) + 2\ PO_4^{3-}(aq) \rightarrow Zn_3(PO_4)_2(s)$

The reactants are the ions that make up the ionic solid product. See Section 17.6, Ion Combinations That Form Precipitates, on Pages 486-491.

18. $NaCHO_2(aq) + HNO_3(aq) \rightarrow NaNO_3(aq) + HCHO_2(aq)$
$Na^+(aq) + CHO_2^-(aq) + H^+(aq) + NO_3^-(aq) \rightarrow Na^+(aq) + NO_3^-(aq) + HCHO_2(aq)$
$CHO_2^-(aq) + H^+(aq) \rightarrow HCHO_2(aq)$

$HNO_3(aq)$ is a strong acid; in solution, it primarily consists of ions.
$HCHO_2(aq)$ is a weak acid; in solution, it primarily consists of molecules.
See Section 17.4, Net Ionic Equations: What They Are and How to Write Them, on Pages 480-482. Also see Section 17.7, Ion Combinations That Form Molecules, on Pages 491-493.

19. $Ca(OH)_2(aq) + 2\ HBr(aq) \rightarrow CaBr_2(aq) + 2\ HOH(\ell)$
$Ca^{2+}(aq) + 2\ OH^-(aq) + 2\ H^+(aq) + 2\ Br^-(aq) \rightarrow Ca^{2+}(aq) + 2\ Br^-(aq) + 2\ HOH(\ell)$
$2\ OH^-(aq) + 2\ H^+(aq) \rightarrow 2\ HOH(\ell)$
$OH^-(aq) + H^+(aq) \rightarrow H_2O(\ell)$

$HBr(aq)$ is a strong acid; in solution, it primarily consists of ions.
See Section 17.4, Net Ionic Equations: What They Are and How to Write Them, on Pages 480-482. Also see Section 17.7, Ion Combinations That Form Molecules, on Pages 491-493.

20. $NaC_4H_7O_2(aq) + HCl(aq) \rightarrow NaCl(aq) + HC_4H_7O_2(aq)$
$Na^+(aq) + C_4H_7O_2^-(aq) + H^+(aq) + Cl^-(aq) \rightarrow Na^+(aq) + Cl^-(aq) + HC_4H_7O_2(aq)$
$C_4H_7O_2^-(aq) + H^+(aq) \rightarrow HC_4H_7O_2(aq)$

HCl(aq) is a strong acid; in solution, it primarily consists of ions.
$HC_4H_7O_2$(aq) is a weak acid; in solution, it primarily consists of molecules.
See Section 17.4, Net Ionic Equations: What They Are and How to Write Them, on Pages 480-482. Also see Section 17.7, Ion Combinations That Form Molecules, on Pages 491-493.

21. $(NH_4)_2SO_3(aq) + 2\ HBr(aq) \rightarrow 2\ NH_4Br(aq) + \text{“}H_2SO_3\text{”}$
$2\ NH_4^+(aq) + SO_3^{2-}(aq) + 2\ H^+(aq) + 2\ Br^-(aq) \rightarrow 2\ NH_4^+(aq) + 2\ Br^-(aq) + SO_2(aq) + H_2O(\ell)$
$SO_3^{2-}(aq) + 2\ H^+(aq) \rightarrow SO_2(aq) + H_2O(\ell)$

HBr(aq) is a strong acid; in solution, it primarily consists of ions.
See Section 17.4, Net Ionic Equations: What They Are and How to Write Them, on Pages 480-482. Also see Section 17.8, Ion Combinations That Form Unstable Products, on Pages 493-494.

22. $2\ LiOH(aq) + (NH_4)_2SO_4(aq) \rightarrow Li_2SO_4(aq) + 2\ \text{“}NH_4OH\text{”}$
$2\ Li^+(aq) + 2\ OH^-(aq) + 2\ NH_4^+(aq) + SO_4^{2-}(aq) \rightarrow 2\ Li^+(aq) + SO_4^{2-}(aq) + 2\ NH_3(aq) + 2\ H_2O(\ell)$
$2\ OH^-(aq) + 2\ NH_4^+(aq) \rightarrow 2\ NH_3(aq) + 2\ H_2O(\ell)$
$OH^-(aq) + NH_4^+(aq) \rightarrow NH_3(aq) + H_2O(\ell)$

See Section 17.4, Net Ionic Equations: What They Are and How to Write Them, on Pages 480-482. Also see Section 17.8, Ion Combinations That Form Unstable Products, on Pages 493-494.

23. $2\ AgNO_3(aq) + (NH_4)_2CO_3(aq) \rightarrow Ag_2CO_3(s) + 2\ NH_4NO_3(aq)$
$2\ Ag^+(aq) + 2\ NO_3^-(aq) + 2\ NH_4^+(aq) + CO_3^{2-}(aq) \rightarrow Ag_2CO_3(s) + 2\ NH_4^+(aq) + 2\ NO_3^-(aq)$
$2\ Ag^+(aq) + CO_3^{2-}(aq) \rightarrow Ag_2CO_3(s)$

See Section 17.4, Net Ionic Equations: What They Are and How to Write Them, on Pages 480-482. Also see Section 17.6, Ion Combinations That Form Precipitates, on Pages 486-491.

24. $Al(NO_3)_3(aq) + K_3PO_4(aq) \rightarrow AlPO_4(s) + 3\ KNO_3(aq)$
$Al^{3+}(aq) + 3\ NO_3^-(aq) + 3\ K^+(aq) + PO_4^{3-}(aq) \rightarrow AlPO_4(s) + 3\ K^+(aq) + 3\ NO_3^-(aq)$
$Al^{3+}(aq) + PO_4^{3-}(aq) \rightarrow AlPO_4(s)$

See Section 17.4, Net Ionic Equations: What They Are and How to Write Them, on Pages 480-482. Also see Section 17.6, Ion Combinations That Form Precipitates, on Pages 486-491.

25. $Zn(s) + 2\ AgNO_3(aq) \rightarrow Zn(NO_3)_2(aq) + 2\ Ag(s)$
$Zn(s) + 2\ Ag^+(aq) + 2\ NO_3^-(aq) \rightarrow Zn^{2+}(aq) + 2\ NO_3^-(aq) + 2\ Ag(s)$
$Zn(s) + 2\ Ag^+(aq) \rightarrow Zn^{2+}(aq) + 2\ Ag(s)$

See Section 17.4, Net Ionic Equations: What They Are and How to Write Them, on Pages 480-482. Also see Section 17.5, Single-Replacement Oxidation-Reduction (Redox) Reactions, on Pages 483-486.

26. $Na_2SO_3(aq) + 2\ HCl(aq) \rightarrow 2\ NaCl(aq) + \text{“}H_2SO_3\text{”}$
$2\ Na^+(aq) + SO_3^{2-}(aq) + 2\ H^+(aq) + 2\ Cl^-(aq) \rightarrow 2\ Na^+(aq) + 2\ Cl^-(aq) + SO_2(aq) + H_2O(\ell)$
$SO_3^{2-}(aq) + 2\ H^+(aq) \rightarrow SO_2(aq) + H_2O(\ell)$

HCl(aq) is a strong acid; in solution, it primarily consists of ions.
Also see Section 17.8, Ion Combinations That Form Unstable Products, on Pages 493-494.

27. $HClO_3(aq) + KOH(aq) \rightarrow HOH(\ell) + KClO_3(aq)$
$H^+(aq) + ClO_3^-(aq) + K^+(aq) + OH^-(aq) \rightarrow H_2O(\ell) + K^+(aq) + ClO_3^-(aq)$
$H^+(aq) + OH^-(aq) \rightarrow H_2O(\ell)$

$HClO_3(aq)$ is a strong acid; in solution, it primarily consists of ions.
See Section 17.4, Net Ionic Equations: What They Are and How to Write Them, on Pages 480-482. Also see Section 17.7, Ion Combinations That Form Molecules, on Pages 491-493.

28. $AgCl(s) + KI(aq) \rightarrow AgI(s) + KCl(aq)$
$AgCl(s) + K^+(aq) + I^-(aq) \rightarrow AgI(s) + K^+(aq) + Cl^-(aq)$
$AgCl(s) + I^-(aq) \rightarrow AgI(s) + Cl^-(aq)$

See Section 17.4, Net Ionic Equations: What They Are and How to Write Them, on Pages 480-482. Also see Section 17.9, Ion-Combination Reactions with Undissolved Reactants, on Pages 494-495.

29. $Pb(s) + CuSO_4(aq) \rightarrow PbSO_4(aq) + Cu(s)$
$Pb(s) + Cu^{2+}(aq) + SO_4^{2-}(aq) \rightarrow Pb^{2+}(aq) + SO_4^{2-}(aq) + Cu(s)$
$Pb(s) + Cu^{2+}(aq) \rightarrow Pb^{2+}(aq) + Cu(s)$

See Section 17.4, Net Ionic Equations: What They Are and How to Write Them, on Pages 480-482. Also see Section 17.5, Single-Replacement Oxidation-Reduction (Redox) Reactions, on Pages 483-486.

30. $NaC_7H_5O_2(aq) + HCl(aq) \rightarrow NaCl(aq) + HC_7H_5O_2(aq)$
$Na^+(aq) + C_7H_5O_2^-(aq) + H^+(aq) + Cl^-(aq) \rightarrow Na^+(aq) + Cl^-(aq) + HC_7H_5O_2(aq)$
$C_7H_5O_2^-(aq) + H^+(aq) \rightarrow HC_7H_5O_2(aq)$

$HCl(aq)$ is a strong acid; in solution, it primarily consists of ions.
$HC_7H_5O_2(aq)$ is a weak acid; in solution, it primarily consists of molecules.
See Section 17.4, Net Ionic Equations: What They Are and How to Write Them, on Pages 480-482. Also see Section 17.7, Ion Combinations That Form Molecules, on Pages 491-493.

31. $NH_4Cl(aq) + KOH(aq) \rightarrow$ "NH_4OH" $+ KCl(aq)$
$NH_4^+(aq) + Cl^-(aq) + K^+(aq) + OH^-(aq) \rightarrow NH_3(aq) + H_2O(\ell) + K^+(aq) + Cl^-(aq)$
$NH_4^+(aq) + OH^-(aq) \rightarrow NH_3(aq) + H_2O(\ell)$

See Section 17.4, Net Ionic Equations: What They Are and How to Write Them, on Pages 480-482. Also see Section 17.8, Ion Combinations That Form Unstable Products, on Pages 493-494.

32. $Na_2CO_3(aq) + Ca(NO_3)_2(aq) \rightarrow 2\ NaNO_3(aq) + CaCO_3(s)$
$2\ Na^+(aq) + CO_3^{2-}(aq) + Ca^{2+}(aq) + 2\ NO_3^-(aq) \rightarrow 2\ Na^+(aq) + 2\ NO_3^-(aq) + CaCO_3(s)$
$CO_3^{2-}(aq) + Ca^{2+}(aq) \rightarrow CaCO_3(s)$

See Section 17.4, Net Ionic Equations: What They Are and How to Write Them, on Pages 480-482. Also see Section 17.6, Ion Combinations That Form Precipitates, on Pages 486-491.

33. $Mg(s) + Zn(NO_3)_2(aq) \rightarrow Mg(NO_3)_2(aq) + Zn(s)$
$Mg(s) + Zn^{2+}(aq) + 2\ NO_3^-(aq) \rightarrow Mg^{2+}(aq) + 2\ NO_3^-(aq) + Zn(s)$
$Mg(s) + Zn^{2+}(aq) \rightarrow Mg^{2+}(aq) + Zn(s)$

See Section 17.4, Net Ionic Equations: What They Are and How to Write Them, on Pages 480-482. Also see Section 17.5, Single-Replacement Oxidation-Reduction (Redox) Reactions, on Pages 483-486.

34. $2\ HNO_3(aq) + Ba(OH)_2(s) \rightarrow 2\ HOH(\ell) + Ba(NO_3)_2(aq)$

$2\ H^+(aq) + 2\ NO_3^-(aq) + Ba(OH)_2(s) \rightarrow 2\ H_2O(\ell) + Ba^{2+}(aq) + 2\ NO_3^-(aq)$

$2\ H^+(aq) + Ba(OH)_2(s) \rightarrow 2\ H_2O(\ell) + Ba^{2+}(aq)$

$HNO_3(aq)$ is a strong acid; in solution, it primarily consists of ions.
See Section 17.4, Net Ionic Equations: What They Are and How to Write Them, on Pages 480-482. Also see Section 17.9, Ion-Combination Reactions with Undissolved Reactants, on Pages 494-495.

35. $2\ HClO_3(aq) + Cu(OH)_2(s) \rightarrow 2\ HOH(\ell) + Cu(ClO_3)_2(aq)$

$2\ H^+(aq) + 2\ ClO_3^-(aq) + Cu(OH)_2(s) \rightarrow 2\ H_2O(\ell) + Cu^{2+}(aq) + 2\ ClO_3^-(aq)$

$2\ H^+(aq) + Cu(OH)_2(s) \rightarrow 2\ H_2O(\ell) + Cu^{2+}(aq)$

$HClO_3(aq)$ is a strong acid; in solution, it primarily consists of ions.
See Section 17.4, Net Ionic Equations: What They Are and How to Write Them, on Pages 480-482. Also see Section 17.9, Ion-Combination Reactions with Undissolved Reactants, on Pages 494-495.

36. $Ag(s) + Mg(NO_3)_2(aq) \rightarrow NR$

Mg is above Ag on the activity series (Table 17.2, Page 485), so Ag will not replace Mg^{2+} in the ionic compound. See Section 17.5, Single-Replacement Oxidation-Reduction (Redox) Reactions, on Pages 483-486.

38. True: a, b, c, h, i, j, k. False: d, e, f, g, l, m.

d) In the solution of a weak acid, almost all of the molecules of the original compound are converted to ions.
e) Only seven "important" acids are strong.
f) Hydrofluoric acid, which is used to etch glass, is a weak acid.
g) Spectators are not included in a net ionic equation.
l) The products of a molecule-formation reaction usually are a salt and water or a weak acid.
m) Ammonium hydroxide is not a possible product of a molecule-formation reaction because it is unstable under normal conditions, yielding ammonia and water.

Chapter 18

Acid–Base (Proton-Transfer) Reactions

1. Acid, H^+; base, OH^-.

   Examples of substances commonly regarded as acids or bases that contain the H^+ ion or the OH^- ion are found throughout Chapter 18. See Section 18.1, The Arrhenius Theory of Acids and Bases, on Pages 502-503.

2. See Section 18.3 summary. The concepts are in agreement regarding acids, but not bases.

   According to the Arrhenius theory, the properties of an acid are the properties of the hydrogen ion. The Brønsted-Lowry theory classifies an acid as a proton donor; a proton is a hydrogen ion. The theories are in agreement about acids. See Section 18.1, The Arrhenius Theory of Acids and Bases, on Pages 502-503, Section 18.2, The Brønsted-Lowry Theory of Acids and Bases, on Pages 503-505, and the Summary: Identifying Features of Acids and Bases box on Page 506.

3. A Lewis acid must have a vacant valence orbital. A Lewis base must have an unshared electron pair.

   See Section 18.3, The Lewis Theory of Acids and Bases, on Pages 505-506.

4. *Please see the Lewis diagrams on textbook Page 525.*
   Aluminum in aluminum chloride is able to accept an electron pair, so it qualifies as a Lewis acid. The chloride ion can contribute the electron pair to the bond, so it is a Lewis base.

   See Section 18.3, The Lewis Theory of Acids and Bases, on Pages 505-506.

5. HOH and H_2CO_3; H_2O and CO_3^{2-}

   To write the formula of a conjugate acid, add an H^+ to the base:
   $H^+ + OH^- \rightarrow HOH$ $\quad H^+ + HCO_3^- \rightarrow H_2CO_3$
   To write the formula of a conjugate base, remove an H^+ from the acid:
   $H_3O^+ \rightarrow H^+ + H_2O$ $\quad HCO_3^- \rightarrow H^+ + CO_3^{2-}$
   See Section 18.4, Conjugate Acid-Base Pairs, on Pages 506-507.

6. Forward: acid, HNO_2; base, CN^-. Reverse: acid, HCN; base, NO_2^-.

   In the forward reaction, HNO_2 donates a proton to CN^-; HNO_2 is the acid and CN^- is the base. In the reverse reaction, HCN donates a proton to NO_2^-; HCN is the acid and NO_2^- is the base. See Section 18.4, Conjugate Acid-Base Pairs, on Pages 506-507.

7. HNO_2 and NO_2^-; CN^- and HCN

   Conjugate acid-base pairs are the species in a Brønsted-Lowry reaction that differ by a H^+ because of an acid losing a proton or a base gaining one. See Section 18.4, Conjugate Acid-Base Pairs, on Pages 506-507.

8. HSO_4^- and SO_4^{2-}; $C_2O_4^{2-}$ and $HC_2O_4^-$

Conjugate acid-base pairs are the species in a Brønsted-Lowry reaction that differ by a H^+ because of an acid losing a proton or a base gaining one. See Section 18.4, Conjugate Acid-Base Pairs, on Pages 506-507.

9. $H_2PO_4^-$ and HPO_4^{2-}; HCO_3^- and H_2CO_3

Conjugate acid-base pairs are the species in a Brønsted-Lowry reaction that differ by a H^+ because of an acid losing a proton or a base gaining one. See Section 18.4, Conjugate Acid-Base Pairs, on Pages 506-507.

10. A strong acid loses protons readily; a weak acid does not lose protons easily. Strong acids are at the top of the left column in Table 18.1 and weak acids are at the bottom.

See Section 18.5, Relative Strengths of Acids and Bases, on Pages 508-510. Table 18.1 is on Page 509.

11. $CO_3^{2-} > H_2PO_4^- > SO_4^{2-} > Br^-$

All species are found in the *Base Formula* column of Table 18.1 on Page 509. The strongest bases are at the bottom of the column, and base strength decreases as you move up the column. See Section 18.5, Relative Strengths of Acids and Bases, on Pages 508-510.

12. $HI > H_2C_2O_4 > HSO_3^- > NH_4^+ > H_2O$

All species are found in the *Acid Formula* column of Table 18.1 on Page 509. The strongest acids are at the top of the column, and acid strength decreases as you move down the column. H_2O is given as HOH in the *Acid Formula* column. See Section 18.5, Relative Strengths of Acids and Bases, on Pages 508-510.

13. $HC_7H_5O_2(aq) + SO_4^{2-}(aq) \rightleftharpoons C_7H_5O_2^-(aq) + HSO_4^-(aq)$; Reverse

weaker acid — weaker base — stronger base — stronger acid

The reaction is completed by transferring a single proton, H^+, from acid to base. Relative strengths of the acids and of the bases are found in Table 18.1 on Page 509. The favored direction is toward the weaker species. See Section 18.6, Predicting Acid-Base Reactions, on Pages 510-512. The Procedure: How to Predict the Favored Direction of an Acid-Base Reaction box on Page 511 summarizes the process.

14. $H_2C_2O_4(aq) + NH_3(aq) \rightleftharpoons HC_2O_4^-(aq) + NH_4^+(aq)$; Forward

stronger acid — stronger base — weaker base — weaker acid

The reaction is completed by transferring a single proton, H^+, from acid to base. Relative strengths of the acids and of the bases are found in Table 18.1 on Page 509. The favored direction is toward the weaker species. See Section 18.6, Predicting Acid-Base Reactions, on Pages 510-512. The Procedure: How to Predict the Favored Direction of an Acid-Base Reaction box on Page 511 summarizes the process.

15. $H_3PO_4(aq) + CN^-(aq) \rightleftharpoons H_2PO_4^-(aq) + HCN(aq)$; Forward
stronger acid stronger base weaker base weaker acid

The reaction is completed by transferring a single proton, H^+, from acid to base. Relative strengths of the acids and of the bases are found in Table 18.1 on Page 509. The favored direction is toward the weaker species. See Section 18.6, Predicting Acid-Base Reactions, on Pages 510-512. The Procedure: How to Predict the Favored Direction of an Acid-Base Reaction box on Page 511 summarizes the process.

16. $H_2BO_3^-(aq) + NH_4^+(aq) \rightleftharpoons H_3BO_3(aq) + NH_3(aq)$; Reverse
weaker base weaker acid stronger acid stronger base

The reaction is completed by transferring a single proton, H^+, from acid to base. Relative strengths of the acids and of the bases are found in Table 18.1 on Page 509. The favored direction is toward the weaker species. See Section 18.6, Predicting Acid-Base Reactions, on Pages 510-512. The Procedure: How to Predict the Favored Direction of an Acid-Base Reaction box on Page 511 summarizes the process.

17. $HPO_4^{2-}(aq) + HC_2H_3O_2(aq) \rightleftharpoons H_2PO_4^-(aq) + C_2H_3O_2^-(aq)$; Forward
stronger base stronger acid weaker acid weaker base

The reaction is completed by transferring a single proton, H^+, from acid to base. Relative strengths of the acids and of the bases are found in Table 18.1 on Page 509. The favored direction is toward the weaker species. See Section 18.6, Predicting Acid-Base Reactions, on Pages 510-512. The Procedure: How to Predict the Favored Direction of an Acid-Base Reaction box on Page 511 summarizes the process.

18. Water ionizes very slightly and does not produce enough ions to light an ordinary conductivity device. With a sufficiently sensitive detector, water displays a very weak conductivity.

See Section 18.8, The Water Equilibrium, on Pages 512-513.

19. Acidic, $[H^+] > [OH^-]$; basic, $[H^+] < [OH^-]$; neutral, $[H^+] = [OH^-]$

See Section 18.8, The Water Equilibrium, on Pages 512-513.

20. *GIVEN:* $[H^+] = 10^{-12}$ M *WANTED:* $[OH^-]$

$$[H^+][OH^-] = 10^{-14} \qquad \frac{[H^+][OH^-]}{[H^+]} = \frac{10^{-14}}{[H^+]} \qquad [OH^-] = \frac{10^{-14}}{[H^+]}$$

EQUATION: $$[OH^-] = \frac{10^{-14}}{[H^+]} = \frac{10^{-14}}{10^{-12}} = 10^{-2}\text{ M}$$

See Section 18.8, The Water Equilibrium, on Pages 512-513.

21. Strongly acidic, pH <4; weakly acidic, $4 \le pH < 6$; strongly basic, $11 \le pH$; weakly basic, $8 \le pH \le 11$; neutral or near neutral, $6 \le pH < 8$

These arbitrary ranges are identified in Table 18.3 on Page 516. See Section 18.9, pH and pOH (Integer Values Only), on Pages 514-518.

22. If pH = x, then $[H^+] = 10^{-x}$

See Section 18.9, pH and pOH (Integer Values Only), on Pages 514-518.

23. *GIVEN:* pOH = 6
$pH + pOH = 14 \quad pH = 14 - pOH = 14 - 6 = 8$
$[OH^-] = 10^{-pOH} = 10^{-6}$
$[H^+] = 10^{-pH} = 10^{-8}$
pH = 8 is near neutral/weakly basic (see Table 18.3, Page 516)

See Section 18.9, pH and pOH (Integer Values Only), on Pages 514-518.

24. *GIVEN:* $[H^+] = 0.1\ M = 10^{-1}\ M$
$pH = -\log [H^+] = -\log (10^{-1}) = -(-1) = 1$
$pH + pOH = 14 \quad pOH = 14 - pH = 14 - 1 = 13$
$[OH^-] = 10^{-pOH} = 10^{-13}$
pH = 1 is strongly acidic (see Table 18.3, Page 516)

See Section 18.9, pH and pOH (Integer Values Only), on Pages 514-518.

25. *GIVEN:* $[OH^-] = 10^{-2}\ M$
$pOH = -\log [OH^-] = -\log (10^{-2}) = -(-2) = 2$
$pH + pOH = 14 \quad pH = 14 - pOH = 14 - 2 = 12$
$[H^+] = 10^{-pH} = 10^{-12}$
pH = 12 is strongly basic (see Table 18.3, Page 516)

See Section 18.9, pH and pOH (Integer Values Only), on Pages 514-518.

26. *GIVEN:* pH = 4
$pH + pOH = 14 \quad pOH = 14 - pH = 14 - 4 = 10$
$[H^+] = 10^{-pH} = 10^{-4}$
$[OH^-] = 10^{-pOH} = 10^{-10}$
pH = 4 is weakly/strongly acidic (see Table 18.3, Page 516)

See Section 18.9, pH and pOH (Integer Values Only), on Pages 514-518.

27. *GIVEN:* pH = 6.62
$pH + pOH = 14.00 \quad pOH = 14.00 - pH = 14.00 - 6.62 = 7.38$
$[H^+] = 10^{-pH} = 10^{-6.62} = 2.4 \times 10^{-7}$
$[OH^-] = 10^{-pOH} = 10^{-7.38} = 4.2 \times 10^{-8}\ M$

See Section 18.10, Noninteger pH-[H⁺] and pOH -[OH⁻] Conversions, on Pages 518-521.

28. *GIVEN:* $[OH^-] = 1.1 \times 10^{-11}\ M$
$pOH = -\log [OH^-] = -\log (1.1 \times 10^{-11}) = 10.96$
$pH + pOH = 14.00 \quad pH = 14.00 - pOH = 14.00 - 10.96 = 3.04$
$[H^+] = 10^{-pH} = 10^{-3.04} = 9.1 \times 10^{-4}\ M$

See Section 18.10, Noninteger pH-[H⁺] and pOH -[OH⁻] Conversions, on Pages 518-521.

29. *GIVEN:* pOH = 5.54
pH + pOH = 14.00 pH = 14.00 – pOH = 14.00 – 5.54 = 8.46
$[OH^-] = 10^{-pOH} = 10^{-5.54} = 2.9 \times 10^{-6}$ M
$[H^+] = 10^{-pH} = 10^{-8.46} = 3.5 \times 10^{-9}$ M

See Section 18.10, Noninteger pH-$[H^+]$ and pOH -$[OH^-]$ Conversions, on Pages 518-521.

30. *GIVEN:* $[H^+] = 7.2 \times 10^{-2}$ M
$pH = -\log[H^+] = -\log(7.2 \times 10^{-2}) = 1.14$
pH + pOH = 14.00 pOH = 14.00 – pH = 14.00 – 1.14 = 12.86
$[OH^-] = 10^{-pOH} = 10^{-12.86} = 1.4 \times 10^{-13}$ M

See Section 18.10, Noninteger pH-$[H^+]$ and pOH -$[OH^-]$ Conversions, on Pages 518-521.

32. True: a, b, c, e, f, g, h. False: d, i, j, k, l.

d) HS^- is the conjugate acid of S^{2-}.
i) A proton-transfer reaction is always favored in the direction that yields the weaker acid.
j) A solution with pH = 9 is less acidic than one with pH = 4.
k) A solution with pH = 3 is 1000 times as acidic as one with pH = 6 (6-3 = 3; 10^3 = 1000).
l) A pOH of 4.65 expresses the hydroxide ion concentration of a solution in two significant figures.

33. $OH^- + NH_3 \rightarrow H_2O + NH_2^-$

A Brønsted-Lowry acid-base reaction requires a proton donor and a proton acceptor. NH_3 has a proton to donate, and OH^- has an unshared electron pair that can accept a proton. See Section 18.2, The Brønsted-Lowry Theory of Acids and Bases, on Pages 503-505.

34. An amphoteric substance can be an acid by losing a proton, or a base by gaining a proton. HX^- is a general formula of an amphoteric substance: $HX^- \rightarrow H^+ + X^{2-}$ (acid reaction); $HX^- + H^+ \rightarrow H_2X$ (base reaction).

Common examples include HSO_4^-, HSO_3^-, $H_2PO_4^-$, HPO_4^{2-}, and HS^-. See Section 18.2, The Brønsted-Lowry Theory of Acids and Bases, on Pages 503-505.

35. *GIVEN:* $[Cl^-] = 7.49 \times 10^{-8}$ M
In general $pQ = -\log[Q]$, so $pCl = -\log[Cl^-] = -\log(7.49 \times 10^{-8}) = 7.126$

See Section 18.10, Noninteger pH-$[H^+]$ and pOH -$[OH^-]$ Conversions, on Pages 518-521.

36. When a proton is removed from a H_3X species, a single positive charge is being pulled away from a particle with a single minus charge, H_2X^-. When a proton is removed from an H_2X^- species, a single positive charge is being pulled away from a particle with a double minus charge, HX^{2-}. The loss of the second proton is more difficult, so H_2X^- is a weaker acid than H_3X.

See Section 18.5, Relative Strengths of Acids and Bases, on Pages 508-510.

37. There can be no proton transfer without a proton—an H^+ ion.

See Section 18.2, The Brønsted-Lowry Theory of Acids and Bases, on Pages 503-505.

38. Carbonate ion is a proton acceptor: $H^+ + CO_3^{2-} \rightarrow HCO_3^-$.

See Section 18.2, The Brønsted-Lowry Theory of Acids and Bases, on Pages 503-505.

Chapter 19

Oxidation–Reduction (Redox) Reactions

1. An electric current through a wire is a one-way movement of electrons. Electrolysis is a two-way movement of charged ions in a solution.

   Electrolysis is explained in the first paragraph of Section 19.1, Electrolytic and Voltaic Cells, on Page 528. Also review Figure 19.1(a) on Page 528.

2. No. An electrolytic cell must be connected to an outside source of electricity for electrolysis to occur. It cannot function as a voltaic cell, which spontaneously produces an electric current when connected to an outside circuit.

   See Section 19.1, Electrolytic and Voltaic Cells, on Pages 528-529. Figure 19.1 on Page 528 illustrates electrolytic and voltaic cells.

3. Oxidation is loss of electrons; reduction is gain of electrons. If one species loses electrons, they must go someplace—there must be another species that will gain the electrons. Therefore, oxidation and reduction must occur simultaneously.

   See Section 19.2, Electron-Transfer Reactions, on Pages 529-532.

4. a, b, c: oxidation; d: reduction

   a) Cl^- loses electrons; loss of electrons is oxidation
   b) Na loses electrons; loss of electrons is oxidation
   c) Sn^{2+} loses electrons; loss of electrons is oxidation
   d) O_2 gains electrons; gain of electrons is reduction
   See Section 19.2, Electron-Transfer Reactions, on Pages 529-532.

5. Oxidation

   Ag loses electrons; loss of electrons is oxidation
   See Section 19.2, Electron-Transfer Reactions, on Pages 529-532.

6. Reduction

   O_2^{2-} gains electrons; gain of electrons is reduction
   See Section 19.2, Electron-Transfer Reactions, on Pages 529-532.

7. Reduction: $Ni^{2+} + 2\,e^- \rightarrow Ni$
   Oxidation: $Mg \rightarrow Mg^{2+} + 2\,e^-$

   Redox: $Ni^{2+} + Mg \rightarrow Ni + Mg^{2+}$

   See Section 19.2, Electron-Transfer Reactions, on Pages 529-532.

8. Reduction: $PbO_2 + SO_4^{2-} + 4\ H^+ + 2\ e^- \rightarrow PbSO_4 + 2\ H_2O$
Oxidation: $Pb + SO_4^{2-} \rightarrow PbSO_4 + 2\ e^-$

Redox: $PbO_2 + SO_4^{2-} + 4\ H^+ + Pb \rightarrow 2\ PbSO_4 + 2\ H_2O$

See Section 19.2, Electron-Transfer Reactions, on Pages 529-532.

9. Mg^{2+}: The oxidation number of a monatomic ion is the same as the charge on the ion: +2
Cl^-: The oxidation number of a monatomic ion is the same as the charge on the ion: –1
$\underline{Cl}O^-$: The oxidation number of combined oxygen is –2; x + (–2) = –1; x = 1; +1
$K\underline{Cl}O_3$: This is an ionic compound, K^+ and ClO_3^-, so we analyze ClO_3^-; the oxidation number of combined oxygen is –2; x + 3(–2) = –1; x = 5; +5

See Section 19.3, Oxidation Numbers and Redox Reactions, on Pages 532-536. The Summary: Oxidation Number Rules box on Page 533 summarizes the oxidation number rules.

10. $\underline{N}_2O_5$: The oxidation number of combined oxygen is –2; 2x + 5(–2) = 0; x = 5; +5
$\underline{N}H_4^+$: The oxidation number of combined hydrogen is +1; x + 4(1) = 1; x = –3; –3
$\underline{Mn}O_4^-$: The oxidation number of combined oxygen is –2; x + 4(–2) = –1; x = 7; +7
$NaH_2\underline{P}O_3$: This is an ionic compound, Na^+ and $H_2PO_3^-$, so we analyze $H_2PO_3^-$; the oxidation number of combined hydrogen is +1; the oxidation number of combined oxygen is –2; 2(1) + x + 3(–2) = –1; x = 3; +3

See Section 19.3, Oxidation Numbers and Redox Reactions, on Pages 532-536. The Summary: Oxidation Number Rules box on Page 533 summarizes the oxidation number rules.

11. a) Copper reduced from +2 to 0. b) Cobalt reduced from +3 to +2.

a) Cu^{2+}: +2; Cu: 0; decrease in oxidation number is reduction
b) Co^{3+}: +3; Co^{2+}: +2; decrease in oxidation number is reduction
See Section 19.3, Oxidation Numbers and Redox Reactions, on Pages 532-536. The Summary: Definitions of Oxidation and Reduction box on Page 535 defines oxidation and reduction in terms of change in oxidation number.

12. a) Sulfur oxidized from +4 to +6. b) Phosphorus oxidized from –3 to 0.

a) H_2O: +1 each hydrogen, -2 oxygen; SO_3^{2-}: +4 sulfur, -2 each oxygen; SO_4^{2-}: +6 sulfur, -2 each oxygen; H+: +1; the oxidation number of sulfur changes from +4 to +6; increase in oxidation number is oxidation
b) PH_3: -3 phosphorus, +1 each hydrogen; P: 0; H^+: +1; the oxidation number of phosphorus changes from -3 to 0; increase in oxidation number is oxidation
See Section 19.3, Oxidation Numbers and Redox Reactions, on Pages 532-536. The Summary: Definitions of Oxidation and Reduction box on Page 535 defines oxidation and reduction in terms of change in oxidation number.

13. a) Fluorine oxidized from –1 to 0. b) Manganese reduced from +6 to +4.

a) HF: +1 hydrogen, -1 fluorine; F_2: 0 each fluorine; H^+: +1; the oxidation number of fluorine changes from -1 to 0; increase in oxidation number is oxidation
b) MnO_4^{2-}: +6 manganese, -2 each oxygen; H_2O: +1 each hydrogen, -2 oxygen; MnO_2: +4 manganese, -2 each oxygen; OH^-: -2 oxygen, +1 hydrogen; the oxidation number of manganese changes from +6 to +4; decrease in oxidation number is reduction
See Section 19.3, Oxidation Numbers and Redox Reactions, on Pages 532-536. The Summary: Definitions of Oxidation and Reduction box on Page 535 defines oxidation and reduction in terms of change in oxidation number.

14. H_2 is the reducing agent, and CuO is the oxidizing agent.

CuO: +2 copper, -2 oxygen; H_2: 0 each hydrogen; Cu: 0; H_2O: +1 each hydrogen, -2 oxygen; copper changes from +2 to 0; hydrogen changes from 0 to +1; CuO accepts electrons, so it is the oxidizing agent; H_2 donates electrons, so it is the reducing agent
See Section 19.4, Oxidizing Agents (Oxidizers) and Reducing Agents (Reducers), on Page 536.

15. BrO_3^- is the oxidizing agent, and HNO_2 is the reducing agent.

BrO_3^-: +5 bromine, -2 each oxygen; HNO_2: +1 hydrogen, +3 nitrogen; -2 each oxygen; Br^-: -1; NO_3^-: +5 nitrogen, -2 each oxygen; H^+: +1; bromine changes from +5 to -1; nitrogen changes from +4 to +5; BrO_3^- accepts electrons, so it is the oxidizing agent; HNO_2 donates electrons, so it is the reducing agent
See Section 19.4, Oxidizing Agents (Oxidizers) and Reducing Agents (Reducers), on Page 536.

16. Ag^+ is a stronger oxidizer than H^+, based on their relative positions in Table 19.2. One species is a stronger oxidizing agent than a second species if the first species attracts electrons more strongly than the second.

Table 19.2 is on Page 537. See Section 19.5, Strengths of Oxidizing Agents and Reducing Agents, on Pages 537-538.

17. Al, H_2, Fe^{2+}, Cl^-

Table 19.2 on Page 537 lists reducing agents in order of relative strength in the *Reducing Agent* column. The stronger reducing agents are at the bottom, and the strength decreases as you move up the column. See Section 19.5, Strengths of Oxidizing Agents and Reducing Agents, on Pages 537-538.

18. $Ni + Zn^{2+} \rightleftharpoons Ni^{2+} + Zn$; reverse
WRA WOA SOA SRA

W = weaker, S = stronger, RA= reducing agent, OA= oxidizing agent; the relative strengths of reducing and oxidizing agents are given in Table 19.2 on Page 537. The reaction is favored in the direction of the weaker species. See Section 19.6, Predicting Redox Reactions, on Pages 538-542.

19. $2\ Fe^{3+} + Co \rightleftharpoons 2\ Fe^{2+} + Co^{2+}$; forward
SOA SRA WRA WOA

W = weaker, S = stronger, RA= reducing agent, OA= oxidizing agent; the relative strengths of reducing and oxidizing agents are given in Table 19.2 on Page 537. The reaction is favored in the direction of the weaker species. See Section 19.6, Predicting Redox Reactions, on Pages 538-542.

20. $O_2 + 4\ H^+ + 2\ Ca \rightleftharpoons 2\ H_2O + 2\ Ca^{2+}$; forward
SOA SRA WRA WOA

W = weaker, S = stronger, RA= reducing agent, OA= oxidizing agent; the relative strengths of reducing and oxidizing agents are given in Table 19.2 on Page 537. The reaction is favored in the direction of the weaker species. See Section 19.6, Predicting Redox Reactions, on Pages 538-542.

21. See the five items listed in Section 19.7.

See Section 19.7, Redox Reactions and Acid-Base Reactions Compared, on Page 542.

22. $SO_4^{2-} + 4\ H^+ + 2\ e^- \rightarrow SO_2 + 2\ H_2O$
$(Ag \rightarrow Ag^+ + e^-) \times 2$

$SO_4^{2-} + 4\ H^+ + 2\ Ag \rightarrow SO_2 + 2\ H_2O + 2\ Ag^+$

$Ag \rightarrow Ag^+$
$Ag \rightarrow Ag^+ + e^-$

$SO_4^{2-} \rightarrow SO_2$
$SO_4^{2-} \rightarrow SO_2 + 2\ H_2O$
$4\ H^+ + SO_4^{2-} \rightarrow SO_2 + 2\ H_2O$
$2\ e^- + 4\ H^+ + SO_4^{2-} \rightarrow SO_2 + 2\ H_2O$

See Section 19.8, Writing Redox Equations, on Pages 543-546. In particular, see the Summary: Writing Redox Equations for Half-Reactions in Acidic Solutions box on Page 543.

23. $NO_3^- + 10\ H^+ + 8\ e^- \rightarrow NH_4^+ + 3\ H_2O$
$(Zn \rightarrow Zn^{2+} + 2\ e^-) \times 4$

$NO_3^- + 10\ H^+ + 4\ Zn \rightarrow NH_4^+ + 3\ H_2O + 4\ Zn^{2+}$

$NO_3^- \rightarrow NH_4^+$
$NO_3^- \rightarrow NH_4^+ + 3\ H_2O$
$10\ H^+ + NO_3^- \rightarrow NH_4^+ + 3\ H_2O$
$8\ e^- + 10\ H^+ + NO_3^- \rightarrow NH_4^+ + 3\ H_2O$

$Zn \rightarrow Zn^{2+}$
$Zn \rightarrow Zn^{2+} + 2\ e^-$

See Section 19.8, Writing Redox Equations, on Pages 543-546. In particular, see the Summary: Writing Redox Equations for Half-Reactions in Acidic Solutions box on Page 543.

24. $Cr_2O_7^{2-} + 14\ H^+ + 6\ e^- \rightarrow 2\ Cr^{3+} + 7\ H_2O)$
$(Fe^{2+} \rightarrow Fe^{3+} + e^-) \times 6$

$Cr_2O_7^{2-} + 14\ H^+ + 6\ Fe^{2+} \rightarrow 2\ Cr^{3+} + 7\ H_2O + 6\ Fe^{3+}$

$Cr_2O_7^{2-} \rightarrow$ Cr3+
$Cr_2O_7^{2-} \rightarrow 2\ Cr^{3+}$
$Cr_2O_7^{2-} \rightarrow 2\ Cr^{3+} + 7\ H_2O$
$14\ H^+ + Cr_2O_7^{2-} \rightarrow 2\ Cr^{3+} + 7\ H_2O$
$6\ e^- + 14\ H^+ + Cr_2O_7^{2-} \rightarrow 2\ Cr^{3+} + 7\ H_2O$

$Fe^{2+} \rightarrow Fe^{3+}$
$Fe^{2+} \rightarrow Fe^{3+} + e^-$

See Section 19.8, Writing Redox Equations, on Pages 543-546. In particular, see the Summary: Writing Redox Equations for Half-Reactions in Acidic Solutions box on Page 543.

25. $(MnO_4^- + 4\ H^+ + 3\ e^- \rightarrow MnO_2 + 2\ H_2O) \times 2$
$(2\ I^- \rightarrow I_2 + 2\ e^-) \times 3$

$2\ MnO_4^- + 8\ H^+ + 6\ I^- \rightarrow 2\ MnO_2 + 4\ H_2O + 3\ I_2$

$I^- \rightarrow I_2$
$2\ I^- \rightarrow I_2$
$2\ I^- \rightarrow I_2 + 2\ e^-$

$MnO_4^- \rightarrow MnO_2$
$MnO_4^- \rightarrow MnO_2 + 2\ H_2O$
$4\ H^+ + MnO_4^- \rightarrow MnO_2 + 2\ H_2O$
$3\ e^- + 4\ H^+ + MnO_4^- \rightarrow MnO_2 + 2\ H_2O$

See Section 19.8, Writing Redox Equations, on Pages 543-546. In particular, see the Summary: Writing Redox Equations for Half-Reactions in Acidic Solutions box on Page 543.

26. $2\ BrO_3^- + 12\ H^+ + 10\ e^- \rightarrow Br_2 + 6\ H_2O$
$(2\ Br^- \rightarrow Br_2 + 2\ e^-) \times 5$

$2\ BrO_3^- + 12\ H^+ + 10\ Br^- \rightarrow 6\ Br_2 + 6\ H_2O$
$BrO_3^- + 6\ H^+ + 5\ Br^- \rightarrow 3\ Br_2 + 3\ H_2O$

$BrO_3^- \rightarrow Br_2$
$2\ BrO_3^- \rightarrow Br_2$
$2\ BrO_3^- \rightarrow Br_2 + 6\ H_2O$
$12\ H^+ + 2\ BrO_3^- \rightarrow Br_2 + 6\ H_2O$
$10\ e^- + 12\ H^+ + 2\ BrO_3^- \rightarrow Br_2 + 6\ H_2O$

$Br^- \rightarrow Br_2$
$2\ Br^- \rightarrow Br_2$
$2\ Br^- \rightarrow Br_2 + 2\ e^-$

See Section 19.8, Writing Redox Equations, on Pages 543-546. In particular, see the Summary: Writing Redox Equations for Half-Reactions in Acidic Solutions box on Page 543.

28. True: a, c, g. False: b, d, e, f.

b) The sum of oxidation numbers in a molecular compound is zero. The sum of oxidation numbers in a formula unit of an ionic compound is zero.
d) The oxidation number of alkali metals (Group 1A/1) is always +1.
e) A substance that gains electrons is reduced.
f) A strong reducing agent releases electrons readily.

29. In the electrolytic cell, the force that moves the charges through the circuit is *outside* the cell. In the voltaic cell, the cell itself is the *source* of the force that moves the charged particles.

Figure 19.1 is on Page 528. See Section 19.1, Electrolytic and Voltaic Cells, on Pages 528-529.

30. Your (1) and (2) numbers may be the reverse of ours. a) (1) $Cu \rightarrow Cu^{2+} + 2\ e^-$. (2) $Cu^{2+} + 2\ e^- \rightarrow Cu$. b) (1) occurs at the "+" electrode and (2) at the "–" electrode. c) "+" is the anode, where oxidation occurs. "–" is the cathode, where reduction occurs. d) Charge is carried through the electrolyte by Cu^{2+} ions moving from anode to cathode. (H^+ and SO_4^{2-} ions also move, but without an identifiable "flow" of charge that is responsible for electrolysis in the solution.) e) The bubbles are hydrogen. They come from the only ion that can be "deposited" as a gas, H^+. The occurs at the cathode, the same electrode at which copper is deposited. Any electrons that are used to reduce H^+ ions instead of copper ions cause the mass of copper deposited to be less than the mass of copper dissolved. [What happens to the concentration of Cu^{2+} ion in the solution? (The concentration of copper increases over time.)] In one sentence: Hydrogen gas, instead of copper, is reduced (deposited) at the cathode according to the half-reaction $2\ H^+(aq) + 2\ e^- \rightarrow H_2(g)$.

Figure 19.1 is on Page 528. See Section 19.1, Electrolytic and Voltaic Cells, on Pages 528-529.

31. The moisture on your tongue becomes an electrolyte for the passage of electric current from one metal to the other. The tingle you "taste" is caused by that current.

See Section 19.1, Electrolytic and Voltaic Cells, on Pages 528-529.

Chapter 20

Chemical Equilibrium

1. Rates of change in opposite directions are equal in an equilibrium.

   Examples of equilibrium systems are found throughout the chapter. See item 4 in the list of four conditions found in every equilibrium on Page 552.

2. Neither matter nor energy may enter or leave a closed system.

   See item 2 in the list of four conditions found in every equilibrium on Page 552.

3. The system is not closed—water enters and leaves—so it is not an equilibrium.

   See Section 20.1, The Character of an Equilibrium, on Page 552. In particular, see item 2 in the list of four conditions found in every equilibrium.

4. Colliding particles must have sufficient energy and proper orientation to react.

   See Section 20.2, The Collision Theory of Chemical Reactions, on Pages 552-554. In particular, review Figure 20.1 on Page 553.

5. ΔE is negative; the reaction is exothermic. $\Delta E = c - b$. Activation energy = $a - b$.

   Figure 20.2 is on Page 554. A delta quantity is final value minus initial value. The final energy value is lower than the initial energy value. Smaller number minus larger number yields a negative value. A reaction that releases energy (negative ΔE) is exothermic (see the Energy in Chemical Change subsection on Pages 33-34). See Section 20.3, Energy Changes during a Molecular Collision, on Pages 554-555.

6. The activation energy is greater for the reverse reaction: $a - c$.

   Figure 20.2 is on Page 554. See Section 20.3, Energy Changes during a Molecular Collision, on Pages 554-555.

7. See text for discussion of activation energy. All other things being equal, the reaction with the lower activation energy will be faster because a larger fraction of reacting particles will be able to engage in reaction-producing collisions.

   See Section 20.3, Energy Changes during a Molecular Collision, on Pages 554-555.

8. The statement refers to the fraction of all possible reacting particles that has enough energy to engage in a reaction-producing collision. In Figure 20.4,

$$\frac{\text{shaded area}}{\text{total area}} = \frac{\text{number of particles with enough energy to react}}{\text{total number of particles}}$$

Figure 20.4 is on Page 556. If all particles in a sample are represented by the area under the curve, those with sufficient energy to engage in a reaction-producing collision are represented by the area under the curve and to the right of the E_a line. This is a small fraction of the total number of particles. See The Effect of Temperature on Reaction Rate subsection on Page 556.

9. A catalyst is a substance that increases reaction rate by lowering activation energy. The catalyst is not permanently change in the reaction.

   See The Effect of a Catalyst on Reaction Rate subsection on Pages 556-557. Figure 20.5 on Page 557 is particularly important.

10. Rate varies directly with reactant concentration. As A concentration increases, rate increases; as B concentration decreases, rate decreases.

    See The Effect of Concentration on Reaction Rate subsection on Pages 557-558.

11. At the beginning, when both reactants are at highest concentration.

    The red curve in Figure 20.7, Page 558 (repeated on Page 559), represents the forward reaction rate. The highest reaction rate occurs at the initial time. See Section 20.5, Development of a Chemical Equilibrium, on Pages 558-559.

12. Nitrogen and hydrogen concentrations decrease; ammonia increases.

    The reactants, $N_2(g)$ and $H_2(g)$, react to form the product, $NH_3(g)$. The concentrations of the reactants decrease as they react, and the concentration of the product increases as it is formed, until equilibrium concentrations are achieved. See Section 20.5, Development of a Chemical Equilibrium, on Pages 558-559.

13. Reverse shift to use up some of the added reactant in the reverse direction.

    The concentration of a species on the product side of the reaction is increased, so the reaction will shift in the reverse direction to reduce that concentration. See The Concentration Effect subsection on Pages 559-560.

14. Forward direction to replenish some of the product that was removed.

    The concentration of a species on the product side of the reaction is reduced, so the reaction will shift in the forward direction to increase that concentration. See The Concentration Effect subsection on Pages 559-560.

15. Reverse shift to use up some of the added $C_2H_3O_2^-(aq)$.

    The concentration of a species on the product side of the reaction is increased, so the reaction will shift in the reverse direction to reduce that concentration. See The Concentration Effect subsection on Pages 559-560.

16. Reverse shift to reduce total number of gaseous molecules and thereby reduce the pressure that had been created by the reduced volume.

The increased pressure that results from the reduced volume is relieved by a shift in the direction of fewer molecules. One mole of reactant yields two moles of products, so the shift is in the reverse direction. See The Volume Effect subsection on Pages 561-562.

17. Forward direction to increase total number of gaseous molecules and thereby increase the pressure that had been reduced by larger volume.

    The decreased pressure that results from the increased volume is relieved by a shift in the direction of greater molecules. One mole of *gaseous* reactant yields two moles of gaseous products, so the shift is in the forward direction. See The Volume Effect subsection on Pages 561-562.

18. Reverse direction to sue up some of the added energy.

    The "concentration" of a species on the product side of the reaction is increased, so the reaction will shift in the reverse direction to reduce that "concentration." See The Temperature Effect subsection on Pages 563-564.

19. Heat the system to produce a shift in forward direction.

    To increase the amount of HI, the reaction must shift in the forward direction. Increasing the "concentration" of a species on the reactant side of the reaction will shift the reaction in the forward direction. See The Temperature Effect subsection on Pages 563-564.

20. (a) and (d), forward; (b) and (c), reverse.

    a) The concentration of a species on the reactant side of the reaction is increased, so the reaction will shift in the forward direction to reduce that concentration. See The Concentration Effect subsection on Pages 559-560.
    b) The "concentration" of a species on the product side of the reaction is increased, so the reaction will shift in the reverse direction to reduce that "concentration." See The Temperature Effect subsection on Pages 563-564.
    c) The increased pressure that results from the reduced volume is relieved by a shift in the direction of fewer molecules. Nine moles of reactants yields ten moles of products, so the shift is in the reverse direction. See The Volume Effect subsection on Pages 561-562.
    d) The concentration of a species on the product side of the reaction is reduced, so the reaction will shift in the forward direction to increase that concentration. See The Concentration Effect subsection on Pages 559-560.

21. $$\frac{[SO_3]^2}{[SO_2]^2[O_2]}$$

    See Section 20.7, The Equilibrium Constant, on Pages 564-567. In particular, see the four-step procedure on Page 565 for writing an equilibrium constant expression.

22. $$\frac{[CH_4][H_2S]^2}{[H_2]^4[CS_2]}$$

    See Section 20.7, The Equilibrium Constant, on Pages 564-567. In particular, see the four-step procedure on Page 565 for writing an equilibrium constant expression.

23. $[Cd^{2+}][OH^-]^2$

Note that $Cd(OH)_2(s)$ is a solid and therefore not included in the equilibrium constant expression. See Section 20.7, The Equilibrium Constant, on Pages 564-567. In particular, see the four-step procedure on Page 565 for writing an equilibrium constant expression.

24. $$\frac{[H^+][NO_2^-]}{[HNO_2]}$$

See Section 20.7, The Equilibrium Constant, on Pages 564-567. In particular, see the four-step procedure on Page 565 for writing an equilibrium constant expression.

25. $$\frac{[Ag^+][CN^-]^2}{[Ag(CN)_2^-]}$$

See Section 20.7, The Equilibrium Constant, on Pages 564-567. In particular, see the four-step procedure on Page 565 for writing an equilibrium constant expression.

26. In a K expression, the species on the right are always in the numerator, and those on the left are in the denominator. As the example equation is written, $K = \frac{[NH_3]^2}{[N_2][H_2]^3}$. If the equation is reversed, $2\ NH_3(g) \rightleftharpoons N_2(g) + 3\ H_2(g)$, $K = \frac{[N_2][H_2]^3}{[NH_3]^2}$.

See Section 20.7, The Equilibrium Constant, on Pages 564-567. In particular, see the paragraph at the top of Page 566.

27. The equilibrium constant is very small, so the reaction is favored in the reverse direction.

$K = 0.000000023$. If $K < 0.01$, the reverse direction is favored. See Section 20.8, The Significance of the Value of K, on Pages 567-568.

28. $HC_2H_3O_2 \rightleftharpoons H^+ + C_2H_3O_2^-$. The equilibrium constant, $K = \frac{[H^+][C_2H_3O_2^-]}{[HC_2H_3O_2]}$, will be small. A weak acid ionizes only slightly, producing very small concentrations of the numerator species compared with the almost unchanged concentration of the denominator species.

The experimentally determined value of K for this equilibrium is 0.000018, verifying the prediction of a small K. See Section 20.8, The Significance of the Value of K, on Pages 567-568.

29. a) Forward, with large K. HCl is a strong acid and ionizes almost completely.
b) Reverse, with small K. $BaSO_4$ is a low-solubility solid and releases few ions.

a) HCl is one of the seven common strong acids. A strong acid ionizes almost completely.
b) Table 17.3 and Table 17.4 on Page 489, Solubilities of Ionic Compounds, show that barium sulfate is insoluble. The product ion concentrations are very small, giving a small K value. See Section 20.8, The Significance of the Value of K, on Pages 567-568.

30. *GIVEN:* $s = 8.8 \times 10^{-14}$ mol/L *WANTED:* K_{sp}

EQUATIONS: $CdS(s) \rightleftharpoons Cd^{2+}(aq) + S^{2-}(aq)$ $K_{sp} = [Cd^{2+}][S^{2-}]$

$[Cd^{2+}] = [S^{2-}] = 8.8 \times 10^{-14}$ $K_{sp} = [Cd^{2+}][S^{2-}] = (8.8 \times 10^{-14})^2 = 7.7 \times 10^{-27}$

See the Solubility Equilibria subsection on Pages 569-573.

31. *GIVEN:* 1.0×10^{-3} g CuBr; 100 mL *WANTED:* M CuBr (mol/L)

EQUATION: $M \equiv \frac{mol}{L} = \frac{1.0 \times 10^{-3} \text{ g CuBr}}{0.100 \text{ L}} \times \frac{1 \text{ mol CuBr}}{143.45 \text{ g CuBr}} = 7.0 \times 10^{-5} \text{ M CuBr}$

GIVEN: $s = 7.0 \times 10^{-5}$ mol/L *WANTED:* K_{sp}

EQUATIONS: $CuBr(s) \rightleftharpoons Cu^{+}(aq) + Br^{-}(aq)$ $K_{sp} = [Cu^{+}][Br^{-}]$

$[Cu^{+}] = [Br^{-}] = 7.0 \times 10^{-5}$ $K_{sp} = [Cu^{+}][Br^{-}] = (7.0 \times 10^{-5})^2 = 4.9 \times 10^{-9}$

See the Solubility Equilibria subsection on Pages 569-573.

32. *GIVEN:* $K_{sp} = 8.7 \times 10^{-9}$ *WANTED:* s (mol/L)

EQUATIONS: $CaCO_3(s) \rightleftharpoons Ca^{2+}(aq) + CO_3^{2-}(aq)$ $K_{sp} = [Ca^{2+}][CO_3^{2-}]$

$[Ca^{2+}] = [CO_3^{2-}] = s$ $K_{sp} = [Ca^{2+}][CO_3^{2-}] = s^2 = 8.7 \times 10^{-9}$ $s = 9.3 \times 10^{-5}$ mol/L

GIVEN: 100 mL (0.100 L); 9.3×10^{-5} mol/L *WANTED:* s (g/100 mL)

PER/PATH: $L \xrightarrow{9.3 \times 10^{-5} \text{ mol } CaCO_3/L} \text{mol } CaCO_3 \xrightarrow{100.09 \text{ g } CaCO_3/\text{mol } CaCO_3} \text{g } CaCO_3$

$$0.100 \text{ L} \times \frac{9.3 \times 10^{-5} \text{ mol } CaCO_3}{L} \times \frac{100.09 \text{ g } CaCO_3}{\text{mol } CaCO_3} = 9.3 \times 10^{-4} \text{ g } CaCO_3$$

Since 9.3×10^{-4} g $CaCO_3$ dissolves in 100 mL, the solubility is $\frac{9.3 \times 10^{-4} \text{ g } CaCO_3}{100 \text{ mL}}$

See the Solubility Equilibria subsection on Pages 569-573.

33. *GIVEN:* $K_{sp} = 1.7 \times 10^{-6}$ *WANTED:* s (mol/L)

EQUATIONS: $BaF_2 \rightleftharpoons Ba^{2+} + 2 F^{-}$ $K_{sp} = [Ba^{2+}][F^{-}]^2$

$K_{sp} = [Ba^{2+}][F^{-}]^2 = 1.7 \times 10^{-6}$ solubility = s = $[Ba^{2+}]$

$[F^{-}] = 2s$ $(s)(2s)^2 = 4s^3 = 1.7 \times 10^{-6}$ $s^3 = \frac{1.7 \times 10^{-6}}{4} = 4.3 \times 10^{-7}$

$s = \sqrt[3]{4.3 \times 10^{-7}}$ $s = 7.5 \times 10^{-3}$ mol/L

See the Solubility Equilibria subsection on Pages 569-573.

34. *GIVEN:* $K_{sp} = 8.1 \times 10^{-9}$; 0.10 M $BaCl_2$ *WANTED:* s (mol/L)

EQUATIONS: $BaCO_3(s) \rightleftharpoons Ba^{2+}(aq) + CO_3^{2-}(aq)$ $K_{sp} = [Ba^{2+}][CO_3^{2-}]$

$BaCl_2(s) \rightarrow Ba^{2+}(aq) + 2 Cl^{-}(aq)$ 1 mol $BaCl_2$ yields 1 mol Ba^{2+}, so $[Ba^{2+}] = 0.10$ M

$K_{sp} = [Ba^{2+}][CO_3^{2-}] = 0.10[CO_3^{2-}] = 8.1 \times 10^{-9}$ $[CO_3^{2-}] = \frac{8.1 \times 10^{-9}}{0.10} = 8.1 \times 10^{-8}$ M

solubility = $[CO_3^{2-}] = 8.1 \times 10^{-8}$ mol/L

See the Solubility Equilibria subsection on Pages 569-573.

35. *GIVEN:* pH = 1.93; 1.0 M $HC_3H_5O_3$ *WANTED:* K; % ionization

EQUATIONS: $HC_3H_5O_3(aq) \rightleftharpoons H^+(aq) + C_3H_5O_3^-(aq)$ $K_a = \frac{[H^+][C_3H_5O_3^-]}{[HC_3H_5O_3]}$

$[H^+] = 10^{-pH} = 10^{-1.93}$ M = 0.012 M

Since 1 mol H^+ is formed for every 1 mol $C_3H_5O_3^-$ formed, $[H^+] = [C_3H_5O_3^-] = 0.012$ M

$$K_a = \frac{[H^+][C_3H_5O_3^-]}{[HC_3H_5O_3]} = \frac{(0.012)^2}{1.0} = 1.4 \times 10^{-4}$$

EQUATION: $\%\ \text{ionization} = \frac{\text{mol ionized}}{\text{total mol}} = \frac{0.012}{1.0} \times 100 = 1.2\%$ ionized

See the Ionization Equilibria subsection on Pages 573-576.

36. *GIVEN:* 0.1 M HNO_2; $K_a = 4.6 \times 10^{-4}$ *WANTED:* pH

EQUATIONS: $HNO_2(aq) \rightleftharpoons H^+(aq) + NO_2^-(aq)$ $K_a = \frac{[H^+][NO_2^-]}{[HNO_2]}$

$K_a [HNO_2] = [H^+][NO_2^-]$

Since ionization is the only source of H^+ and NO_2^-, $[H^+] = [NO_2^-]$, so $K_a [HNO_2] = [H^+][H^+]$

$[H^+] = \sqrt{K_a[HNO_2]} = \sqrt{(4.6 \times 10^{-4})(0.1)} = 6.8 \times 10^{-3}$ M

$pH = -\log [H^+] = -\log (6.8 \times 10^{-3}) = 2.17$

pH = 2.2 (answer rounded off to one significant figure to match 0.1)

See the Ionization Equilibria subsection on Pages 573-576.

37. *GIVEN:* 0.25 M $NaNO_2$; 0.75 M HNO_2; $K_a = 4.6 \times 10^{-4}$ *WANTED:* pH

EQUATIONS: $HNO_2(aq) \rightleftharpoons H^+(aq) + NO_2^-(aq)$ $K_a = \frac{[H^+][NO_2^-]}{[HNO_2]}$

$NaNO_2(s) \rightarrow Na^+(aq) + NO_2^-(aq)$; 1 mol $NaNO_2$ yields 1 mol NO_2^-, so $[NO_2^-] = 0.25$ M

$$K_a \times \frac{[HNO_2]}{[NO_2^-]} = \frac{[H^+][NO_2^-]}{[HNO_2]} \times \frac{[HNO_2]}{[NO_2^-]} \qquad K_a \times \frac{[HNO_2]}{[NO_2^-]} = [H^+]$$

$$[H^+] = K_a \times \frac{[HNO_2]}{[NO_2^-]} = 4.6 \times 10^{-4} \times \frac{0.75}{0.25} = 1.4 \times 10^{-3}\ \text{M}$$

$pH = -\log [H^+] = -\log (1.4 \times 10^{-3}) = 2.85$

See the Ionization Equilibria subsection on Pages 573-576.

38. *GIVEN:* pH = 4.80; $K_a = 6.5 \times 10^{-5}$ *WANTED:* $[HC_7H_5O_2]/[C_7H_5O_2^-]$

EQUATIONS: $HC_7H_5O_2(aq) \rightleftharpoons H^+(aq) + C_7H_5O_2^-(aq)$ $K_a = \frac{[H^+][C_7H_5O_2^-]}{[HC_7H_5O_2]}$

$$\frac{K_a}{[H^+]} = \frac{[H^+][C_7H_5O_2^-]}{[H^+][HC_7H_5O_2]} \qquad \frac{K_a}{[H^+]} = \frac{[C_7H_5O_2^-]}{[HC_7H_5O_2]} \qquad \frac{[H^+]}{K_a} = \frac{[HC_7H_5O_2]}{[C_7H_5O_2^-]}$$

$$\frac{[HC_7H_5O_2]}{[C_7H_5O_2^-]} = \frac{[H^+]}{K_a} = \frac{10^{-4.80}}{6.5 \times 10^{-5}} = 0.24$$

See the Ionization Equilibria subsection on Pages 573-576.

39.

	$PCl_3(g)$ +	$Cl_2(g)$ $\rightleftharpoons$	$PCl_5(g)$
mol/L at start	0.069	0.058	
mol/L change, + or –	–0.027	–0.027	+0.027
mol/L at equilibrium	0.042	0.031	0.027

$$K = \frac{[PCl_5]}{[PCl_3][Cl_2]} = \frac{0.027}{(0.042)(0.031)} = 21$$

The given data allow you to fill in the mol/L at start line and the Cl_2 mol/L at equilibrium value. (Start) - (Change) = (Equilibrium), so the Cl_2 (Change) = (Start) - (Equilibrium) = 0.058 - 0.031 = 0.027. The change line follows the reaction stoichiometry, which is 1:1:1, so the change for PCl_3 is -0.027 and the change for PCl_5 is +0.027. 0.069 - 0.027 gives 0.042 for the equilibrium value of PCl_3. 0 + 0.027 gives 0.027 for the equilibrium value of PCl_5. Once you have the equilibrium values, write the K expression based on the reaction equation, substitute the values, and calculate the value of K. See the Gaseous Equilibria subsection on Pages 576-578.

41 True: b, d, f, g, i, j, l, m, n. False: a, c, e, h, k, o.

a) All equilibria must be closed systems. If a species in the reaction has a steady supply, it is not an equilibrium reaction.
c) Not all collisions between reactant molecules result in a reaction; in fact, most do not.
e) An activated complex is an intermediate molecular species presumed to be formed during the collision of reacting molecules in a chemical change. It has no "properties."
h) An increase in temperature speeds both the forward reaction and the reverse reaction.
k) An increase in the concentration of a substance in an equilibrium increases the reaction rate in which the substance is a reactant.
o) A large K indicates that an equilibrium is favored in the forward direction.

42. Increase [AB], heat the system, and introduce catalyst.

Figure 20.2 is on Page 554. It shows an exothermic reaction, $A_2 + B_2 \rightleftharpoons 2\ AB$ + heat. Increasing the concentration of a species on the right side of the reaction will speed the reverse reaction. A catalyst speeds both forward and reverse reaction rates. See Section 20.4, Conditions that Affect the Rate of a Chemical Reaction, on Pages 556-558.

43. a) High pressure to force reaction to the smaller number of gaseous product molecules.
b) High temperature, at which all reaction rates are faster.

a) The question is solved by applying Le Chatelier's Principle. 4 moles of reactant yields 2 moles of product. See The Volume Effect subsection on Pages 561-562.
b) All reaction rates are higher at higher temperatures.

44. A manufacturer cannot use an equilibrium, which is a closed system from which no product can be removed.

An equilibrium requires a closed system (Section 20.1, Page 552).

45. The higher temperature is used to speed the reaction rate to an acceptable level. Lower pressure is dictated by limits of mechanical design and safety.

Manufacturing decisions are not exclusively based on chemistry! Cost efficiency must also be considered.

46. Temperature affects reaction rates in forward and reverse directions differently. Therefore, the value of an equilibrium constant depends on temperature. If you change the temperature, the value of K changes.

See Section 20.7, The Equilibrium Constant, on Pages 564-567.

Chapter 21

Nuclear Chemistry

1. A stable nucleus is one that will last forever under normal conditions. Radioactive nuclides have unstable nuclei, which change into other nuclides.

   See Section 21.2, Radioactivity, on Pages 586-588.

2. Alpha, beta, and gamma emissions are produced in radioactive decay. Alpha particles ($^{4}_{2}He$) are helium nuclei. Beta particles ($^{0}_{-1}e$) are electrons. Gamma rays are a form of electromagnetic radiation.

   See Section 21.2, Radioactivity, on Pages 586-588.

3. Gamma rays are different from alpha and beta particles because gamma rays are a form of energy, and alpha and beta particles are forms of matter.

   See Section 21.2, Radioactivity, on Pages 586-588.

4. Ionizing radiation refers to radioactive emissions energetic enough to cause particles to ionize. Not all radioactive emissions have sufficient energy to cause the target particles to lose an electron.

   See Section 21.3, The Detection and Measurement of Radioactivity, on Pages 588-589.

5. Radiation can be detected by its ability to expose photographic film, ionize a gas, or cause a substance to emit light.

   See Figure 21.2 on Page 588 and Figure 21.3 on Page 589. Also see Section 21.3, The Detection and Measurement of Radioactivity, on Pages 588-589.

6. A scintillation counter is a device used to measure radiation. Particles in a transparent solid absorb energy from the radiation, and subsequently release some of the energy in the form of light. This light is detected and measured by the counter.

   See Section 21.3, The Detection and Measurement of Radioactivity, on Pages 588-589.

7. A gamma camera is stationary, taking a single image, and a scanner moves, taking multiple images of the area to be observed. Their principal advantage is that they allow medical personnel to "see" inside the body without surgery.

   See Section 21.3, The Detection and Measurement of Radioactivity, on Pages 588-589.

8. Background radiation is potentially dangerous, but most of it is unavoidable. You can avoid voluntary exposure to radiation primarily by avoiding tobacco smoke.

   See Section 21.4, The Effects of Radiation on Living Systems, on Pages 589-591.

9. The half-life of a radioactive substance is the time it takes for one-half of the radioactive atoms in a sample to decay.

   See Section 21.5, Half-Life, on Pages 591-595.

10. $(1/2)^4 = 1/16$ of the original radioactivity remains, thus 15/16 is lost.

    $R = S\,(1/2)^n$, and the given 15/16 is neither R nor S. Amount lost = S - R, so R = S - amount lost = 1 - 15/16 = 1/16. Thus $1/16 = 1 \times (1/2)^n$, and n = 4. See Section 21.5, Half-Life, on Pages 591-595.

11. *GIVEN:* S = 188 clicks/s; R = 47 clicks/s *WANTED:* n
    R/S = 47 clicks per s/188 clicks per s = 0.25. From Figure 21.5, this is 2 half-lives.
    Also, if R/S = 0.25 = 1/4, you know n =2 because $(1/2)^2 = 1/4$.
    GIVEN: 2 half-lives; 132 min *WANTED:* half-life, $t_{1/2}$

    $$\textit{EQUATION: } \text{half-life} = \frac{\text{time}}{\text{number of half - lives}} = \frac{132 \text{ min}}{2 \text{ half - lives}} = 66 \text{ min/half-life} = t_{1/2}$$

    Figure 21.5 is on Page 591. See Section 21.5, Half-Life, on Pages 591-595.

12. a) *GIVEN:* 1 yr *WANTED:* number of half-lives

    $$\textit{PER/PATH: } \text{yr} \xrightarrow{\text{5.2 yr/half - life}} \text{half-life}$$

    $$1 \text{ yr} \times \frac{1 \text{ half - life}}{5.2 \text{ yr}} = 0.19 \text{ half-life; One year is 0.19 half-life.}$$

    GIVEN: S = 100 g [assume 100 g (100%) to start]; n = 0.19 *WANTED:* R
    EQUATION: $R = S \times (1/2)^n = 100 \times (1/2)^{0.19} = 88$ g
    Amount lost = S – R; R = S – amount lost = 100 – 88 = 12 g or 12% lost annually

    b) *GIVEN:* 13 yr; 5.2 yr/half-life *WANTED:* number of half-lives

    $$\textit{PER/PATH: } \text{yr} \xrightarrow{\text{5.2 yr/half - life}} \text{half-life}$$

    $$13 \text{ yr} \times \frac{1 \text{ half - life}}{5.2 \text{ yr}} = 2.5 \text{ half-lives}$$

    GIVEN: S = 105 g; n = 2.5 half-lives *WANTED:* R
    EQUATION: $R = S \times (1/2)^n = 105 \text{ g} \times (1/2)^{2.5} = 19$ g

    c) A significant proportion of the radioisotope would decay before it could be administered.

    See Section 21.5, Half-Life, on Pages 591-595.

13. *GIVEN:* R/S = 0.92 *WANTED:* number of half-lives
    From Figure 21.5, this is 0.12 half-life
    GIVEN: 24 hr; 0.12 half-life *WANTED:* half-life, $t_{1/2}$

    $$\textit{EQUATION: } \text{half-life} = \frac{\text{time}}{\text{number of half - lives}} = \frac{24 \text{ hr}}{0.12 \text{ half} - \text{life}} = 2.0 \times 10^2 \text{ hr/half-life} = t_{1/2}$$

    Figure 21.5 is on Page 591. See Section 21.5, Half-Life, on Pages 591-595.

14. *GIVEN:* R/S = 0.163 *WANTED:* number of half-lives
From Figure 21.5, this is 2.6 half-lives
GIVEN: 17.2 min/half-life; 2.6 half-lives *WANTED:* time (assume min)
PER/PATH: half-lives $\xrightarrow{\text{17.2 min/half - life}}$ min

$$2.6 \text{ half-lives} \times \frac{17.2 \text{ min}}{\text{half - life}} = 45 \text{ min}$$

Figure 21.5 is on Page 591. See Section 21.5, Half-Life, on Pages 591-595.

15. *GIVEN:* R = 19.7 g; S = 50.0 g *WANTED:* number of half-lives
$R/S = \frac{19.7 \text{ g}}{50.0 \text{ g}} = 0.394$; From Figure 21.5, this is 1.34 half-lives
GIVEN: 198 days/half-life, 1.34 half-lives *WANTED:* time (assume days)
PER/PATH: half-lives $\xrightarrow{\text{198 days/half - life}}$ days

$$1.34 \text{ half-lives} \times \frac{198 \text{ days}}{\text{half - life}} = 265 \text{ days}$$

Figure 21.5 is on Page 591. See Section 21.5, Half-Life, on Pages 591-595.

16. Beta decay is the ejection of an electron from the nucleus. The final nuclide has one fewer neutron and one more proton than the original. This is a transmutation because the number of protons changes.

See Section 21.2, Radioactivity (Pages 586-587), for a description of the beta particle. The term *nuclide* is also defined in this section. See Section 21.6, Natural Radioactive Decay Series—Nuclear Equations, on Pages 595-598, for a description of *transmutation*.

17. $^{212}_{82}\text{Pb} \rightarrow {}^{212}_{83}\text{Bi} + {}^{0}_{-1}\text{e}$ $\qquad$ $^{231}_{90}\text{Th} \rightarrow {}^{231}_{91}\text{Pa} + {}^{0}_{-1}\text{e}$

Beta decay of lead-212: reactant is $^{212}_{82}\text{Pb}$, one product is a beta particle, $^{0}_{-1}\text{e}$; for other product, 212 = A + 0, A = 212; 82 = Z + -1, Z = 83; from the periodic table, Z = 83 is Bi
Beta decay of thorium-231: reactant is $^{231}_{90}\text{Th}$; one product is a beta particle, $^{0}_{-1}\text{e}$; for other product, 231 = A + 0, A = 231; 90 = Z + -1, Z = 91; from the periodic table, Z = 91 is Pa
See Section 21.6, Natural Radioactive Decay Series—Nuclear Equations, on Pages 595-598.

18. $^{238}_{90}\text{Th} \rightarrow {}^{234}_{88}\text{Ra} + {}^{4}_{2}\text{He}$ $\qquad$ $^{212}_{84}\text{Po} \rightarrow {}^{208}_{82}\text{Pb} + {}^{4}_{2}\text{He}$

Ejection of alpha particle from thorium-238: reactant is $^{238}_{90}\text{Th}$, one product is an alpha particle, $^{4}_{2}\text{He}$; for other product, 238 = A + 4, A = 234; 90 = Z + 2, Z = 88; from the periodic table, Z = 88 is Ra
Ejection of alpha particle from polonium-212: reactant is $^{212}_{84}\text{Po}$, one product is an alpha particle, $^{4}_{2}\text{He}$; for other product, 212 = A + 4, A = 208; 84 = Z + 2, Z = 82; from the periodic table, Z = 82 is Pb
See Section 21.6, Natural Radioactive Decay Series—Nuclear Equations, on Pages 595-598.

19. The chemical properties of all isotopes of an element are essentially the same.

See item 1 on the list of areas of comparison of nuclear reactions and ordinary chemical reactions on Page 598.

20. The number of uranium atoms is decreasing, and the number of lead atoms is increasing. The end product of the natural decay of uranium is lead.

    See the radioactive decay series illustrated in Figure 21.7 on Page 596, where uranium-238 decays to lead-206. Also see Section 21.7, Nuclear Reactions and Ordinary Chemical Reactions Compared, on Page 598.

21. Induced radioactivity is a result of man-made radionuclides.

    See Section 21.8, Nuclear Bombardment and Induced Radioactivity, on Pages 598-600.

22. A particle accelerator increases the kinetic energy of charged particles that bombard nuclei. It is used to create new isotopes.

    See Section 21.8, Nuclear Bombardment and Induced Radioactivity, on Pages 598-600.

23. The transuranium elements are those with $Z \geq 93$. They are all radioactive. Americium ($Z = 95$) is used in smoke detectors.

    The photographs in the margin on Page 600 show americium-241 in a smoke detector. See Section 21.8, Nuclear Bombardment and Induced Radioactivity, on Pages 598-600.

24. All elements with $Z \geq 93$ are "transuranium elements." As with the other transuranium elements, $Z = 118$ would be expected to be radioactive. Chemically, it should be similar to the other Group 8A/18 elements because it would be in Group 8A/18.

    See Section 21.8, Nuclear Bombardment and Induced Radioactivity, on Pages 598-600.

25. a) ${}^{99}_{43}\text{Tc}$ b) ${}^{239}_{94}\text{Pu}$ c) ${}^{59}_{27}\text{Co}$

    a) 98 + 2 = A + 1, A = 99; 42 + 1 = Z + 0, Z = 43; from the periodic table, Z = 43 is Tc
    b) 238 + 4 = (3 × 1) + A, A = 239; 92 + 2 = (3 × 0) + Z, Z = 94; from the periodic table, Z = 94 is Pu
    c) A + 2 = 60 + 1, A = 59; Z + 1 = 27 + 1, Z = 27; from the periodic table Z = 27 is Co
    See Section 21.8, Nuclear Bombardment and Induced Radioactivity, on Pages 598-600.

26. Radioactive decay is the *spontaneous* breaking up of a large nucleus into two smaller nuclei. Nuclear fission is almost the same thing, but fission is initiated by neutron bombardment.

    See Section 21.10, Nuclear Fission, on Pages 601-602.

27. ${}^{235}_{92}\text{U} + {}^{1}_{0}\text{n} \rightarrow {}^{144}_{55}\text{Cs} + {}^{90}_{37}\text{Rb} + 2\ {}^{1}_{0}\text{n}$

    Equation 21.7 is on Page 601. The "same reactants" are ${}^{235}_{92}\text{U} + {}^{1}_{0}\text{n}$. The problem statement gives the product species ${}^{144}_{55}\text{Cs} + {}^{90}_{37}\text{Rb}$. You then need to figure out how many neutrons, ${}^{1}_{0}\text{n}$, are needed to balance the nuclear reaction. 235 + 1 = 144 + 90 + (x × 1), x = 2. See Section 21.10, Nuclear Fission, on Pages 601-602.

28. A chain reaction requires the product neutrons to be captured within the sample. The minimum quantity of matter for this purpose is called the critical mass.

    See Section 21.10, Nuclear Fission, on Pages 601-602.

29. Perhaps the primary advantage of nuclear power is avoidance of the consumption of our dwindling supply of fossil fuels and the pollution associated with burning them. The primary disadvantages are the radioactive waste and the threat of an accident.

See Section 21.11, Electrical Energy from Nuclear Fission, on Pages 603-604.

31. True: a, d, f, g, j, k. False: b, c, e, h, i, m. The answer to l is left to you.

b) The chemical properties of all isotopes of an element, radioactive or nonradioactive, are essentially the same.
c) β-rays have more penetrating power than α-rays.
e) When alpha, beta, or gamma radiation emitted from one source collides with a nearby atom, it can change the electron arrangement in the target or even knock an electron completely out of the target.
h) Isotopes with higher atomic numbers generally have shorter half-lives than isotopes with lower atomic numbers.
i) The first transmutations were achieved before humans were on earth.
m) The main obstacle to developing nuclear fusion as a source of electrical energy is the extremely high temperature needed to start and sustain the reaction.

32. Presumably it takes an infinite time for all of a sample of radioactive matter to decay.

If half of a sample decays in a given quantity of time, and half of that decays in the next quantity of time, etc., a fraction will theoretically always exist. See Section 21.5, Half-Life, on Pages 591-595. In particular, review the Figure 21.5 decay curve on Page 591. The curve will get closer and closer to the x-axis, but, theoretically, it will never touch it.

33. Rate of decay is best measured by Geiger counters or other devices described in Section 21.3, whose measurements are independent of the mass of the sample or the compound in which the radionuclide is found.

Note that half-life is defined as the time it takes for one-half the radioactive atoms in a sample to *decay*. When an atom decays, it changes to product particles that have mass, so change in mass is an indirect measure of the number of atoms that have decayed that can be difficult to measure in a sample that contains many different atoms. See Section 21.5, Half-Life, on Pages 591-595.

34. *GIVEN:* R = 22.7 units; S = 29.0 units *WANTED:* number of half-lives

$$R/S = \frac{22.7 \text{ units}}{29.0 \text{ units}} = 0.783$$; From Figure 21.5, this is 0.35 half-life.

GIVEN: 5.73×10^3 years/half-life; 0.35 half-lives *WANTED:* time (assume years)

PER/PATH: half-lives $\xrightarrow{5.73 \times 10^3 \text{ years/half-life}}$ years

$$0.35 \text{ half-life} \times \frac{5.73 \times 10^3 \text{ years}}{\text{half-life}} = 2.0 \times 10^3 \text{ years}$$

Radiocarbon dating indicates that the cloth is about 2000 years old, which places it at the beginning of the Christian era.

Figure 21.5 is on Page 591. See Section 21.5, Half-Life, on Pages 591-595.

35. US is $\frac{238}{(238 + 32)} \times 100 = 88\%$ uranium by mass. US_2 is 79% uranium by mass. Thus for samples of equal mass, there are more uranium atoms in US than in US_2. Only the radioactive element, uranium, contributes to radioactivity, so US will exhibit the greater amount of radioactivity. The radioactivity of 0.5 mole of US will be the same as 0.5 mole of US_2 because both samples contain the same number of uranium atoms.

Note that half-life is defined as the time it takes for one-half the radioactive *atoms* in a sample to decay. See Section 21.5, Half-Life, on Pages 591-595.

Chapter 22

Organic Chemistry

1. Originally, organic chemistry was defined as the chemistry of compounds found in living organisms. The modern definition describes organic chemistry as the chemistry of carbon compounds. The definition was changed because it was discovered that is was possible to synthesize organic compounds from non-living sources. The original definition includes the modern definition.

   See Section 22.1, The Nature of Organic Chemistry, on Pages 612-613.

2. 120°; trigonal planar.

   A double bond and two single bonds are three regions of electron density. Table 12.2 on Page 327 shows that this arrangement yields a trigonal planar electron-pair and molecular geometry. You may wish to review Section 12.3, Molecular Geometry, on Pages 326-332. Table 22.1, Bonding in Organic Compounds, on Page 613, summarizes the important bonding types found in organic compounds. See Section 22.2, The Molecular Structure of Organic Compounds, on Pages 613-614.

3. The everyday definition of *saturated* is completely full. Often applied to moisture, it means holding all the water something can hold. In chemistry, the term is used for solutions to indicate that the solution has the maximum amount of solute that can dissolve at a given temperature. With respect to hydrocarbons, the term means that the bonding capacity of a carbon atom is filled by being bonded to four other atoms.

   Section 22.3, Saturated Hydrocarbons: The Alkanes and Cycloalkanes, on Pages 615-620, shows a number of examples of structural diagrams of saturated hydrocarbons. Table 22.3, Unsaturated Hydrocarbons, on Page 622 shows examples of structural diagrams of unsaturated hydrocarbons.

4. The difference from one alkane to the next is a $—CH_2—$ structural unit. A homologous series is one where the compounds differ by the same structural unit.

   See Section 22.3, Saturated Hydrocarbons: The Alkanes and Cycloalkanes, on Pages 615-620.

5. See Section 22.3, which shows the two isomers of butane, C_4H_{10}.

   The two structural formulas are shown on Page 616.

6. $C_{12}H_{26}$; $CH_3CH_2CH_2CH_2CH_2CH_2CH_2CH_2CH_2CH_2CH_2CH_3$; $CH_3(CH_2)_{10}CH_3$

   The molecular formula of an alkane has the form C_nH_{2n+2}, where n is the number of carbon atoms. See Page 615. Line and condensed formulas are defined on Page 616.

7. An alkyl is an alkane from which one hydrogen has been removed. The symbol R is used to indicate a general alkyl.

   See the Alkyl Groups subsection on Page 617.

8. Pentane; C_9H_{20}.

Prefixes for the alkane series are given in Table 22.2 on Page 615. These are normal alkanes that have the general formula C_nH_{2n+2}.

9. *Please see the answer on textbook Page 650.*

A heptane is a 7-carbon parent alkane. 3-ethyl is a 2-carbon alkyl group on carbon 3, and 4-methyl is a 1-carbon alkyl group on carbon 4. See the Naming the Alkanes by the IUPAC System subsection on Pages 618–619.

10. 2,2,4-trimethylpentane

The longest continuous chain is 5 carbons, a pentane. There are 2 one-carbon alkyl groups, named methyl, on carbon 2, and an additional one-carbon group on carbon 4. Three methyl groups are indicated by the tri- prefix. See the Naming the Alkanes by the IUPAC System subsection on Pages 618–619.

11. *Please see the answer on textbook Page 650.*

A ethane is a 2-carbon parent alkane. There are 2 (di-) chlorine atoms, one on carbon 1 and one on carbon 2, and there are 4 (tetra-) fluorine atoms, 2 on carbon 1 and 2 on carbon 2. See the Naming the Alkanes by the IUPAC System subsection on Pages 618–619.

12. Carbon atoms in a cycloalkane, C_nH_{2n}, are arranged in a ring. Carbon atoms in a normal alkane, C_nH_{2n+2}, are in an unbroken chain.

See the Cycloalkanes subsection on Pages 619–620.

13. Cyclobutane

A square has 4 corners, so this cycloalkane has 4 carbon atoms. The name for a 4-carbon alkane is butane, and these atoms are in a ring, so the molecule is named with the cyclo- prefix. See the Cycloalkanes subsection on Pages 619–620.

14. *Please see the answer on textbook Page 650.*

The parent cycloalkane is a cyclohexane, 6 carbon atoms (hexane) in a ring (cyclo-). A bromine atom is attached to carbon 1, and a methyl group, a one carbon alkyl, is attached to carbon 3. Each atom in the cyclohexane parent ring needs two additional atoms, so either the specified atom or group is attached or a hydrogen atom is attached to saturate the ring. See the Cycloalkanes subsection on Pages 619–620.

15. 5-ethyl-2-methyloctane

The longest continuous chain is 8 carbons, an octane. There is a single carbon (methyl) attached to carbon 2, and there is a 2-carbon group (ethyl) attached to carbon 5. See the Naming the Alkanes by the IUPAC System subsection on Pages 618–619.

16. 4,4-dichloro-1,2-difluorohexane

The longest continuous chain is 6 carbons, a hexane. Carbon 1 and carbon 2 have fluorine (fluoro-) atoms attached. Two fluorines are indicated with the di- prefix. Carbon 4 has two chlorine (chloro-) atoms attached, again using a di- (2) prefix. See the Naming the Alkanes by the IUPAC System subsection on Pages 618-619.

17. *Please see the answer on textbook Page 650.*

The parent chain is a cyclohexane, 6 carbon atoms in a ring. A bromine (bromo-) atom is on carbon 1, and a methyl group, one carbon atom, is on carbon 3 of the parent chain. See the Naming the Alkanes by the IUPAC System subsection on Pages 618-619 and the Cycloalkanes subsection on Pages 619-620.

18. 1-fluoro-2-propylcyclopentane

The parent chain is 5 carbon atoms in a ring, which is cyclopentane. A fluorine (fluoro-) atom is on carbon 1, and a 3-carbon alkyl group, propyl-, is attached to carbon 2. See the Naming the Alkanes by the IUPAC System subsection on Pages 618-619 and the Cycloalkanes subsection on Pages 619-620.

19. An alkyne has a triple bond; an alkane has only single bonds. The general formula for an alkane is C_nH_{2n+2}, for an alkyne, C_nH_{2n-2}.

See Section 22.4, Unsaturated Hydrocarbons: The Alkenes and the Alkynes, on Pages 620-623.

20. *Please see the answer on textbook Page 650.*

The parent chain is two (eth-) double-bonded (-ene) carbons. Carbon 1 has two (di-) chlorine (chloro-) atoms attached. See the Structure and Nomenclature subsection of Section 22.4, Unsaturated Hydrocarbons: The Alkenes and the Alkynes, on Pages 620-622.

21. *Please see the answer on textbook Page 650.*

The parent chain is a 7-carbon (hept-) chain that includes a triple bond (-yne) next to carbon 3. See the Structure and Nomenclature subsection of Section 22.4, Unsaturated Hydrocarbons: The Alkenes and the Alkynes, on Pages 620-622. Cis-trans isomerism is discussed in the Isomerism among the Unsaturated Hydrocarbons subsection on Pages 622-623. In an alkyne, the bond angle between the triple-bonded carbons is 180°, so there is nothing "on this side" or "across."

22. 2-methyl-7-octadecene

The parent chain contains 18 carbons and a double bond, which is named in the problem statement as an octadecene. The lowest number possible for the location of the double bond is at carbon 7. This places the 1-carbon alkyl group, a methyl group, at carbon 2. See the Structure and Nomenclature subsection of Section 22.4, Unsaturated Hydrocarbons: The Alkenes and the Alkynes, on Pages 620-622.

23. 3-ethyl-1-hexyne

The parent chain has 6 carbons, giving a hex- prefix, and it contains a triple bond, yielding a -yne suffix. The lowest number possible for the triple bond is at carbon 1. A 2-carbon alkyl group, a ethyl group, is attached at carbon 3. See the Structure and Nomenclature subsection of Section 22.4, Unsaturated Hydrocarbons: The Alkenes and the Alkynes, on Pages 620-622.

24. *Please see the answers on textbook Page 650.*

The parent name benzene indicates a substituted benzene ring. The substituents are three (tri-) methyl groups, which are 1-carbon alkyl groups. To find all possible isomers, proceed in a systematic fashion. The 3 methyl groups can be on adjacent carbons (1,2,3-), they can skip one carbon (1,2,4-), or they can have a carbon between each methyl group (1,3,5-). See Section 22.5, Aromatic Hydrocarbons, on Pages 623-624.

25. 1-bromo-3-fluorobenzene

The parent ring is benzene. The substituents are fluorine, fluoro-, and bromine, bromo-. These are named in alphabetical order, so the bromine atom is assigned to carbon 1. The lowest number possible for the fluorine atom to be attached is therefore at carbon 3. See Section 22.5, Aromatic Hydrocarbons, on Pages 623-624.

26. *Please see the answers on textbook Page 651.*

The parent chain is butane, a 4-carbon alkane. "Dichloro substitution products" indicates that two (di-) hydrogens in butane have been replaced by chlorine atoms (chloro-). To find all possible isomers, proceed in a systematic manner. One chlorine can be on carbon 1, and the other can be on carbons 1, 2, 3, or 4, giving the 1,1-, 1,2-, 1,3-, and 1,4- isomers. One chlorine can be on carbon 2, and the other can be on carbons 2 or 3, giving the 2,2- and 2,3- isomers. See Section 22.8, Chemical Reactions of Hydrocarbons, on Pages 626-628, for a discussion of substitution products. See the Naming the Alkanes by the IUPAC System subsection on Pages 618-619 for a review of nomenclature.

27. *Please see the answers on textbook Page 651.*

"Dibromo addition products" indicates that two (di-) bromine atoms (from a bromine molecule, Br_2) have been added across the double bond. The pentene reactant is a 5-carbon (pent-) parent chain with a double bond (-ene). The original double bond could have been between carbon 1 and carbon 2, giving the 1,2- product, or it could have been between carbon 2 and carbon 3, giving the 2,3- product. See Section 22.8, Chemical Reactions of Hydrocarbons, on Pages 626-628, for a discussion of addition products.

28. Both products are butane.

Hydrogenation is the addition of two hydrogen atoms (a hydrogen molecule, H_2) across a multiple bond. In this question, the process is the hydrogenation of an alkene (-ene), a double bond. Both reactants are butenes, so when the double bond is opened and hydrogen atoms are added, the product is butane. No matter whether the hydrogen atoms are added to carbons 1 and 2 or to carbons 2 and 3, the resulting product is saturated butane. See Section 22.8, Chemical Reactions of Hydrocarbons, on Pages 626-628, for a discussion of addition products.

29. See the structural diagrams at the beginning of Section 22.10.

The structural diagrams at the top of Page 630 show how water, H-O-H, is structurally similar to an alcohol, R-O-H, and an ether, R-O-R'. See the Structures and Names of Alcohols and Ethers subsection on Pages 629-630.

30. Both the alcohol and water have hydrogen bonding, and substances with similar intermolecular forces tend to dissolve in one another. The ether is polar and can accept but not donate a hydrogen bond to water.

The ether, CH_3-O-CH_3, is a polar molecule that does not have the -O-H group necessary to donate a hydrogen bond. The isomeric alcohol, CH_3-CH_2-O-H, can donate and accept hydrogen bonds. Water, H-O-H, can donate and accept hydrogen bonds. Like dissolves like, in terms of similar intermolecular forces. You may want to review Section 15.3, Types of Intermolecular Forces, on Pages 404-408.

31. 2-pentanol

The parent alkane is a 5-carbon chain, a propane. To name an alcohol, the -e is replaced with an -ol. The hydroxyl group is on carbon 2. See the Structures and Names of Alcohols and Ethers subsection on Pages 629-630.

32. *Please see the answer on textbook Page 651.*

The compound is a ether, which means that it has the general structure R-O-R'. One R group is an ethyl, a 2-carbon alkyl, and the other R group is a propyl, a 3-carbon alkyl. See the Structures and Names of Alcohols and Ethers subsection on Pages 629-630.

33. *Please see the answer on textbook Page 651.*

The product ether has the general structure R-O-R'. Both (di-) R groups are isopropyl groups, branched 3-carbon alkyl groups. See the Structures and Names of Alcohols and Ethers subsection on Pages 629-630. An ether can be formed from the dehydration of—removal of a water molecule from—two alcohols. To produce an ether with an isopropyl group on each side of the oxygen atom, the reactant alcohols must be isopropyl alcohol. See the reaction at the bottom of Page 630.

34. Formaldehyde is a one-carbon aldehyde. With only one carbon, it is not possible to attach two alkyl groups to the carbonyl carbon, so "formanone" is not possible.

Formaldehyde is HCHO. The -one ending indicates a ketone, which has the general structure R-CO-R'. Two alkyl groups are required to form a ketone. See Section 22.11, Aldehydes and Ketones, on Pages 632-634.

35. *Please see the answer on textbook Page 651.*

The product butanal is a 4-carbon (butan-) aldehyde (-al). An aldehyde is characterized by a carbonyl group attached to an alkyl group, R-CHO. Thus the target product is CH_3-CH_2-CH_2-CHO. This is to be prepared by oxidation of an alcohol, R-CH(-R')-OH $1/2\ O_2 \rightarrow$ R-C(-R')=O + H_2O. In the product aldehyde, R is CH_3-CH_2-CH_2- and R' is H-. These are the same in the reactant alcohol, giving R-CH(-R')-OH = CH_3-CH_2-CH_2-CH(-H)-OH, or CH_3-CH_2-CH_2-CH_2-OH. See Section 22.11, Aldehydes and Ketones, on Pages 632-634.

36. *Please see the answer on textbook Page 651.*

See Section 22.12, Carboxylic Acids and Esters, on Pages 634–636. The general structure of a carboxylic acid is shown at the bottom of page 634.

37. *Please see the answer on textbook Page 651.*

Acetic acid is a 2-carbon carboxylic acid, CH_3-COOH. 1-propanol is a 3-carbon (propan-) alcohol (-ol) with the hydroxyl group on carbon 1, HO-$CH_2CH_2CH_3$. The reaction between an acid and an alcohol yields a water molecule, formed from the acid hydroxyl group and the alcohol hydrogen atom, and an ester. The resulting ester is CH_3-COO-$CH_2CH_2CH_3$. It is named by giving the name of the alkyl group associated with the alcohol, -$CH_2CH_2CH_3$, propyl, then the name of the anion derived from the acid, CH_3-COO-, acetate. If the reactant alcohol is 2-propanol, the ester group attaches to the second carbon, making the alkyl group isopropyl. See Section 22.12, Carboxylic Acids and Esters, on Pages 634–636.

38. *Please see the answers on textbook Page 651.*

An amine is formed by replacing one, two, or all three hydrogens in an ammonia molecule, NH_3, with an alkyl group. All 3 carbon atoms in C_3H_9N can be attached in a straight chain, $CH_3CH_2CH_2$-NH_2, or a branched chain, $(CH_3)_2$-CH-NH_2, to form propylamine and isopropylamine, respectively. An amine is named by identifying the alkyl groups attached to the nitrogen atom, followed by the suffix -amine. A third possibility is to attach 2 carbon atoms as one alkyl group (ethyl) and 1 carbon atom as a second alkyl group (methyl), to form CH_3CH_2-NH-CH_3, ethylmethlyamine. Finally, three (tri) 1-carbon groups (methyl) can be attached to the nitrogen atom to form $(CH_3)_3N$, trimethylamine. See Section 22.13, Amines and Amides, on Pages 636–638.

39. Propylamine and isopropylamine are primary amines; ethylmethylamine is a secondary amine; trimethylamine is a tertiary amine.

The number of hydrogens replaced in NH_3 determines whether an amine is primary (1 H replaced), secondary (2), or tertiary (3). See Section 22.13, Amines and Amides, on Pages 636–638.

40. *Please see the answer on textbook Page 651.*

Propanoic acid is a 3-carbon carboxylic acid, CH_3CH_2-COOH. Diethyl amine is an amine with two ethyl groups attached, $(CH_3CH_2)_2$-NH. The -OH part of the hydroxyl group in an acid reacts with a hydrogen atom on an amine to form water and a new molecule. See Section 22.13, Amines and Amides, on Pages 636–638.

41. *Please see the answer on textbook Page 651.*

An addition polymer is formed from alkene monomers by breaking the double bonds in the monomers and forming single bonds to neighboring molecules. See Section 22.15, Addition Polymers, on Pages 639–641.

42. *Please see the answer on textbook Page 652.*

The addition polymer was formed from a alkene monomer. Note that the pattern in the polymer is a carbon with two hydrogens followed by a carbon with two methyl groups, followed by a carbon with two hydrogens and a carbon with two methyl groups, etc. Thus the alkene monomer must be a carbon with two hydrogens double bonded to a carbon with two methyl groups. See Section 22.15, Addition Polymers, on Pages 639-641.

43. *Please see the answer on textbook Page 652.*

The parent chain of 4-methyl-1-pentene is a 5-carbon chain (pent-) with a double bond (-ene) between carbons 1 and 2. A 1-carbon alkyl group (methyl) is attached to carbon 4. An addition polymer is formed from alkene monomers by breaking the double bonds in the monomers and forming single bonds to neighboring molecules. See Section 22.15, Addition Polymers, on Pages 639-641.

44. *Please see the answer on textbook Page 652.*

A condensation polymer can be formed from a dicarboxylic acid and a dialcohol. The reaction between an acid and an alcohol yields a water molecule, formed from the acid hydroxyl group and the alcohol hydrogen atom, and an ester. See Section 22.16, Condensation Polymers, on Pages 641-644.

45. *Please see the answer on textbook Page 652.*

The reactants are a dicarboxylic acid and a diamine. The -OH part of the hydroxyl group in an acid reacts with a hydrogen atom on an amine to form water and a new molecule. See Section 22.16, Condensation Polymers, on Pages 641-644.

46. *Please see the answer on textbook Page 652.*

It is critical to recognize the group of atoms in the polymer where there is a carbon double bonded to an oxygen atom and single bonded to a nitrogen atom. This is where the two monomers were joined. Split the polymer at this location and add a water molecule in the form of a -OH group to the carbon end and add an -H to the nitrogen on the other end. Repeat to form the dicarboxylic acid and diamine monomers. See Section 22.16, Condensation Polymers, on Pages 641-644.

47. *Please see the answer on textbook Page 652.*

It is critical to recognize the group of atoms in the polymer where there is a carbon double bonded to an oxygen atom and single bonded to a nitrogen atom. This is where the two monomers were joined. Split the polymer at this location and add a water molecule in the form of a -OH group to the carbon end and add an -H to the nitrogen on the other end. Repeat to form the dicarboxylic acid and diamine monomers. See Section 22.16, Condensation Polymers, on Pages 641-644.

49. True: b, c, d, g, h, j, k, n, o, p, s. False: a, e, f, i, l, m, q, r.

a) To be classified as organic, a compound must contain carbon atoms (a tiny number of carbon compounds are commonly classified as inorganic).
e) Alkanes and cycloalkanes are saturated hydrocarbons. Alkenes and alkynes are unsaturated hydrocarbons.
f) An alkyl group is an alkane from which an end hydrogen atom has been removed.
i) An aliphatic hydrocarbon is an alkane, an alkene, or an alkyne. Alkanes are saturated; alkenes and alkynes are unsaturated.
l) Carbonyl groups are found in aldehydes and ketones.
m) An ester is neither a hydrocarbon nor aromatic, and it is made by the reaction of an alcohol with a carboxylic acid.
q) Forming cross-links in a polymer is the most effective way of making a polymer with mechanical strength.
r) Addition polymers are made from monomers that bond to each other without forming any other product.

50. An alkene and a cycloalkane with the same number of carbon atoms have the same molecular formula, C_nH_{2n}. The alkene has a double bond between two carbon atoms, and there is no closed loop of carbon atoms. The cycloalkane has only single bonds, and the carbon atoms are assembled in a closed ring.

Alkenes are discussed in Section 22.4, Unsaturated Hydrocarbons: The Alkenes and the Alkynes, on Pages 620-623. Cycloalkanes are discussed in a Cycloalkanes subsection on Pages 619-620.

51. Experimentally, there is only one type of carbon-carbon bond in benzene, not two. Benzene also does *not* undergo the addition reactions typical of alkenes. The delocalized structure reminds us that benzene is different from both alkanes and alkenes.

See Section 22.5, Aromatic Hydrocarbons, on Pages 623-624.

52. The number of monomers in different polymeric molecules varies, so they have no definite molecular mass.

See Section 22.15, Addition Polymers, on Pages 639-641, and Section 22.16, Condensation Polymers, on Pages 641-644.

53. For an aldehyde to be oxidized to a carboxylic acid, an oxygen atom must be inserted between the carbonyl carbon and the hydrogen bonded to it. If there is no hydrogen atom bonded to the carbonyl atom, as in a ketone, there can be no oxidation.

See Section 22.11, Aldehydes and Ketones, on Pages 632-634.

54. *Please look at the equation describing the formation of a peptide linkage on textbook Page 652.* This condensation reaction requires loss of a water molecule, one hydrogen of which must come from the amine. A tertiary amine has no hydrogens to lose. Trimethylamine (or any tertiary amine) can *not* form an amide.

See Section 22.13, Amines and Amides, on Pages 636-638.

Chapter 23

Biochemistry

1. Nonpolar R groups, polar but neutral R groups, acidic R groups, basic R groups.

   See Table 23.1 on Page 655.

2. Proline is the only amino acid in which the amine group is bonded to two carbons (a secondary amine).

   Table 23.1 is on Page 655.

3. S–F–G–Y is serylphenylalanylglycyltyrosine; serine is the N-terminal amino acid. Y–F–G–S is tyrosylphenylalanylglycylserine; tyrosine is the N-terminal amino acid.

   S = serine; F = phenylalanine; G = glycine; Y = tyrosine. To name the peptide, drop the -ine suffix of the amino acid name and add a -yl suffix. List the -yl versions of the amino acid names in order from N-terminal acid to C-terminal acid. For the C-terminal acid, use the unaltered amino acid name with the -ine ending. See the examples in Section 23.1, Amino Acids and Proteins, on Pages 654-660.

4. *Please see the answer on textbook Page 677.*

   A = alanine; C = cysteine; F = phenylalanine. To write the Lewis diagram of the peptide, draw each amino acid diagram (from Table 23.1 on Page 655), working from N-terminal end to C-terminal end, left to right. To connect amino acid molecules, split out a water molecule in the form of the -OH group from the acid end and the H- group from the amine end to form peptide linkages. See Section 23.1, Amino Acids and Proteins, on Pages 654-660.

5. Secondary structure refers to regularities in local conformations; tertiary structure refers to the bends in the chain.

   See the Secondary Protein Structure subsection on Pages 657-658 and the Tertiary Protein Structure subsection on Page 658.

6. β-pleated sheets have hydrogen bonds between the carbonyl oxygens and the amide hydrogens on adjacent chains.

   See the Secondary Protein Structure subsection on Pages 657-658.

7. Most enzymes catalyze only one reaction. Many enzymes are needed for the many chemical reactions that occur in living systems.

   See Section 23.2, Enzymes, on Pages 660-661.

8. Enzyme specificity refers to the fact that only a specific substrate molecule can bind to the enzyme.

   See Section 23.2, Enzymes, on Pages 660-661.

9. Enzymes and inorganic catalysts are both catalysts because they speed up a reaction without participating in the reaction itself. They differ in the mechanism by which they speed up the reaction.

   See Section 23.2, Enzymes, on Pages 660-661.

10. Enzyme feedback inhibition is similar to a furnace thermostat in that when a specified temperature is reached (high concentration of reaction product), the furnace is stopped (the enzyme stops catalyzing the reaction).

    See Section 23.2, Enzymes, on Pages 660-661.

11. The example of a monosaccharide ketose sugar presented in the textbook is fructose.

    See the Monosaccharides subsection on Pages 661-663.

12. Saccharides can be classified as monosaccharides, disaccharides, and polysaccharides. They differ in the number of monomer units comprising the sugar.

    See Section 23.3, Carbohydrates, on Pages 661-666.

13. Cellulose is a polysaccharide, ribose is a monosaccharide, and lactose is a disaccharide.

    See Section 23.3, Carbohydrates, on Pages 661-666.

14. Glucose and fructose.

    See the Disaccharides subsection on Pages 663-664.

15. Ketoses give negative Tollens' tests: fructose and sucrose.

    See Section 23.3, Carbohydrates, on Pages 661-666.

16. *Please see the answer on textbook Page 677.*

    See the Haworth projection of a single glucose molecule on the top of Page 662. It has numbered carbon atoms, and you want to connect carbon 1 of one molecule to carbon 4 of another. The molecules are bonded in a condensation reaction, splitting out a water molecule. See Section 23.3, Carbohydrates, on Pages 661-666.

17. $\alpha(1 \rightarrow 4)$ bonds can be broken by human enzymes; $\beta(1 \rightarrow 4)$ bonds cannot.

    See the Polysaccharides subsection on Pages 664-666.

18. Fats, oils, and phospholipids; waxes; and those without ester groups.

    See Section 23.4, Lipids, on Pages 666-669.

19. Fats are solids at room temperature; oils are liquids at room temperature.

    See the Fats and Oils subsection on Pages 666-667.

20. Fatty acids are built from 2-carbon acids.

    See the Fats and Oils subsection on Pages 666-667.

21. *Please see the answer on textbook Page 677.*

    See the Waxes subsection on Pages 667-668.

22. Deoxyribonucleic acid (DNA) and ribonucleic acid (RNA).

    See Section 23.5, Nucleic Acids, on Pages 670-674.

23. *Please see the answer on textbook Page 677.*

    The Lewis diagram for thymine is given in Figure 23.6 on Page 670.

24. *Please see the answers on textbook Page 677.*

    The Lewis diagrams for guanine and cytosine are given in Figure 23.6 on Page 670.

25. *Please see the answer on textbook Page 678.*

    The Lewis diagram for deoxyribose is shown in the Monosaccharides subsection of Section 23.3, Carbohydrates, on Page 663.

26. *Please see the answer on textbook Page 678.*

    The Lewis diagram for adenine is given in Figure 23.6 on Page 670. The Lewis diagram for deoxyribose is shown in the Monosaccharides subsection of Section 23.3, Carbohydrates, on Page 663. To combine them into a single molecule, you split out a water molecule. The -OH group attached to the carbon 1 position in the deoxyribose molecule combines with the H-atom from the N-H nitrogen in the adenine molecule.

27. *Please see the answer on textbook Page 678.*

    The Lewis diagrams for guanine, thymine, and adenine are given in Figure 23.6 on Page 670. The sugar-phosphate chain that forms the backbone DNA is shown in Figure 23.7 on Page 671. To attach the bases to the sugar-phosphate backbone, you split out a water molecule. The -OH group attached to the carbon 1 position in the deoxyribose molecule combines with the H-atom from the N-H nitrogen in the base molecule.

28. *Please see the answer on textbook Page 678.*

    The Lewis diagrams for uracil, cytosine, and guanine are given in Figure 23.6 on Page 670. The sugar-phosphate chain that forms the backbone of *deoxy*ribonucleic acid is shown in Figure 23.7 on Page 671. The question asks for the RNA, ribonucleic acid, fragment, however. Thus, the sugar molecules must be modified. Compare the ribose and deoxyribose molecules on the top of Page 663. Note that at the carbon 2 position, ribose has an -H and a -OH, but deoxyribose has two -H's. To attach the bases to the sugar-phosphate backbone, you split out a water molecule. The -OH group attached to the carbon 1 position in the ribose molecule combines with the H- atom from the N-H nitrogen in the base molecule.

29. In transcription, DNA transfers its information to mRNA. The mRNA travels to a ribosome where it is translated.

See Section 23.5, Nucleic Acids, on pages 670-674.

31. *GIVEN:* 64,500 g/mole hemoglobin *WANTED:* amino acids/hemoglobin

PER/PATH: g/mole hemoglobin $\xrightarrow{\text{120 g/ mole amino acid}}$ amino acids/hemoglobin

$$\frac{64{,}500 \text{ g}}{\text{mole hemoglobin}} \times \frac{1 \text{ mole amino acid}}{120 \text{ g}} = 538 \frac{\text{amino acids}}{\text{hemoglobin}}$$

See Section 23.1, Amino Acids and Proteins, on Pages 654-660, to review amino acids and proteins. A model of the hemoglobin molecule is illustrated in Figure 23.3 on Page 659.

32. All are polymers.

Cellulose is described on Page 664. Proteins are discussed in Section 23.1, Amino Acids and Proteins, on Pages 654-660. See Section 23.5, Nucleic Acids, for a discussion of both DNA and RNA. Starch is discussed in the Polysaccharides subsection on Pages 664-666.

33. Nitrogen in found in all proteins; if you picked sulfur (from cysteine or methionine), that's also true.

If you analyze Table 23.1 on Page 655, you see that the elements that make up the 20 amino acids commonly found in proteins are C, H, O, N, and S. Carbohydrates, which were originally thought to be hydrates of carbon, $(C{\cdot}H_2O)_n$, contain C, H, and O. See Page 661. Fats and oils contain C, H, and O, as shown on Pages 666-667.

34. The coil of the α-helix and the strong hydrogen bonding within the protein chains keep water from soaking into the nylon; there is no further hydrogen bonding to be made by the water to the nylon. In cotton, however, there is little hydrogen bonding between the cellulose chains, so water molecules can form hydrogen bonds with (and therefore soak into) the hydroxyl groups on the sugars that make up cellulose.

The structure of nylon 66 is shown on Page 643. See the Secondary Protein Structure subsection on Pages 657-658 for a discussion of hydrogen bonding in proteins. See the Polysaccharides subsection on Pages 664-666 for a discussion of hydrogen bonding in cellulose.

35. Because guanine and cytosine are complementary base pairs in DNA, there must also be 21% guanine. If guanine + cytosine are 42%, then adenine and thymine must equal 58%. Adenine is then 29%, as is thymine.

See Section 23.5, Nucleic Acids, on Pages 670-674, for a discussion of complementary base pairs. Adenine and thymine are a complementary pair, as are guanine and cytosine. 21% × 2 = 42%; 58% ÷ 2 = 29%.

Active Learning Workbook Answers

Chapter 2

Matter and Energy

1. Macroscopic: c. Microscopic: a, e. Particulate: b, d.

2. Studying matter at the particulate level is a characteristic of chemistry that sets it apart from the other sciences. An advantage is that understanding the behavior of particles allows the prediction of the macroscopic behavior of samples of matter made from those particles. Chemists can then design particles to exhibit the macroscopic characteristics desired, as seen with drug design and synthesis, for example.

3. Your illustration should resemble the particulate view in Figure 2.5.

4. A dense gas that is concentrated at the bottom of a container can be poured because its particles can move relative to each other. Chunks of solids, such as sugar crystals, can be poured.

5. Gases are most easily compressed because of the large spaces between molecules.

6. Chemical: b, c. Physical: a, d, e.

7. a, c, d.

8. Physical properties: a, b, d; chemical property: c.

9. Physical properties: a, b; chemical properties: c, d.

10. It is a pure substance, one kind of matter. However, it is heterogeneous, consisting of two visibly different forms or phases of carbon.

11. The everyday definition of the term *solution* is a liquid that appears the same throughout. The scientific definition of the term is a homogeneous mixture, whether solid, liquid, or gaseous.

12. Your sketch should show one type of particle (a pure substance), but in more than one state or form (heterogeneous).

13. Your sketch should include two or more different types of particles mixed together.

14. b, c, d.

15. Ice cubes from a home refrigerator are usually heterogeneous, containing trapped air. Homogeneous cubes in liquid water are heterogeneous, having visible solid and liquid phases.

16. Examples include glass products, plastic products, aluminum foil, cleaning and grooming solutions, and the air.

17. Natural milk from a cow separates into layers, with the cream on top and skim milk on the bottom. Milk is homogenized to give it consistency in properties, notably taste.

18. Homogeneous: c. Heterogeneous: a, b.

19. Pick out the ball bearings (no change); use the magnetic property of steel to pick up the ball bearings with a magnet (no change); dissolve the salt in water and filter or pick out the ball bearings (physical change).

20. The original liquid must be a mixture because the freezing point changed when some of the liquid was removed. The freezing point of a pure substance is the same no matter how much of the substance you have.

21. Compounds: a, b, c. Elements: d, e, f.

22. Elements: c, d. Compounds: a, b, e.

23. Yes. For example, limestone decomposes into lime and carbon dioxide.

24. On the macroscopic level, a compound can be decomposed chemically; an element cannot. On the particulate level, an element is one type of atom; a compound is more than one type of atom.

25. (a) elements: 2, 3; compounds: 1, 4, 5. (b) In general, if there are two or more words in the name, the substance is a compound. However, many compounds are known by one-word common names, and one-word names for many carbon-containing compounds are assembled from prefixes, suffixes, and special names for recurring groups. Chloromethane is such a compound. The name of an element is always a single word.

26. There is no evidence that A can be broken down into two or more other pure substances by a chemical or physical change, but only two methods have been tried. A is most likely an element, but the evidence is not conclusive.

27.	**G, L, S**	**P, M**	**Hom, Het**	**E, C**
Factory smokestack emissions	All, but mostly G	M	Het	
Concrete	S	M	Het	
Helium in a steel cylinder	G	P	Hom	E
Hummingbird feeder solution	L	M	Hom	
Table salt	S	P	Hom	C

28. Gravitational forces are attractive only; electrostatic forces can be attractive or repulsive. Magnetic forces can be attractive or repulsive also. All three can be acting simultaneously.

29. Reactants: $AgNO_3$, $NaCl$. Products: $AgCl$, $NaNO_3$.

30. The reactant Ni is an element; the product $Ni(NO_3)_2$ is a compound.

31. a, b, c.

32. Kinetic energy is greatest when the swing moves through its lowest point. Potential energy is at a maximum when the swing is at its highest point.

33. A gaseous substance was driven off by the heating process.

34. Examples include electrical energy being converted to mechanical energy (washing machine), light energy (light bulb), or heat energy (oven). These changes are useful because they are advantageous to you, but they are wasteful because they are not 100% efficient and thus an imperfect use of energy.

35. Nothing.

36. Examples include salt, sugar, and pencil "lead" (graphite, a form of carbon).

37. Air, certainly, and perhaps tap water, and/or some glass or plastic products.

38. Mercury, water, ice, carbon.

39. Nitrogen oxides, for example, form at least six different compounds. You also may have thought of carbon monoxide and carbon dioxide.

40. Rainwater is more pure. Ocean water is a solution of salt and other substances. Ocean water is distilled by evaporation and condensed into rain.

41. The substance is a mixture. If it was a pure substance, its density would not change.

42. If the objects have opposite charge, there is an attraction between them. An increase in separation will be an increase in potential energy. If they have the same charge (repulsion), the greater distance will be lower in potential energy.

43. Because each object contains particles with both positive and negative charges, there are both attractive and repulsive forces between the objects. If the net force is one of attraction, the particles move toward each other; if the net force is one of repulsion, the particles separate. If the two forces are balanced, the net force is zero and the particles remain separated at a constant distance.

Active Learning Workbook Answers

Chapter 3

Measurement and Chemical Calculations

1. a) 3.22×10^{-4} b) 6.03×10^{9} c) 6.19×10^{-12}

2. a) 5,120,000 b) 0.000000840 c) 1,920,000,000,000,000,000,000

3. a) 7.29×10^{-3} b) $5.13 \times 10-^{12}$ c) 2.98×10^{5} d) 4.02×10^{-6}

4. a) 75.6 b) 9.41×10^{-12} c) 3.24×10^{10} d) 1.49×10^{3}

5. a) 3.77 b) 1.97×10^{12}

6. a) 4.65×10^{8} b) 2.96×10^{-7}

7. 8.5 hr

8. 2.9 min

9. 11 dollars

10. 8.94 dollars

11. 5.2×10^{2} weeks

12. Her weight is the same when the elevator is standing still as when it moves at a constant rate. It decreases when the elevator slows; increases when the elevator accelerates. Her mass is constant no matter what the elevator is doing.

13. The meter.

14. A kilobuck is $1000. A megabuck is $1,000,000.

15. 1 mL = 0.001 L.

16. Megagrams because a gram is a very small unit of mass when compared to the mass of an automobile.

17. 0.0574 g; 1.41×10^{3} g; 4.54×10^{9} mg

18. 2.17×10^{3} cm; 0.517 km; 6.66×10^{4} cm

19. 494 mL; 1.91×10^{3} mL; 0.874 L

20. 711 g; 5.27×10^{5} pm; 3.63×10^{5} dag

21. 3, 5, 3, uncertain—2 to 5, 2, 3, 5, 4

22. 6.40×10^{-3} km; 0.0178 g; 7.90×10^{4} m; 4.22×10^{4} tons; 6.50×10^{2} dollars

23. 147 lb + 67.7 lb + 3.6×10^{2} lb + 135.43 lb = 7.1×10^{2} lb

24. 22.93 mL – 19.4 mL = 3.5 mL

25. 171.2 g; 262 g

26. 1.1 g/mL

27. 271 cm^3; 6.00 gal

28. 1.12 lb

29. 6.6×10^{2} kg

30. 2.53×10^{-4} lb

31. 152 lb, a middleweight

32. a) 436 yd b) 1.31×10^{3} ft

33. 443 m

34. 8.84 km

35. 2.34×10^{3} L

36.

Celsius	Fahrenheit	Kelvin
69	156	342
– 34	– 29	239
–162	–260	111
2	36	275
85	185	358
–141	–222	132

37. 37.0°C

38. 26°C

39. 136°F

40. $Q \propto m$; $Q = \Delta H_{fus} \times m$; cal/g; heat energy lost or gained per gram

41. 8.0×10^{1} cal/g

42. $P \propto \frac{1}{V}$; $P = k' \times \frac{1}{V}$; atm · L

43. 0.883 g/mL

44. 1.18 g/L

45. 6.40×10^2 cm^3

46. 1.6×10^3 g

47. Sample calculation for a 150-lb person:

$$150 \text{ lb} \times \frac{1 \text{ kg}}{2.20 \text{ lb}} = 68.2 \text{ kg} = 68{,}200 \text{ g} = 68{,}200{,}000 \text{ mg}$$

Kilograms are best because the number of g and mg are inconveniently large.

48. 2.1×10^3 cm^3

49. 0.4 ft; 1.6 m

50. 30.1 kg

51. 280 C degrees

52. 8.3 lb

Active Learning Workbook Answers

Chapter 4

Introduction to Gases

1. All gas properties relate in some way to the kinetic character. Specifically, particle motion explains why gases fill their containers. Also, pressure results from the large numbers of particle collisions with the container walls.

2. Gas molecules move independently of each other in all directions. Therefore, they exert pressure on the container walls, including the top, uniformly in all directions. Liquid and solid pressures are the result of gravity, so they exert a downward pressure on the bottom of the container. Liquid particles can move amongst themselves with the body of the liquid, so they also exert horizontal pressures against the walls of their container. Solid particles lack this freedom of horizontal movement, so they exert no horizontal pressure on the walls of a container.

3. Because gas molecules are widely spaced, air is compressible, which makes for a soft and comfortable surface to lie on. The uniform pressure of air in the mattress keeps the person on the mattress off the ground. The low density of a gas allows the entire volume inside the mattress to be filled with a small mass of air, allowing the mattress to be filled by mouth rather than by a pump.

4. Gas particles collide with the walls of the container, exerting pressure uniformly in all directions, including the top of the tank and its sides.

5. Gas particles collide with the walls of the container, exerting pressure uniformly in all directions.

6. Gas particles are very widely spaced, which leaves room for the addition of more.

7. Gas particles are always in motion, "pushing" back the surrounding water. As the bubble rises, there is less liquid pushing on the bubble, so the gas volume increases because the gas particles are pushing against less force. Gas particles are widely spaced, thus gases have lower densities than liquids, so the bubble rises.

8. Pressure measures force per unit area. Temperature measures the average kinetic energy of the particles in a sample.

9. Manometer operation is explained in Figure 4.7.

10.

atm	1.03	0.163	1.16	0.902	0.964
psi	15.2	2.40	17.0	13.3	14.2
in. Hg	30.9	4.88	34.7	27.0	28.9
cm Hg	78.5	12.4	88.5	68.6	73.3
mm Hg	785	124	885	686	733
torr	785	124	885	686	733
Pa	1.05×10^5	1.65×10^4	1.18×10^5	9.14×10^4	9.77×10^4
kPa	105	16.5	118	91.4	97.7
bar	1.05	0.165	1.18	0.914	0.977

11. 9.20×10^2 torr

12. It is impossible to have a negative kelvin temperature. 0 K, absolute zero, is the lowest temperature that can be approached. The student probably meant to record –18°C = 255 K.

13. 304 K

14. 20 K; 14 K

15. –14°C; 26°C

16. 21.5 L

17. 38°C

18. Squeezing bulb reduces gas volume, increasing pressure and forcing some air bubbles out of pipet. Releasing bulb increases volume and decreases pressure. Liquid pressure is then greater than gas pressure, so liquid enters pipet to reduce gas volume until liquid and gas pressures are equal.

19. 7.20×10^2 mL

20. 2.46 atm

21. 0.0395 L

22. 701 L

23. STP conditions have been established so that all data are reported at the same conditions. One atmosphere pressure is easily achieved in a laboratory, but 0°C is not a convenient temperature.

24. 25.5 L

25. 54.9 mL

26. 3.11 atm

27. -32°C

28. Dust particles are being pushed around by moving gas particles, which are too small to be seen.

29. 8.50×10^2 torr

30. 3.45 atm

31. 72°C

32. 159 psi gauge

Active Learning Workbook Answers

Chapter 5

Atomic Theory: The Nuclear Model of the Atom

1. Yes, see Figure 5.2.

2. The Law of Definite Composition says that any compound is always made up of elements in the same proportion by mass. Dalton's atomic theory explains this by stating that atoms of different elements combine to form compounds.

3. The calcium atoms and perhaps some of the oxygen atoms are in the calcium oxide; the carbon atoms and perhaps some of the oxygen atoms are in the carbon dioxide.

4. The Law of Multiple Proportions in this case states that the same mass of sulfur combines with masses of fluorine in the ratio of simple whole numbers, 4 to 6.

5. See Table 5.1.

6. Alpha particles and atomic nuclei are positively charged. As an alpha particle approached a nucleus, the repulsion between the positive charges deflected the alpha particle from its path.

7. The nucleus.

8. The electrons were thought to travel in circular orbits around the nucleus.

9. No, the atomic number is the number of protons, and all atoms of an element have the same number of protons.

10. Mass number is the sum of protons plus neutrons. Isotopes of the same element have different numbers of neutrons, but the same number of protons. The sums must be different. Atoms of different elements *must* have different numbers of protons and they *may* have different numbers of neutrons. An atom with one less proton than another may have one more neutron than the other, so their mass numbers would be the same. Example: carbon-14 (6 protons, 8 neutrons) and nitrogen-14 (7 protons, 7 neutrons).

11.

Name of Element	Nuclear Symbol	Atomic Number	Mass Number	Protons	Neutrons	Electrons
scandium	$^{45}_{21}Sc$	21	45	21	24	21
germanium	$^{76}_{32}Ge$	32	76	32	44	32
tin	$^{122}_{50}Sn$	50	122	50	72	50
chlorine	$^{37}_{17}Cl$	17	37	17	20	17
sodium	$^{23}_{11}Na$	11	23	11	12	11

12. The mass of a proton and neutron is close to 1 amu, which is $\frac{1}{6.02 \times 10^{23}}$ g. It is more convenient to use the amu.

13. 79.9 amu, Br

14. The atomic mass of neon is less than the average of the three atomic masses, so the isotope with the lowest mass must be present in the greatest abundance.

15. 69.7 amu, Ga, gallium

16. 107.9 amu, Ag, silver

17. 121.8 amu, Sb, antimony

18. 140.1 amu, Ce, cerium

19. 18; 21, 39, 57, 89

20. (a) Period 4, Group 2A/2; (b) Period 3, Group 4A/14; (c) Period 5, Group 7B/7

21. Z = 29, 63.55 amu; Z = 55, 132.9 amu; Z = 82, 207.2 amu

22. He, 4.003 amu; Al, 26.98 amu

23.

Name of Element	Atomic Number	Symbol of Element
Magnesium	12	Mg
Oxygen	8	O
Phosphorus	15	P
Calcium	20	Ca
Zinc	30	Zn
Lithium	3	Li
Nitrogen	7	N
Sulfur	16	S
Iodine	53	I
Barium	56	Ba
Potassium	19	K
Neon	10	Ne
Helium	2	He
Bromine	35	Br
Nickel	28	Ni
Tin	50	Sn
Silicon	14	Si

24. The ratio of the masses of mercury is 402/201 = 2, a simple ratio of whole numbers.

25. The atomic mass will be closer to the atomic mass of the more abundant isotope, whose atomic mass is 63 amu—about 1/4 of the way between them. The difference in masses is 2, and 1/4 of 2 is 0.5. One-quarter of the way between 63 amu and 65 amu is therefore 63.5 amu. The element is copper, Cu.

26. $x \times 6.10512$ amu + $(1 - x) \times 7.01600$ amu = 6.941 amu; $x = 0.08234$; 8.234% at 6.10512 amu; $1 - 0.08234 = 0.91766$; 91.77% at 7.01600 amu

27. The planetary model of the atom is similar to the solar system in that electrons orbit the nucleus as planets orbit the sun. Both models are similar in terms of the size of the nucleus/sun being large compared with the electrons/planets and the vast amount of empty space in the atom/solar system.

28. Chemical properties of isotopes of an element are identical.

29. a) 1.0×10^1 g/cm^3
b) In packing carbon atoms into a crystal there are spaces between the atoms. There are no voids in a single atom. (In fact, voids in diamond account for 66% of the total volume, and in graphite, 78%.)
c) 2×10^{-39} cm^3
d) 1×10^{16} g/cm^3
e) 4×10^5 tons

Active Learning Workbook Answers

Chapter 6

Chemical Nomenclature

1. He, Ne, Ar, Kr, Xe, Rn

2. F_2, B, Ni, S or S_8

3. Chromium, chlorine, beryllium, iron

4. Kr, Cu, Mn, N_2

5. Cl_2O, Br_3O_8, hydrogen bromide, diphosphorus trioxide

6. When an atom gains one, two, or three electrons, the particle that remains is a monatomic anion. The ion has a negative charge because it has more electrons than protons.

7. Copper(I) ion, iodide ion, potassium ion, mercury(I) ion, sulfide ion

8. Fe^{3+}, H^+, O^{2-}, Al^{3+}, Ba^{2+}

9. The formula of an acid usually begins with H.

10. Monoprotic, 1; diprotic, 2; triprotic, 3

11.

Acid Name	Acid Formula	Ion Name	Ion Formula
Sulfuric	H_2SO_4	Sulfate	SO_4^{2-}
Carbonic	H_2CO_3	Carbonate	CO_3^{2-}
Chloric	$HClO_3$	Chlorate	ClO_3^-
Hydrofluoric	HF	Fluoride	F^-
Bromic	$HBrO_3$	Bromate	BrO_3^-
Sulfurous	H_2SO_3	Sulfite	SO_3^{2-}
Arsenic	H_3AsO_4	Arsenate	AsO_4^{3-}
Periodic	HIO_4	Periodate	IO_4^{3-}
Selenous	H_2SeO_3	Selenite	SeO_3^{2-}
Tellurous	H_2TeO_3	Tellurite	TeO_3^{2-}
Hypoiodous	HIO	Hypoiodite	IO^-
Hypobromous	HBrO	Hypobromite	BrO^-
Telluric	H_2TeO_4	Tellurous	TeO_4^{2-}
Perbromic	$HBrO_4$	Perbromate	BrO_4^-
Hydrobromic	HBr	Bromide	Br^-

12. An anion or cation that contains an ionizable hydrogen, such as HSO_4^- and NH_4^+, can lose the hydrogen as thus behave as an acid.

13. HSO_3^-, HCO_3^-

14. Hydrogen selenite ion, hydrogen telluride ion

15. Ammonium ion, cyanide ion

16. OH^-, Cd^{2+}

17. $Ca(OH)_2$, NH_4Br, K_2SO_4

18. MgO, $AlPO_4$, Na_2SO_4, CaS

19. $BaSO_3$, Cr_2O_3, KIO_4, $CaHPO_4$

20. Lithium phosphate, magnesium carbonate, barium nitrate

21. Potassium fluoride, sodium hydroxide, calcium iodide, aluminum carbonate

22. Copper(II) sulfate, chromium(III) hydroxide, mercury(I) iodide

23. Hydrates: $NiSO_4 \cdot 6\ H_2O$, $Na_3PO_4 \cdot 12\ H_2O$; anhydrate: KCl

24. 7; magnesium sulfate heptahydrate

25. $(NH_4)_3PO_4 \cdot 3\ H_2O$, $K_2S \cdot 5\ H_2O$

26. ClO_4^-, $BaCO_3$, ammonium iodide, phosphorus trichloride

27. Hydrogen sulfide ion, beryllium bromide, $Al(NO_3)_3$, OF_2

28. Hg_2^{2+}, $CoCl_2$, silicon dioxide, lithium nitrite

29. Nitride ion, calcium chlorate, $Fe_2(SO_4)_3$, PCl_5

30. SnF_2, K_2CrO_4, lithium hydride, iron(II) carbonate

31. Nitrous acid, zinc hydrogen sulfate, KCN, CuF

32. Mg_3N_2, $LiBrO_2$, sodium hydrogen sulfite, potassium thiocyanate

33. nickel hydrogen carbonate, copper(II) sulfide, $Cr(IO_3)_3$, K_2HPO_4

34. SeO_2, $Mg(NO_2)_2$, iron(II) bromide, silver oxide

35. Tin(II) oxide, ammonium dichromate, NaH, $H_2C_2O_4$

36. $Co_2(SO_4)_3$, FeI_3, copper(II) phosphate, manganese(II) hydroxide

37. Aluminum selenide, magnesium hydrogen phosphate, $KClO_4$, $HBrO_2$

38. $Sr(IO_3)_2$, $NaClO$, rubidium sulfate, diphosphorus pentoxide

39. Iodine monochloride, silver acetate, $Pb(H_2PO_4)_2$, GaF_3

40. $MgSO_4$, $Hg(BrO_2)_2$, sodium oxalate, manganese(III) hydroxide

Active Learning Workbook Answers

Chapter 7

Chemical Formula Relationships

1. $Al(NO_3)_3$: 1 aluminum atom, 3 nitrogen atoms, 9 oxygen atoms

2. Molecular mass is a term properly applied only to substances that exist as molecules. Sodium nitrate is an ionic compound.

3. Atomic mass, Ba; molecular mass, NH_3 and Cl_2 because these are molecular substances; formula mass, all substances, but particularly CaO and Na_2CO_3 because neither of the other two terms technically fit these ionic compounds.

4. a) 6.941 amu Li + 35.45 amu Cl = 42.39 amu LiCl
   b) 2(26.98) amu Al + 3(12.01 amu C) + 9(16.00 amu O) = 233.99 amu $Al_2(CO_3)_3$
   c) 2(14.01 amu N) + 8(1.008 amu H) + 32.07 amu S + 4(16.00 amu O) = 132.15 amu $(NH_4)_2SO_4$
   d) 4(12.01 amu C) + 10(1.008 amu H) = 58.12 amu C_4H_{10}
   e) 107.9 amu Ag + 14.01 amu N + 3(16.00 amu O) = 169.9 amu $AgNO_3$
   f) 54.94 amu Mn + 2(16.00 amu O) = 86.94 amu MnO_2
   g) 3(65.39 amu Zn) + 2(30.97 amu P) + 8(16.00 amu O) = 386.11 amu $Zn_3(PO_4)_2$

5. The quantity of particles of each is the same.

6. By definition, the mole is the amount of any substance that contains the same number of units as the number of atoms in exactly 12 grams of carbon-12. The definition doesn't say what that number is. Through experiment, we have found that, to three significant figures, there are 6.02×10^{23} atoms in 12 grams of carbon-12.

7. a) 4.67×10^{24} molecules CH_4
   b) 5.35×10^{22} molecules CO
   c) 3.48×10^{25} atoms Fe

8. a) 0.407 mol C_2H_2
   b) 11.6 mol Na

9. Numerically.

10. a) 44.09 g/mol C_3H_8; b) 266.32 g/mol C_6Cl_5OH
    c) 366.01 g/mol $Ni_3(PO_4)_2$; d) 189.41 g/mol $Zn(NO_3)_2$

11. a) 0.212 mol O_2
    b) 0.0610 mol $Mg(NO_3)_2$
    c) 0.00755 mol Al_2O_3
    d) 14.3 mol C_2H_5OH
    e) 0.00410 mol $(NH_4)_2CO_3$
    f) 0.740 mol Li_2S

12. a) 0.00372 mol KIO_3
 b) 0.858 mol $BeCl_2$
 c) 1.65 mol $Ni(NO_3)_2$

13. a) 32.60 g LiCl
 b) 3.43×10^3 g $HC_2H_3O_2$
 c) 4.7 g Li
 d) 213 g $Fe_2(SO_4)_3$
 e) 678 g $NaC_2H_3O_2$

14. a) 41.7 g Li_2SO_4
 b) 801 g $K_2C_2O_4$
 c) 43.7 g $Pb(NO_3)_2$

15. a) 2.58×10^{23} units $LiNO_3$
 b) 1.98×10^{21} units Li_2S
 c) 6.88×10^{23} units $Fe_2(SO_4)_3$

16. a) 5.5×10^{18} molecules I_2
 b) 1.11×10^{24} molecules $C_2H_4(OH)_2$
 c) 1.51×10^{22} units $Cr_2(SO_4)_3$

17. a) 2.22 g $C_{19}H_{37}COOH$
 b) 274 g F
 c) 156 g $NiCl_2$

18. \$ 5.03×10^{-21}

19. 1.1×10^{21} $C_{12}H_{22}O_{11}$ molecules

20. a) 95.9 g N
 b) 192 g N_2
 c) 1.77×10^{23} atoms N
 d) 8.85×10^{22} N_2 molecules
 e) 1.77×10^{23} atoms N

21. a) NH_4NO_3: 35.00% N, 5.037% H, 59.96% O
 b) $Al_2(SO_4)_3$: 15.77% Al, 28.12% S, 56.11% O
 c) $(NH_4)_2CO_3$: 29.16% N, 8.392% H, 12.50% C, 49.95% O
 d) CaO: 71.47% Ca, 28.53% O
 e) MnS_2: 46.14% Mn, 53.86% S

22. 121 g Li

23. 25.8 g K

24. 262 g $Zn(CN)_2$

25. 769 kg $PbMoO_4$

26. 12.17 g $Ca(ClO_3)_2$

27. C_6H_{10} must be a molecular formula because both 6 and 10 and divisible by 2. Its empirical formula is C_3H_5. There is no common divisor for 7 and 10, so it can be an empirical formula. An empirical formula can also be a molecular formula.

	Element	Grams	Moles	Mole Ratio	Formula Ratio	Empirical Formula	Molecular Formula
28.	C	52.2	4.35	2.00	2		
	H	13.0	12.9	5.92	6	C_2H_6O	
	O	34.8	2.18	1.00	1		
29.	Fe	11.89	0.213	1.00	2		
	O	5.10	0.319	1.50	3	Fe_2O_3	
30.	C	17.2	1.43	1.00	1		
	H	1.44	1.43	1.00	1	CHF_3	
	F	81.4	4.28	2.99	3		
31.	C	38.7	3.22	1.00	1		
	H	9.7	9.6	3.0	3	CH_3O	$\frac{62.0}{31.0} = 2$
	O	51.6	3.23	1.00	1		$C_2H_6O_2$
32.	Cl	71.3	2.01	1.00	1		
	C	24.8	2.06	1.02	1	$ClCH_2$	$\frac{97}{49} = 2$
	H	3.9	3.9	1.9	2		$Cl_2C_2H_4$

33. 4.13×10^{23} P_4 molecules; 1.65×10^{24} P atoms

34. 2 oz is about the capacity of a typical salt shaker. 12 oz is a little large for a sugar bowl, but not by much.
1 mol NaCl = 2 oz NaCl
1 mol $C_{12}H_{22}O_{11}$ = 12 oz $C_{12}H_{22}O_{11}$

35. 107 g C_8H_{18}

36.	Element	Grams	Moles	Mole Ratio	Formula Ratio	EF
a)	Co	42.4	0.719	1.00	1	
	S	23.0	0.717	1.00	1	$CoSO_3$
	O	34.6	2.16	3.01	3	
b)	$CoSO_3$	26.1	0.188	1.00	1	
						$CoSO_3 \cdot 5\ H_2O$
	H_2O	16.9	0.938	4.99	5	

43.0 g hydrate – 26.1 g anhydrate = 16.9 g H_2O

Active Learning Workbook Answers

Chapter 8

Reactions and Equations

1. Particulate: Two ozone molecules react to form three oxygen molecules. Molar: Two moles of ozone molecules react to form three moles of oxygen molecules.

2. $4\ Li(s) + O_2(g) \rightarrow 2\ Li_2O(s)$

3. $4\ B(s) + 3\ O_2(g) \rightarrow 2\ B_2O_3(s)$

4. $Ca(s) + Br_2(\ell) \rightarrow CaBr_2(s)$

5. $2\ HI(g) \rightarrow H_2(g) + I_2(s)$

6. $2\ BaO_2(s) \rightarrow 2\ BaO(s) + O_2(g)$

7. $C_3H_8(\ell) + 5\ O_2(g) \rightarrow 3\ CO_2(g) + 4\ H_2O(g)$

8. $C_2H_5OH(\ell) + 3\ O_2(g) \rightarrow 2\ CO_2(g) + 3\ H_2O(g)$

9. $Ca(s) + 2\ HBr(aq) \rightarrow CaBr_2(aq) + H_2(g)$

10. $Cl_2g) + 2\ KI(aq) \rightarrow 2\ KCl(aq) + I_2(s)$

11. $CaCl_2(aq) + 2\ KF(aq) \rightarrow CaF_2(s) + 2\ KCl(aq)$

12. $2\ NaOH(aq) + MgBr_2(aq) \rightarrow 2\ NaBr(aq) + Mg(OH)_2(s)$

13. $H_2SO_4(aq) + Ba(OH)_2(aq) \rightarrow 2\ H_2O(\ell) + BaSO_4(s)$

14. $3\ NaOH(s) + H_3PO_4(aq) \rightarrow Na_3PO_4(aq) + 3\ H_2O(\ell)$

15. $Pb(NO_3)_2(aq) + 2\ NaI(aq) \rightarrow PbI_2(s) + 2\ NaNO_3(aq)$

16. $2\ C_4H_{10}(\ell) + 13\ O_2(g) \rightarrow 8\ CO_2(g) + 10\ H_2O(g)$

17. $H_2SO_3(aq) \rightarrow SO_2(aq) + H_2O(\ell)$

18. $2\ K(s) + 2\ HOH(\ell) \rightarrow 2\ KOH(aq) + H_2(g)$

19. $Zn(s) + 2\ AgClO_3(aq) \rightarrow Zn(ClO_3)_2(aq) + 2\ Ag(s)$

20. $(NH_4)_2S(s) + Cu(NO_3)_2 \rightarrow 2\ NH_4NO_3(aq) + CuS(s)$

21. $2\ P(s) + 3\ Br_2(\ell) \rightarrow 2\ PBr_3(\ell)$ *or* $P_4(s) + 6\ Br_2(\ell) \rightarrow 4\ PBr_3(\ell)$

22. $Ca(OH)_2(s) \rightarrow CaO(s) + H_2O(g)$

23. $2\ C_3H_8O_3(\ell) + 7\ O_2(g) \rightarrow 6\ CO_2(g) + 8\ H_2O(g)$

24. $2\ Sb(s) + 3\ Cl_2(g) \rightarrow 2\ SbCl_3(s)$

25. $2\ KOH(aq) + ZnCl_2(aq) \rightarrow 2\ KCl(aq) + Zn(OH)_2(s)$

26. $4\ Al(s) + 3\ C(s) \rightarrow Al_4C_3(s)$

27. $3\ Li_2SO_3(aq) + 2\ Na_3PO_4(aq) \rightarrow 2\ Li_3PO_4(s) + 3\ Na_2SO_3(aq)$

28. $2\ Cr(s) + 3\ Sn(NO_3)_2(aq) \rightarrow 2\ Cr(NO_3)_3 + 3\ Sn(s)$

29. $SO_3(g) + H_2O(\ell) \rightarrow H_2SO_4(aq)$

30. $3\ Fe(s) + 4\ H_2O(g) \rightarrow Fe_3O_4(s) + 4\ H_2(g)$

31. $Al_4C_3(s) + 12\ H_2O(\ell) \rightarrow 4\ Al(OH)_3(s) + 3\ CH_4(g)$

32. $3\ Mg(s) + 2\ NH_3(g) \rightarrow Mg_3N_2(s) + 3\ H_2(g)$

33.

a)	Double replacement precipitation	$NH_4Cl + AgNO_3 \rightarrow NH_4NO_3 + AgCl$
b)	Combination	$Ca + Cl_2 \rightarrow CaCl_2$
c)	Single replacement redox	$F_2 + 2\ NaI \rightarrow 2\ NaF + I_2$
d)	Double replacement precipitation	$Zn(NO_3)_2 + Ba(OH)_2 \rightarrow Zn(OH)_2 + Ba(NO_3)_2$
e)	Single replacement redox	$Cu + NiCl_2 \rightarrow CuCl_2 + Ni$

34.

(1)	$S + O_2 \rightarrow SO_2$	Combination
(2)	$2\ SO_2 + O_2 \rightarrow 2\ SO_3$	Combination
(3)	$SO_3 + H_2O \rightarrow H_2SO_4$	Combination

35. $4\ Ag + 2\ H_2S + O_2 \rightarrow 2\ Ag_2S + 2\ H_2O$

36. $WO_3 + 3\ H_2 \rightarrow W + 3\ H_2O$; single replacement redox

37. $Ba(OH)_2(aq) + 2\ HNO_3(aq) \rightarrow Ba(NO_3)_2(aq) + 2\ H_2O(\ell)$

38. $HNH_2SO_3(aq) + KOH(aq) \rightarrow H_2O(\ell) + KNH_2SO_3(aq)$

Active Learning Workbook Answers

Chapter 9

Quantity Relationships in Chemical Reactions

1. a) 76.2 mol NH_3 b) 2.89 mol NO c) 2.23 mol NO

2. $MgO + H_2O \rightarrow Mg(OH)_2$; 0.884 mol MgO

3. $2\ SO_2 + O_2 \rightarrow 2\ SO_3$; 3.99 mol SO_3

4. a) 6.70 mol NH_3 b) 857 g H_2O c) 229 g NH_3 d) 7.12 g NO

5. 12.2 g CO_2

6. 42.2 g NaOH

7. 152 g Na_2CO_3

8. 6.10×10^2 mg $Ca(C_{18}H_{35}O_2)_2$

9. 1.50 g Fe_2O_3

10. 61.4 g NaCl

11. 202 g $CaCl_2$

12. 39.9 g NaOH

13. 1.09 g $Mg(OH)_2$

14. 0.360 g H_2SO_4

15. 135 g Zn

16. 22.4 L/mol is the molar volume of a gas at standard temperature and pressure, 0°C and 1 atm.

17. 94.3 L CO

18. 1.13×10^3 g FeS_2

19. 15.7 L H_2

20. 3.14×10^3 mL O_2

21. 308 L SO_2

22. 16.7 g H_2O

23. 75.5 L N_2

24. 43.8 g NH_4NO_3

25. 98.1% yield

26. 2.86 g NH_3 (act)

27. 1.00×10^2 kg H_2

28. 81.7% yield

29. 61.7 kg CH_3COOH

Questions 30–32: The comparison-of-moles and smaller-amount methods may yield slight differences in answers caused by round-offs in calculations.

30. 1.98 g $BaCrO_4$; 1.13 g Na_2CrO_4 left

31. 3.58 kg I_2; 0.42 kg $NaIO_3$ left

32. 236 g P_4S_3; 23 g S_8 left

33. a) 3.06 kJ b) 156 cal c) 1.49 kcal

34. 118 kcal ; 1.18×10^5 cal

35. 8.9×10^5 kJ/year

36. $2.82 \times 10^3 \text{ kJ} + 6\ CO_2(g) + 6\ H_2O(\ell) \rightarrow C_6H_{12}O_6(s) + 6\ O_2(g)$
$6\ CO_2(g) + 6\ H_2O(\ell) \rightarrow C_6H_{12}O_6(s) + 6\ O_2(g) \quad \Delta H = +2.82 \times 10^3 \text{ kJ}$

37. $286 \text{ kJ} + H_2O(\ell) \rightarrow H_2(g) + \frac{1}{2}\ O_2(g)$
$H_2O(\ell) \rightarrow H_2(g) + \frac{1}{2}\ O_2(g) \quad \Delta H = +286 \text{ kJ}$

38. $C_2H_2(g) + \frac{5}{2}\ O_2(g) \rightarrow 2\ CO_2(g) + H_2O(\ell) + 1.31 \times 10^3 \text{ kJ}$
$C_2H_2(g) + \frac{5}{2}\ O_2(g) \rightarrow 2\ CO_2(g) + H_2O(\ell) \quad \Delta H = -1.31 \times 10^3 \text{ kJ}$

39. 1.03×10^4 kJ

40. 2.50×10^2 g CaO

41. 1.99×10^4 g C_8H_{18}

42. 3.77 g $Ca_3(PO_4)_2$

43. a) 7.87 g $NaHCO_3$; 4.12 g CO_2; b) 12.1 g $NaHCO_3$ unreacted; c) $Na_3C_6H_5O_7$ is sodium citrate

44. 0.519 kJ

45. 719 kg SiC

46. a) 50% from Pb and 50% from PbO_2 b) 11.7 g PbO_2

47. 0.86 ton Cl_2 (act)

48. All the silver from AgCl is recovered.
1×10^1 g additional AgCl could have been treated

49. 19.1 lb $C_3H_5(NO_3)_3$

50. 95.59% Cu

51. 1.94 g $AgNO_3$

Active Learning Workbook Answers

Chapter 10

Atomic Theory: The Quantum Model of the Atom

1. Radio waves, TV waves, microwaves, x-rays.

2. Wavelength, frequency, and velocity.

3. d and e

4. All colors of light could be emitted because of the infinite number of energy differences possible. White light consists of all visible colors. Infrared and ultraviolet light would also be emitted.

5. An atom must absorb energy before it can release that energy in the form of light.

6. An atom with its electron(s) in the ground state cannot emit light because there is no lower energy level for the electron to fall to. Only an atom with electrons in excited states can emit light.

7. Individual lines result when an electron jumps from $n = 2$ to $n = 1$, $n = 3$ to $n = 2$, $n = 3$ to $n = 1$, etc.

8. The Bohr model provided an explanation for atomic line spectra in terms of electron energies. It also introduced the idea of quantized electron energy levels in the atom.

9. In general, energies increase as the principal energy level increases: $n = 1 < n = 2 < n = 3 \ldots < n = 7$.

10. The total number of sublevels within a given principal energy level is equal to n, the principal quantum number.

11. *s*: 1; *p*: 3; *d*: 5; *f*: 7

12. Each *p* sublevel contains three orbitals, each of which can hold two electrons, for a maximum of six electrons in the *p* sublevel.

13. This statement is true.

14. An orbital may hold one or two electrons, or it may be empty.

15. See Figure 10.6.

16. 6; 10; 14

17. For elements other than hydrogen, the energy of each principal energy level spreads over a range related to the sublevels.

18. (b) Bohr's quantized electron energy levels appear in quantum theory as principal energy levels.

19. $n = 2$ has 2 sublevels, s and p.

20. Sulfur, which is in Period 3, Group 6A/16.

21. [Ar] substitutes for $1s^2 2s^2 2p^6 3s^2 3p^6$.

22. a) selenium, b) oxygen, c) magnesium

23. a) nitrogen, b) aluminum, c) cobalt

24. Mg: $1s^2 2s^2 2p^6 3s^2$; Ni: $1s^2 2s^2 2p^6 3s^2 3p^6 4s^2 3d^8$

25. Cr: $1s^2 2s^2 2p^6 3s^2 3p^6 4s^1 3d^5$; Se: $1s^2 2s^2 2p^6 3s^2 3p^6 4s^2 3d^{10} 4p^4$

26. Mg: [Ne]$3s^2$; Ni: [Ar]$4s^2 3d^8$; Cr: [Ar]$4s^1 3d^5$; Se: [Ar]$4s^2 3d^{10} 4p^4$

27. [Ar]$4s^2 3d^3$

28. Many of the similar chemical properties of elements in the same column of the periodic table are related to the valence electrons.

29. $ns^2 np^1$ or $\dot{\text{Al}}:$

30. $ns^2 np^4$

31. Ionization energies of elements in the same chemical family decrease as atomic number increases.

32. Within a period, the valence electrons are in the same principal energy level. As the number of protons increases across a period, the positive nuclear charge increases, exerting a greater pull on the electrons.

33. They both have the electron configuration ns^2.

34. Halogens

35. $ns^2 np^6$

36. (a) noble gases; (b) alkaline earths

37. Both chlorine and iodine have seven valence electrons: $ns^2 np^5$.

38. The transition elements are those in the B groups (3–12) of the periodic table. They are metals.

39. (a) Z = 19, 20, 31, 32. (b) Z = 34, 35, 36.

40. As you go from left to right across a row in the periodic table, the valence electrons are all in the same principal energy level. As the number of protons in an atom increases, the positive charge in the nucleus increases. This pulls the valence electrons closer to the nucleus, so the atom becomes smaller.

41. Germanium has $n = 4$ valence electrons; silicon has $n = 3$ valence electrons. The number of occupied energy levels is more important than nuclear charge in determining size.

42. Most of the elements next to the stair-step line in the periodic table have some properties of both metals and nonmetals. These elements are the metalloids or semimetals.

43. (a) G, A, (b) R, T

44. X, Q, R, Z, M, T

45. Q < X < Z

46. The other lines are outside the visible spectrum, in the infrared or ultraviolet regions.

47. I: $[Kr]5s^24d^{10}5p^5$; W: $[Xe]6s^24f^{14}5d^4$ or $[Xe]6s^14f^{14}5d^5$

48. The number of protons identifies an element. A neutral atom can gain or lose electrons, forming ions of that element.

49. We expect magnesium to have a greater ionization energy than sodium and calcium because of the trends in ionization energy explained in the text. Aluminum has a greater nuclear charge than magnesium, but the additional electron is by itself in a 3*p* orbital, which makes it easier to remove.

50. (a) More metallic. Example: In Group 4A/14, carbon is a nonmetal, silicon and germanium are metalloids, and tin and lead are metals. (b) Less metallic. Example: In Period 3, sodium, magnesium, and aluminum are metals, silicon is a metalloid, and phosphorus, sulfur, chlorine, and argon are nonmetals.

51. All species have a single electron. Species with two or more electrons are far more complex.

52. To form a monatomic ion, carbon would have to lose four electrons. The fourth ionization energy of any atom is very high.

53. Xenon has the lowest ionization energy of the noble gases and apparently the greatest reactivity. This is the same ionization energy trend seen in all groups in the periodic table.

54. (a) Ionization energy increases across a period of the periodic table because of increasing nuclear charge and the same principal energy level of the outermost electrons. (b) The breaks in ionization energy trends across Periods 2 and 3 occur just after the *s* orbital is filled and just after the *p* orbitals are half-filled.

Active Learning Workbook Answers

Chapter 11

Chemical Bonding

1. All have the electron configuration of Ne (cations) or Ar (anions): $1s^22s^22p^6$ or $1s^22s^22p^63s^23p^6$.

2. Any two of K^+, Ca^{2+}, Sc^{3+}.

3. Any two of Te^{2-}, I^-, Cs^+.

4.

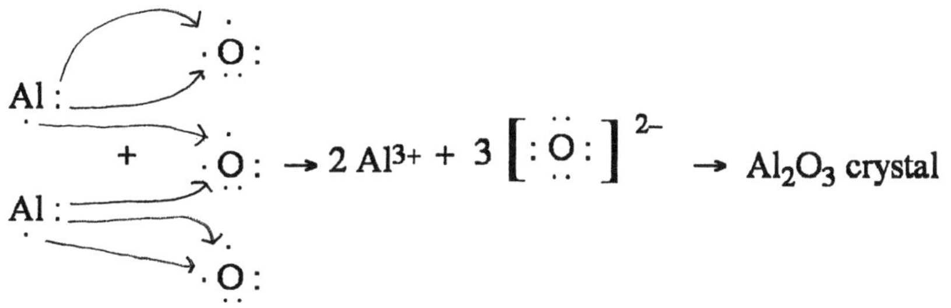

5. A potassium atom forms a K^+ ion by losing one electron. A chlorine atom can accept one electron to form a Cl^- ion, so it takes one potassium atom to donate the one electron to a single chlorine atom.

6. Orbital overlap refers to the covalent bond formed when the atomic orbitals of individual atoms extend over one another so that electrons are shared between the two nuclei in the bond.

7.

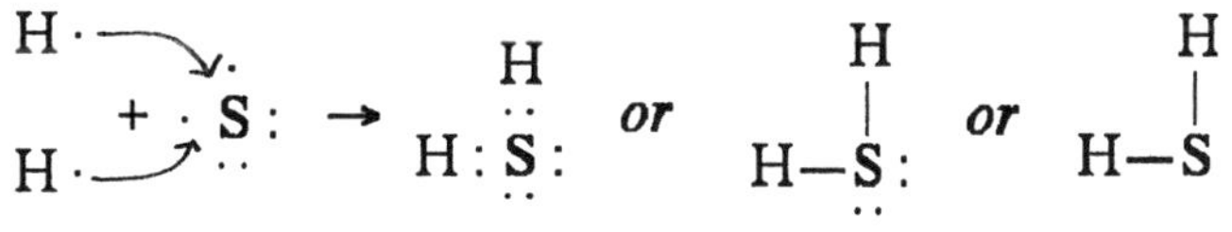

8. You should circle the three lone pairs (··) around the chlorine atom.

9. The energy of a system is reduced when bonds form in which the noble gas electron configuration is reached. In a covalent, or electron-sharing, bond, atoms share electrons to achieve the noble gas configuration. An example is shown in the answer to Question 7 above. The hydrogen atoms obtain the electron configuration of helium and the sulfur atom obtains the electron configuration of argon.

10. The charge density of the electron cloud formed by a bonding electron pair in a nonpolar bond is centered in the region between the bonded atoms. In a polar bond, this charge density is shifted toward the more electronegative atom.

11. S—O > N—Cl ≈ C—C

12. S in S—O

13. Electronegativity values increase from left to right across any row of the table, and they increase from the bottom to top of any column.

14. Atoms bonded by a quadruple bond can conform to the octet rule if they have no unshared electron pairs.

15. X can be bonded to the maximum of two additional atoms by single bonds. X can be bonded to the minimum of no additional atoms if it has two lone pairs.

16. The molecule can theoretically have a maximum of an infinite number of atoms: A—B≡C—D . . . . The minimum number is two: : A≡B :

17. NO_2 and NO have an odd number of electrons and thus cannot satisfy the octet rule.

18. Localized electrons are those that stay near a single atom or pair of atoms; delocalized electrons do not.

19. The calcium ions in the crystal have a 2+ charge, and there are two electrons for each ion.

20. Alloys are mixtures. They are neither pure substances nor compounds. They lack constant composition and have variable physical properties.

21. (a) The electrostatic attractions between positive and negative ions results in a minimization of energy in an ionic bond. (b) The stability of a noble gas electron configuration results in a minimization of energy in the formation of a covalent bond.

22. A bond between sodium and sulfur is most likely to be ionic because it is between a metal and a nonmetal. The bonds between fluorine and chlorine and between oxygen and sulfur are likely to be covalent because they are between two nonmetals.

23. (c) Mg^{2+} is isoelectronic with Ne, not Ar.

24. A bond between identical atoms is completely nonpolar. Their attractions for the bonding electrons are equal.

25. According to the octet rule, a atom achieves a noble gas electron configuration in forming a bond. There is no noble gas electron configuration in metallic bonding, which consists of positive ions in a sea of electrons.

26. In either case, ionic or covalent bonding, the total energy of the system is minimized by the formation of the bond. Atoms have noble gas electron configurations when these bonds are formed.

27. K—O; Ca—O or Na—O; Al—O; S—O. The group-to-group spread predicts Na—O is more polar than Ca—O, but the period-to-period spread predicts Ca—O is more polar than Na—O.

28. Five or six electron pairs may surround a central atom if *d* orbitals are involved in bonding.

29. One explanation is that a boron atom is smaller than an aluminum atom. The valence electrons in boron are therefore closer to the nucleus than they are in aluminum. That makes it more difficult to remove the electrons to form an ion.

Active Learning Workbook Answers

Chapter 12

Structure and Shape

1.

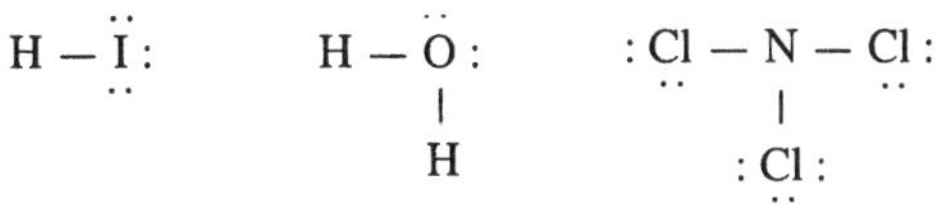

```
       ..              ..              ..       ..
H – I :         H – O :          : Cl – N – Cl :
    ..              |              ..   |   ..
                    H                 : Cl :
                                        ..
```

2.

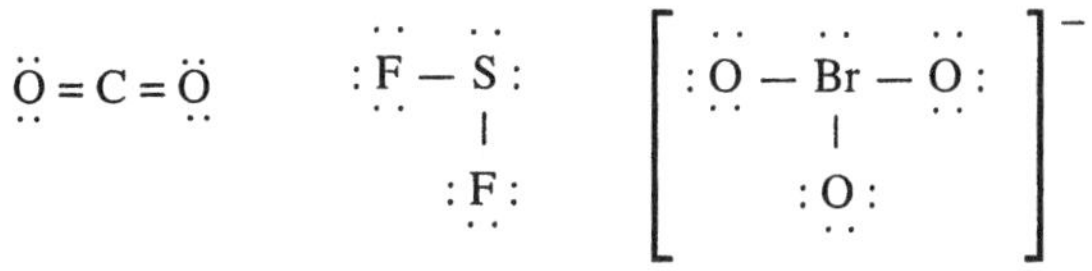

```
..     ..       ..  ..        ..    ..    ..      –
O = C = O     : F – S :     [ : O – Br – O : ]
..     ..       ..  |           ..   |    ..
                  : F :            : O :
                    ..               ..
```

3.

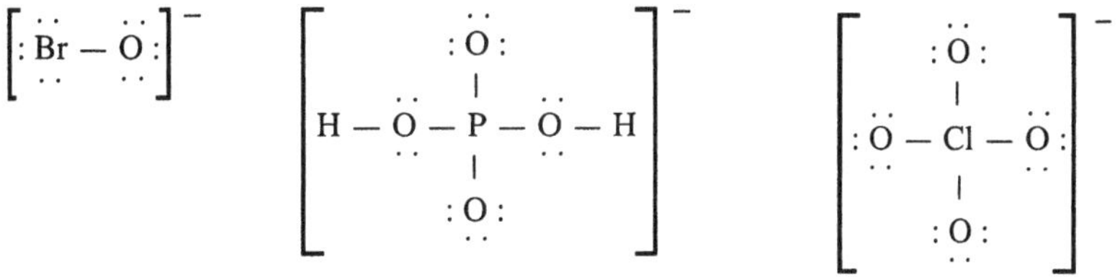

```
  ..    ..   –              ..        –               ..      –
[ : Br – O : ]            : O :                     : O :
    ..   ..                 |                         |
                    ..      |     ..            ..          ..
                [ H – O  –  P  –  O – H ]     [ : O – Cl – O : ]
                    ..      |     ..            ..    |     ..
                          : O :                     : O :
                            ..                        ..
```

4.

```
        H                  H                   ..
        |                  |                 : Cl :
  ..          ..     ..         ..       ..    |    ..
: Cl – C – F :     : F – C – F :       : I – C – I :
  ..    |     ..     ..    |    ..       ..    |    ..
      : Cl :             : F :               : I :
        ..                 ..                  ..
```

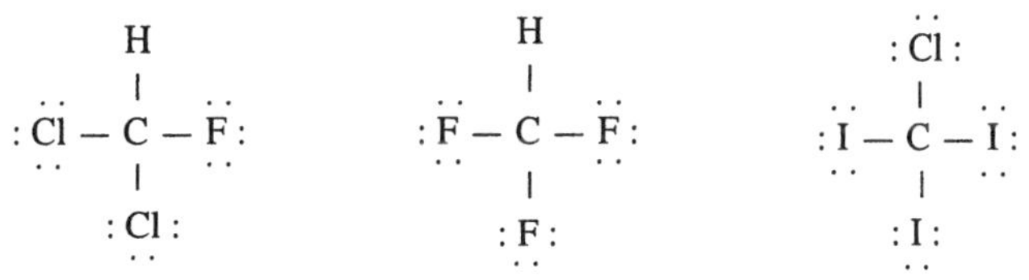

Questions 5-9: Where two or more acceptable diagrams are possible, only one representative diagram is shown.

5.

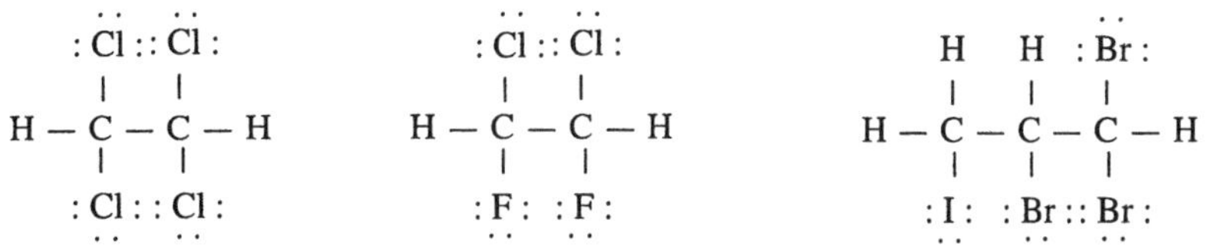

```
   ..    ..              ..    ..                        ..
 : Cl :: Cl :          : Cl :: Cl :          H    H   : Br :
    |     |               |     |            |    |     |
H – C  –  C – H       H – C  –  C – H    H – C  – C  –  C – H
    |     |               |     |            |    |     |
 : Cl :: Cl :           : F : : F :        : I : : Br :: Br :
   ..    ..               ..    ..           ..     ..    ..
```

6.

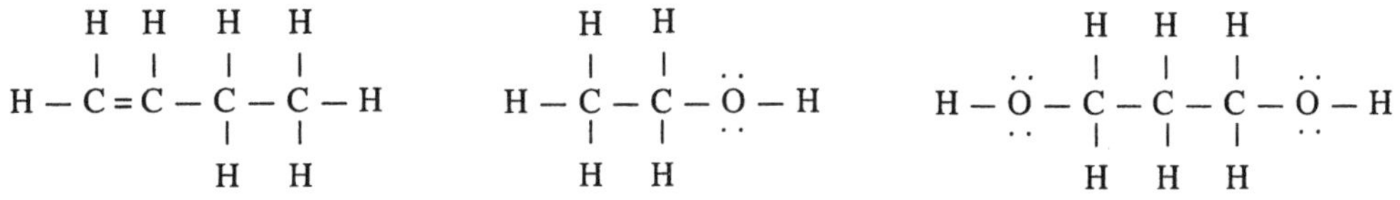

```
    H   H   H   H                 H   H                      H   H   H
    |   |   |   |                 |   |   ..             ..  |   |   |   ..
H – C = C – C – C – H         H – C – C – O – H      H – O – C – C – C – O – H
            |   |                 |   |   ..             ..  |   |   |   ..
            H   H                 H   H                      H   H   H
```

7.

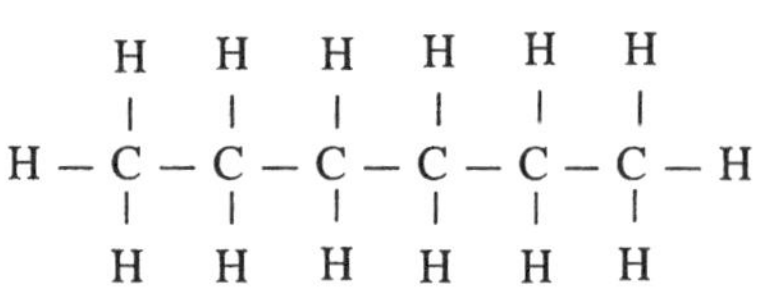

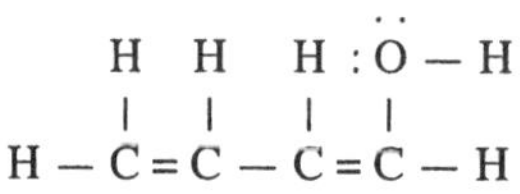

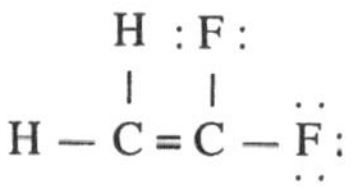

```
     H   H   H   H   H   H              H   H   H :Ö – H            H :F:
     |   |   |   |   |   |              |   |   |   |               |  |    ..
H – C – C – C – C – C – C – H       H – C = C – C = C – H       H – C = C – F:
     |   |   |   |   |   |                                                 ..
     H   H   H   H   H   H
```

8.

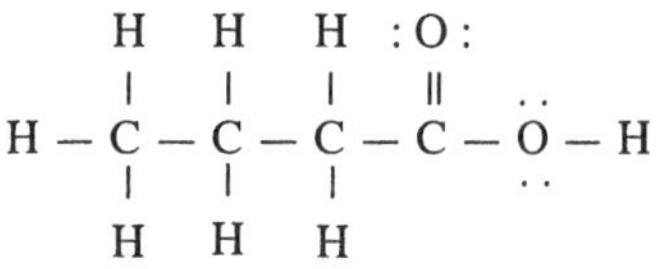

```
     H   H   H  :O:
     |   |   |   ||   ..
H – C – C – C – C – O – H
     |   |   |        ..
     H   H   H
```

9.

```
..     ..        ..      ..
O = N = O        N = N = O        :N ≡ O:
..     ..        ..      ..
```

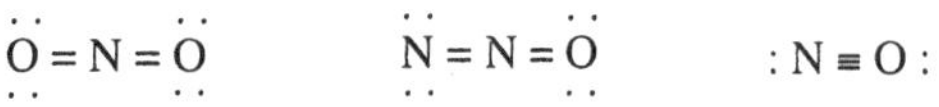

	Substance	Electron Pair Geometry	Molecular Geometry	3D Ball-and-Stick Representation
10.	BCl_3	Trigonal planar	Trigonal planar	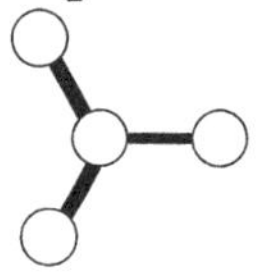
	PH_3	Tetrahedral	Trigonal pyramidal	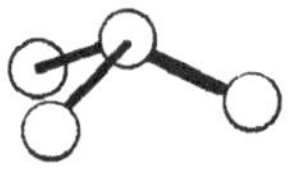
	H_2S	Tetrahedral	Bent	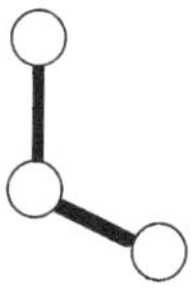
11.	BrO^-	Linear	Linear	
	ClO_3^-	Tetrahedral	Trigonal pyramidal	see above
	PO_4^{3-}	Tetrahedral	Tetrahedral	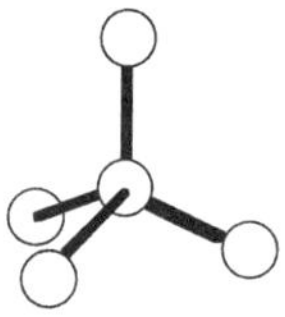
12.	C in C_3H_7OH	Tetrahedral	Tetrahedral	
13.	N in $C_2H_5NH_2$	Tetrahedral	Trigonal pyramidal	
14.	C in C_2H_2	Linear	Linear	
15.	C in HCN	Linear	Linear	

16. Trigonal pyramidal

17. Molecular and Electron-pair: Trigonal planar

18. Molecular: Tetrahedral; Electron-pair: Trigonal pyramidal

19. Molecular and Electron-pair: Tetrahedral

20. b, d, and e.

21. CCl_4 molecules are nonpolar because the polar bonds are symmetrically arranged, which makes the molecule itself net nonpolar.

22. Both molecules are linear. HF is more polar than HBr. The H end of each is more positive.

23.

```
                        H
H → O:                  |   ..
    ↑               H — C → O ← H
    H                   |   ..
                        H
```

Both molecules are polar because of the bent structure around the oxygen atom and the concentration of negative charge near the highly electronegative oxygen atom. The carbon-oxygen electronegativity difference is slightly less than the hydrogen-oxygen electronegativity difference, so CH_3OH is slightly less polar than H_2O.

24. a and d

25. b and c

26. Organic compounds are those that contain carbon (with some exceptions); inorganic compounds do not contain carbon.

27. $CH_3(CH_2)_6CH_3$, C_6H_6, and $C_{18}H_8$.

28. The structure of a molecule simply describes the relative positions of its atoms. The three-dimensional shape of a molecule results from its structure.

29. A straight line of carbon atoms in a Lewis diagram simply indicates that the atoms are bonded to one another. Single-bonded carbon atoms form tetrahedral bond angles, which form a zig-zag chain when in the same plane.

30. Alcohols, ethers, and water molecules all include an oxygen atom with two lone pairs of electrons that is single-bonded to two other atoms. In an alcohol, one of the bonds to the oxygen atom is to a hydrogen atom and the other is to a carbon atom. In an ether, both bonds to the oxygen atom are to carbon atoms.

31. Carboxylic acids contain the carboxyl group, —COOH.

32.

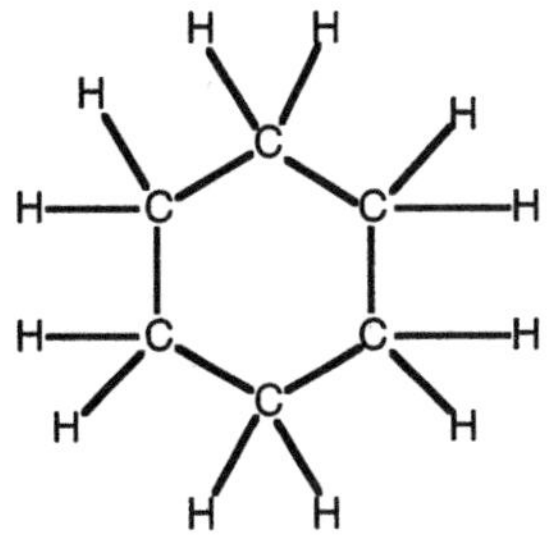

33. 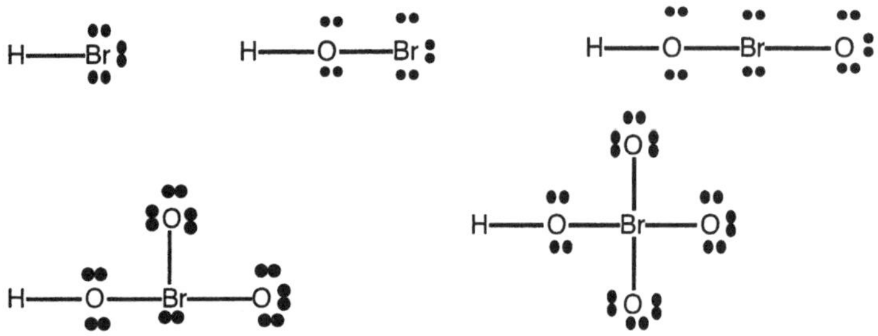 H—Br H—O—Br H—O—Br—O H—O—Br(—O)—O H—O—Br(—O)(—O)—O

34. Bond angles in C_2H_6 are all tetrahedral, which causes the molecule to be three-dimensional. The geometry around both carbon atoms in C_2H_4 is trigonal planar, and the double bond between the carbon atoms keeps all four hydrogen atoms in the same plane.

35. Any two from:

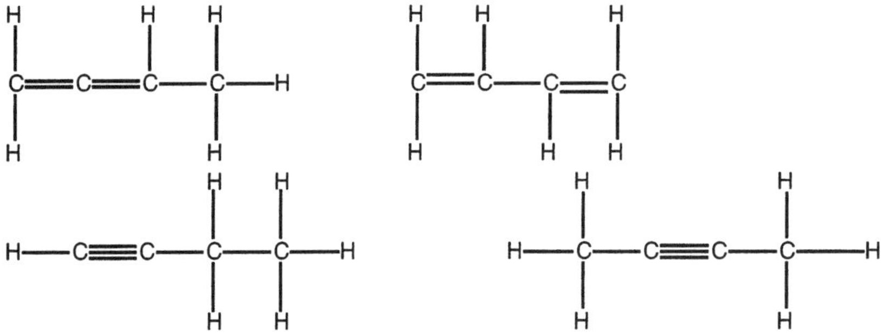

36. All species in the question, including SeO_4^{2-} and CI_4, have 32 valence electrons and four atoms to be distributed around a central atom. Consequently, they all have the same tetrahedral shape.

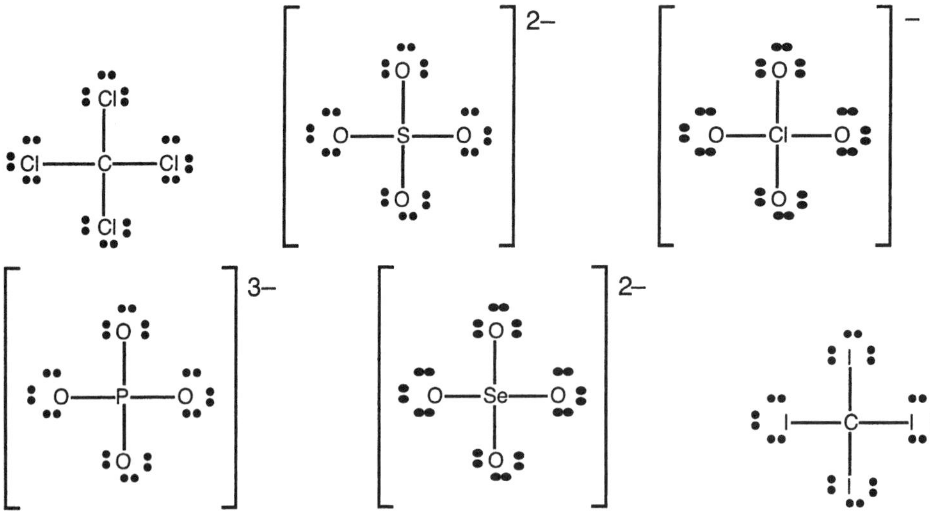

Active Learning Workbook Answers

Chapter 13

The Ideal Gas Law and Its Applications

1. $n = \frac{PV}{RT}$, so since temperature and pressure are the same for all three samples, n is directly proportional to V. The volume of 1×10^{23} hydrogen molecules is the same as the volume of 1×10^{23} oxygen molecules and half the volume of 2×10^{23} nitrogen molecules.

2. 342 torr

3. 8.39 L

4. 0.0320 mol

5. –38°C

6. 2.13 mol

7. 15.9 L

8. 52.0 g/mol

9. a) 1.3 g/L b) 1.2 g/L

10. 15.9 g/mol

11. 30.0 g/mol

12. $D = \frac{m}{V} = \frac{(MM)P}{RT}$, so when temperature and pressure are constant, density is directly proportional to molar mass. N_2O_4 is more dense than NO_2.

13. 1.72×10^4 L/mol; Molar volume is independent of the identity of the gas.

14. 26.6 L/mol

15. 1.57 L O_2

16. 27.7 g $KMnO_4$

17. 308 L SO_2

18. 16.7 g H_2O

19. 75.5 L N_2

20. 43.8 g NH_4NO_3

21. a) 17.6 L O_2 b) 15.0 L O_2

22. 1.78×10^3 L ClO_2

23. 459 L H_2S

24. 216 g

25. 28.5 L/mol
At a given temperature and pressure, molar volume is the same for all gases. The identity of the gas, the volume, and the mass are not needed to solve the problem.

26. 312 g/mol

27. MM = 28.0 g/mol
The molar mass of nitrogen is twice its atomic mass. Therefore, there must be two atoms in the molecule, N_2.

28. a) 334 g/mol
b) The difference between the expected molar mass and the molar mass that was found experimentally is 346 – 334 = 12 g/mol. This is the atomic mass of carbon. Perhaps she has one too many carbon atoms in her expected molecule.

Active Learning Workbook Answers

Chapter 14

Combined Gas Law Applications

1. 22.4 L/mol is the molar volume of a gas at standard temperature and pressure, 0°C and 1 atm.

2. 94.3 L CO

3. 1.72×10^4 L/mol
Molar volume is independent of the identity of the gas.

4. 26.6 L/mol

5. 15.0 L CO

6. 5.27 g

7. a) 0.179 g/L b) 0.22 g/L

8. 0.557 mol

9. 15.9 g/mol

10. 0.950 g/L

11. 30.0 g/mol

12. 1.32 g/L

13. $D = \frac{MM}{MV}$, so if temperature and pressure are the same for two gases, their molar volumes are the same, and therefore density is directly proportional to molar mass. N_2O_4 is more dense than NO_2.

14. 1.57 L O_2

15. 27.7 g $KMnO_4$

16. 308 L SO_2

17. 16.7 g H_2O

18. 75.5 L N_2

19. 43.8 g NH_4NO_3

20. The Volume–Amount (Avogadro's) Law tells us that equal volumes of all gases at the same temperature and pressure contain the same number of molecules. Reasoning in reverse, gases with the same number of molecules at the same temperature and pressure must have equal volumes. The volume of 1×10^{23} hydrogen molecules is the same as the volume of 1×10^{23} oxygen molecules and half the volume of 2×10^{23} nitrogen molecules.

21. a) 17.6 L O_2 b) 14.9 L O_2

22. 1.77×10^3 L ClO_2

23. 0.133 L H_2S

24. 217 g

25. 28.5 L/mol
At a given temperature and pressure, molar volume is the same for all gases. The identity of the gas, the volume, and the mass are not needed to solve the problem.

26. 312 g/mol

27. MM = 28.0 g/mol
The molar mass of nitrogen is twice its atomic mass. Therefore, there must be two atoms in the molecule: N_2.

28. (a) 334 g/mol
(b) The difference between the expected molar mass and the molar mass that was found experimentally is 346 – 334 = 12 g/mol. This is the atomic mass of carbon. Perhaps she has one too many carbon atoms in her expected molecule.

29. 1.2 g/L

30. 33.2 g/mol

Active Learning Workbook Answers

Chapter 15

Gases, Liquids, and Solids

1. Each gas fills the volume of the container, 1×10^2 L. A gas, either pure or a mixture, is made up of tiny particles which are widely separated from each other so that they occupy the whole volume of the container that holds them.

2. 2 torr

3. 717 mm Hg

4. Gas particles are very widely spaced; liquid particles are "touchingly close." When gases are mixed, the particles of one gas can occupy the empty spaces between the particles of the other. When liquids are mixed, particles must be "pushed aside" to make room for others.

5. Intermolecular attractions are stronger in liquids because of the lack of space between molecules.

6. The stronger the intermolecular attractions, the more motion is needed to separate the particles within the liquid, and the higher the boiling point.

7. Energy is required to overcome intermolecular attractions, separate liquid particles from one another, and keep them apart. The more energy required, the higher the molar heat of vaporization.

8. The ball bearing will reach the bottom of the water cylinder first. Molecules in syrup have higher intermolecular attractions than those in water, so the syrup is more viscous. Syrup molecules are being pulled apart so the ball bearing can pass through the liquid. This slows the rate of fall.

9. Intermolecular attractions are stronger in mercury than in water. Mercury therefore has a higher surface tension and clings to itself rather than spreading or penetrating paper.

10. The wetting agent reduces surface tension water, overcoming its intermolecular attractions and allowing it to penetrate the duck's feathers. The duck's buoyancy is reduced, and it sinks.

11. NO_2 has the highest boiling point, which suggests strong intermolecular attractions. It should also have the highest molar heat of vaporization.

12. Only N_2O is a liquid at –90°C, so it alone has a measurable equilibrium vapor pressure as that term is used in this unit. NO is a gas, and its vapor pressure is its gas pressure. NO_2, a solid at –90°C, probably has a very small vapor pressure.

13. Other things being equal, dipole forces are stronger than induced dipole forces. Induced dipole forces are likely to be larger than dipole forces when the molecules are very large.

14. $NH(CH_3)_2$, hydrogen bonding; CH_2F_2, dipole; C_3H_8, induced dipole.

15. Dipole forces are attractive forces between polar molecules; hydrogen bonds are stronger dipole-like forces between polar molecules in which hydrogen is bonded to a highly electronegative element, usually nitrogen, oxygen, or fluorine.

16. NH_3 has the higher boiling point because it has relatively strong hydrogen bonds, versus relatively weak induced dipoles in CH_4.

17. Argon has the higher boiling point because its atoms are larger than those of neon.

18. See the discussion on hydrogen bonds in Section 15.3.

19. a) Induced dipoles b) Hydrogen bonding

20. Induced dipole forces are present in all molecules. In addition, dipole forces are present in (b), whereas (a), (c), and (d) have hydrogen bonding.

21. CS_2 should have the higher melting and boiling points because its molecules are larger than otherwise similar CO_2 molecules.

22. CH_4 should have the higher vapor pressure as a liquid at a given temperature because only weak induced dipoles are present. Relatively stronger dipole forces are present in CH_3F.

23. See the discussion in Section 15.4.

24. At higher temperatures, a greater percentage of molecules in the liquid state have enough energy to vaporize into the gaseous state.

25. The vapor pressure of the compound exceeds the external pressure, so it is a gas. A compound boils, or changes from a liquid to a gas, when its vapor pressure equals the external pressure.

26. The water is delivered at very high pressure, pressure greater than the vapor pressure at the temperature of delivery.

27. Low-boiling liquids and a low heat of vaporization are both characteristic of relatively weak intermolecular attractions. A liquid with weak attractions would therefore exhibit both properties.

28. N should have both the lower boiling point and lower molar heat of vaporization.

29. Ice is a crystalline solid. Ice crystals have a definite geometric order and ice melts at a definite, constant temperature.

30. C: Network solid. D: Ionic solid.

31. 0.732 kJ/g

32. 969 kJ

33. 90.2 g

34. 13.1 kJ

35. 76 kJ

36. 68.0 J/g

37. 44.5 g

38. A. From $q = m \times c \times \Delta T$, when q and m are equal for two objects, c is inversely proportional to ΔT.

39. –6.6 kJ

40. -2.0×10^1 kJ

41. 6°C

42. J (boiling point), K (freezing point)

43. H, I

44. C

45. Gas condenses at boiling point, J; liquid cools from boiling point, J, to freezing point K.

46. P – O

47. q (heat solid) = 2.9 kJ q (melt solid) = 42.5 kJ q (heat liquid) = 11 kJ
Total q = 2.9 kJ + 42.5 kJ + 11 kJ = 56 kJ

48. q (cool liquid) = –46.9 kJ q (freeze liquid) = –68.9 kJ q (cool solid) = -1.0×10^2 kJ
Total q = –46.9 kJ + (–68.9 kJ) + (-1.0×10^2 kJ) = -2.2×10^2 kJ

49. q (heat solid) = 6.0×10^4 kJ q (melt solid) = 2.7×10^4 kJ q (heat liquid) = 1.5×10^4 kJ
Total q = 6.0×10^4 kJ + 2.7×10^4 kJ + 1.5×10^4 kJ = 10.2×10^4 kJ = 1.02×10^5 kJ

50. Large molecules having strong induced dipole forces may have stronger intermolecular attractive forces than small molecules with hydrogen bonding and therefore exhibit greater viscosity.

51. The final ether vapor pressure in Containers A and C is the greatest—the equilibrium vapor pressure. Container B has the lowest vapor pressure, having all evaporated before reaching equilibrium.

52. 779 torr

53. The liquid can be boiled by reducing the pressure and thus the boiling temperature.

54. q_{Al} = –1.2 kJ; q (cool water) = –29 kJ; q (freeze water) = –137 kJ; q (cool ice) = –8 kJ
Total q = –1.2 kJ + (–29 kJ) + (–137 kJ) + (–8 kJ) = –175 kJ

55. Without a regular and uniform structure in an amorphous solid, some intermolecular forces are stronger than others. The weak forces are more easily overcome than the stronger forces, and thus a lower temperature is needed for melting in some parts of the solid than in other parts.

56. 35 g

Active Learning Workbook Answers

Chapter 16

Solutions

1. Provided that there is no chemical reaction, gases will always combine to form a homogeneous mixture, which is, by definition, a solution.

2. If particles were visible, there would be distinctly different phases and the mixture would not be homogeneous and therefore not a solution.

3. a) Salt is solute; water is solvent. When a solid is dissolved in a liquid, the solid is the solute and the liquid is the solvent.
   b) Copper is solute; silver is solvent.
   c) Oxygen is solute; nitrogen is solvent.
   For (b) and (c), the solute is the substance present in the smallest amount.

4. A concentrated solution contains a relatively large amount of solute; a saturated solution cannot hold more solute.

5. a) All salt added will be solid in beaker.
   b) Some or all salt added will dissolve; little or no salt will be solid in beaker.
   c) All added salt and some previously dissolved salt will be solid in beaker.

6. Temperature. Usually, the higher the temperature, the higher the solubility.

7. Acetic acid is soluble in water because it is dispersed uniformly throughout the solution. It is also miscible, a term usually used to express the solubility of liquids in each other. Sugar would not be called miscible in water.

8. Cations are attracted to the negative portion of the water molecule; anions are attracted to the positive portion.

9. As dissolved solute particles move through the solution, they come into contact with undissolved solute and return to the solid state.

10. When dissolving begins, crystallization rate is zero. Dissolving rate remains constant, and as solution becomes more concentrated, crystallization rate increases. Concentration continues to increase until crystallization rate equals dissolving rate.

11. Never.

12. A supersaturated solution is unstable, and a physical disturbance such as stirring will start crystallization.

13. A finely divided solid offers more surface area per unit of mass. Stirring or agitating the solution prevents concentration build-up at the solute surface, which minimizes crystallization rate. All physical processes speed up at higher temperatures because particle movement is more rapid.

14. (a) formic acid; (c) methylamine. Both compounds match the hydrogen bonding found in water. Hexane and tetrafluoromethane are nonpolar.

15. Glycerine exhibits hydrogen bonding, as does water, so they are miscible. Hexane is nonpolar.

16. Carbon dioxide, CO_2.

17. 2.78%

18. 241 g NH_4NO_3; 174 g H_2O

19. 0.290 M KI

20. The anhydrate has more moles and thus has the higher concentration: 1.54 M $NiCl_2$

21. 2.5 g $AgNO_3$

22. 196 g KOH

23. 0.29 L

24. 1.08 L

25. 0.1114 mol $AgNO_3$

26. 0.00210 mol $KMnO_4$

27. 29.0% KNO_3

28. 2.00 m

29. 138 g $(CH_3CH_2)_2NH$

30. 597 mL H_2O

31. Equivalent mass is the mass of a substance that reacts with one mole of hydrogen or hydroxide ions. LiOH has one mole of OH^- ions, so equivalent mass = molar mass. H_2SO_4 can release one or two moles of H^+ ions, so equivalent mass = molar mass or $\frac{1}{2}$ of molar mass.

32. 1 eq/mol HNO_2; 1 eq/mol H_2SeO_4

33. 2 eq/mol $Cu(OH)_2$; 3 eq/mol $Fe(OH)_3$

34. 47.02 g HNO_2/eq; 144.98 g H_2SeO_4/eq

35. 48.78 g $Cu(OH)_2$/eq; 35.62 g $Fe(OH)_3$/eq

36. 0.160 N KOH

37. 0.372 N $H_2C_2O_4$

38. 18.0 g $NaHSO_4$

39. (a) 0.965 M NaOH; (b) 0.474 M H_3PO_4

40. 0.610 eq NaOH

41. 1.60 L

42. 0.51 M $HC_2H_3O_2$

43. 32 mL HNO_3

44. 2.4×10^2 mL H_2SO_4

45. 1.8 N H_3PO_4

46. 0.580 g $Mg(OH)_2$

47. 1.34 g $Ca_3(PO_4)_2$

48. 114 mL

49. 107 mL

50. 0.3976 M Na_2CO_3

51. 0.500 M KOH

52. 0.3829 M H_2SO_4

53. At 2 eq/mol, 0.3976 M Na_2CO_3 = 0.7952 N Na_2CO_3

54. At 1 eq/mol, 0.500 M KOH = 0.500 N KOH

55. 0.645 N Na_2CO_3

56. 0.339 N $H_2C_4H_4O_6$

57. 0.170 N H_3PO_4; The normality of the acid depends on how many hydrogens react.

58. 232 g/eq

59. Partial pressure is a colligative property because it depends on the number of particles and is independent of their identity (for an ideal gas).

60. $T_b = 101.3°C$; $T_f = -4.52°C$

61. 2.87°C

62. 0.64 m

63. 5.0×10^1 g/mol

64. 52 g/mol

65. 5.0°C/m

66. The bubbles are dissolved air (mostly nitrogen and oxygen) that becomes less soluble at higher temperatures.

67. No

68. Distillation is one method. It is used to separate petroleum into its components, which include natural gas, gasoline, lubricating oil, asphalt, and many other product.

69. Dissolve more than 1.02 g of silver acetate at a temperature greater than 20°C, then cool the solution without disturbance. Crystallization odes not occur at the solubility limit because solute particles are not properly organized for crystal formation.

70. 5.38 M HCl

71. 7.88 g $H_2C_2O_4 \cdot 2\,H_2O$

72. 7.15 g $Na_2CO_3 \cdot 10\,H_2O$

73. 0.362 g $Ni(OH)_2$ will precipitate

74. 0.366 M OH^-

75. 78.8% NaH_2PO_4; 21.2% Na_2HPO_4

76. A small sample of pure air is a homogeneous mixture and is therefore a solution. The "atmosphere," even if it were pure air, is a very tall sample that becomes less dense at higher elevations. The atmosphere is therefore not homogeneous, and consequently it is not a solution.

77. The density of a solution must be known in order to convert concentrations based on mass only (percentage, molality) to those based in volume (molarity, normality).

Active Learning Workbook Answers

Chapter 17

Net Ionic Equations

1. Solutions of weak electrolytes conduct electricity poorly because only a small quantity of ions are formed when the electrolyte is dissolved. Solutions of nonelectrolytes do not conduct electricity because essentially no ions exist in the solution.

2. It is the movement of ions that makes up an electric current in solution. Soluble molecular compounds are generally neutral, and are usually nonelectrolytes. Some molecular compounds can react with water, forming ions as a product, and thus act as electrolytes.

3. $NH_4^+(aq)$, $SO_4^{2-}(aq)$; $Mn^{2+}(aq)$, $Cl^-(aq)$

4. $Ni^{2+}(aq)$, $SO_4^{2-}(aq)$; $K^+(aq)$, $PO_4^{3-}(aq)$

5. $H^+(aq)$, $NO_3^-(aq)$; $H^+(aq)$, $Br^-(aq)$

6. $H_2C_4H_4O_4(aq)$, $HF(aq)$

7. $3\ Zn^{2+}(aq) + 2\ PO_4^{3-}(aq) \rightarrow Zn_3(PO_4)_2(s)$

8. $2\ Fe(s) + 6\ H^+(aq) \rightarrow 3\ H_2(g) + 2\ Fe^{3+}(aq)$

9. $Na_2C_2O_4(s) + 2\ H^+(aq) \rightarrow H_2C_2O_4(aq) + 2\ Na^+(aq)$

10. $Cu(s) + Li_2SO_4(aq) \rightarrow NR$

11. $Ba(s) + 2\ H^+(aq) \rightarrow H_2(g) + Ba^{2+}(aq)$

12. $Ni(s) + CaCl_2(aq) \rightarrow NR$

13. $Pb^{2+}(aq) + 2\ I^-(aq) \rightarrow PbI_2(s)$

14. $KClO_3(aq) + Mg(NO_2)_2(aq) \rightarrow NR$

15. $Ag^+(aq) + Br^-(aq) \rightarrow AgBr(s)$

16. $Zn^{2+}(aq) + SO_3^{2-}(aq) \rightarrow ZnSO_3(s)$

17. $Pb^{2+}(aq) + CO_3^{2-}(aq) \rightarrow PbCO_3(s)$; $Ca^{2+}(aq) + 2\ OH^-(aq) \rightarrow Ca(OH)_2(s)$

18. $H^+(aq) + NO_2^-(aq) \rightarrow HNO_2(aq)$

19. $H^+(aq) + C_3H_5O_3^-(aq) \rightarrow HC_3H_5O_3(aq)$

20. $OH^-(aq) + HF(aq) \rightarrow H_2O(\ell) + F^-(aq)$

21. $2\ H^+(aq) + SO_3^{2-}(aq) \rightarrow H_2O(\ell) + SO_2(aq)$

22. $2\ H^+(aq) + SO_3^{2-}(aq) \rightarrow H_2O(\ell) + SO_2(aq)$

23. $Ba^{2+}(aq) + SO_3^{2-}(aq) \rightarrow BaSO_3(s)$

24. $Cu^{2+}(aq) + 2\ OH^-(aq) \rightarrow Cu(OH)_2(s)$

25. $2\ H^+(aq) + MgCO_3(s) \rightarrow Mg^{2+}(aq) + H_2O(\ell) + CO_2(g)$

26. $2\ H^+(aq) + Pb(OH)_2(s) \rightarrow Pb^{2+}(aq) + 2\ H_2O(\ell)$

27. $HC_7H_5O_2(s) + OH^-(aq) \rightarrow C_7H_5O_2^-(aq) + H_2O(\ell)$

28. $Ni(s) + 2\ H^+(aq) \rightarrow Ni^{2+}(aq) + H_2(g)$

29. $H^+(aq) + HSO_3^-(aq) \rightarrow H_2O(\ell) + SO_2(aq)$

30. $MgSO_4(aq) + NH_4Br(aq) \rightarrow NR$

31. $Mg(s) + 2\ H^+(aq) \rightarrow Mg^{2+}(aq) + H_2(g)$

32. $Ni(OH)_2(s) + 2\ H^+(aq) \rightarrow Ni^{2+}(aq) + 2\ H_2O(\ell)$

33. $F^-(aq) + H^+(aq) \rightarrow HF(aq)$

34. $Ag(s) + HCl(aq) \rightarrow NR$

35. $2\ Li(s) + 2\ H_2O(\ell) \rightarrow 2\ Li^+(aq) + 2\ OH^-(aq) + H_2(g)$

36. $2\ Al(s) + 3\ Cu^{2+}(aq) \rightarrow 2\ Al^{3+}(aq) + 3\ Cu(s)$

Active Learning Workbook Answers

Chapter 18

Acid–Base (Proton-Transfer) Reactions

1. The classical properties of acids and bases are listed in the introduction to the chapter. As an example of how a property relates to the ion associated with it, an acid-base neutralization is $H^+ + OH^- \rightarrow H_2O$.

2. An Arrhenius base is a source of OH^- ions, whereas a Brønsted-Lowry base is a proton receiver. The two are in agreement, as the OH^- ion is an excellent proton receiver. Other substances, however, can also receive protons, so there are other bases according to the Brønsted-Lowry concept.

3. In the reaction shown below, $AlCl_3$, a Lewis acid, accepts an electron pair from Cl^-, a Lewis base, in a Lewis acid-Lewis base neutralization reaction.

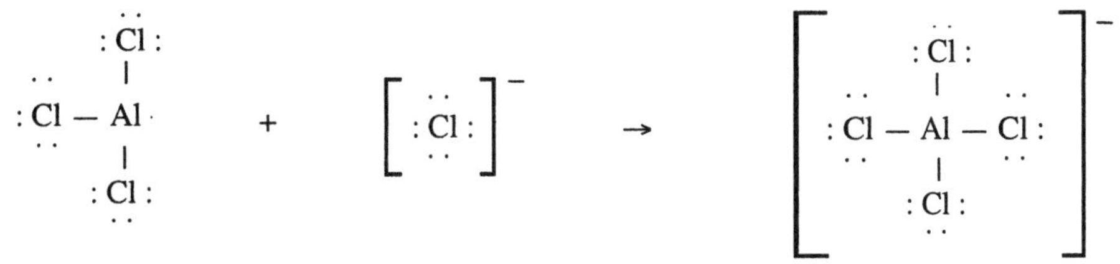

```
        ..                                     ┌      ..       ┐ -
       : Cl :                                  │     : Cl :    │
   ..    |                 ..   -              │  ..   |    .. │
 : Cl — Al ·      +     [ : Cl : ]      →      │: Cl — Al — Cl:│
   ..    |                 ..                  │  ..   |    .. │
       : Cl :                                  │     : Cl :    │
        ..                                     └      ..       ┘
```

4. BF_3 is a Lewis acid because the empty valence orbital in the boron atom accepts an electron pair from the oxygen atom in $C_2H_5OC_2H_5$, a Lewis base because it donates the electron pair.

5. F^-; HPO_4^{2-}; HNO_2; H_3PO_4

6. Acids: HSO_4^- (forward) and $HC_2O_4^-$ (reverse); bases: $C_2O_4^{2-}$ (forward) and SO_4^{2-} (reverse)

7. HSO_4^- and SO_4^{2-}; $HC_2O_4^-$ and $C_2O_4^{2-}$

8. HNO_2 and NO_2^-; $HC_3H_5O_2$ and $C_3H_5O_2^-$

9. NH_4^+ and NH_3; $H_2PO_4^-$ and HPO_4^{2-}

10. A strong base has a strong attraction for protons, while a weak base has little attraction for protons. Strong bases are at the bottom of the right column in Table 18.1 and weak bases are at the top.

11. $H_2O < HClO < HC_2O_4^- < H_2SO_3$

12. $CN^- > ClO^- > HSO_3^- > H_2O > Cl^-$

13. $HC_3H_5O_2 + PO_4^{3-} \rightleftharpoons C_3H_5O_2^- + HPO_4^{2-}$ Forward

14. $HSO_4^- + CO_3^{2-} \rightleftharpoons SO_4^{2-} + HCO_3^-$ Forward

15. $H_2CO_3 + NO_3^- \rightleftharpoons HCO_3^- + HNO_3$ Reverse

16. $NO_2^- + H_3O^+ \rightleftharpoons HNO_2 + H_2O$ Forward

17. $HSO_4^- + HC_2O_4^- \rightleftharpoons H_2SO_4 + C_2O_4^{2-}$ Reverse

$HSO_4^- + HC_2O_4^- \rightleftharpoons SO_4^{2-} + H_2C_2O_4$ Reverse

18. The very small value for K_w indicates that water ionizes to a very small extent.

19. An acidic solution has a higher H^+ concentration than OH^- concentration. The solution is therefore acidic: $10^{-5} > 10^{-9}$.

20. 10^{-12} M

21. (a) neutral; (b) weakly basic; (c) strongly basic

22. Basic. Water solutions with pH = 7 ($[H^+] = [OH^-] = 10^{-7}$) are neutral, pH > 7 ($[H^+] < 10^{-7}$ and $[OH^-] > 10^{-7}$) are basic, and those with pH < 7 ($[H^+] > 10^{-7}$ and $[OH^-] < 10^{-7}$) are acidic.

	pH	pOH	$[H^+]$	$[OH^-]$	
23.	5	9	10^{-5}	10^{-9}	weakly acidic
24.	13	1	10^{-13}	10^{-1}	strongly basic
25.	10	4	10^{-10}	10^{-4}	moderately basic
26.	9	5	10^{-9}	10^{-5}	weakly basic
27.	4.40	9.60	4.0×10^{-5}	2.5×10^{-10}	
28.	4.06	9.94	8.7×10^{-5}	1.1×10^{-10}	
29.	0.55	13.45	2.8×10^{-1}	3.5×10^{-14}	
30.	6.60	7.40	2.5×10^{-7}	4.0×10^{-8}	

31. No. A Brønsted-Lowry acid is a proton donor. A proton is a hydrogen ion. If there are no hydrogen atoms in a substance, it cannot donate a hydrogen ion to another species.

32. Chemists generally prefer the relative simplicity of the pH scale. Most people find that working with numbers such as pH = 4 is much more convenient than the equivalent hydrogen ion concentration 1×10^{-4} M or 0.0001 M.

33. $[Br^-]$ = antilog (–7.2) = 6×10^{-8} M

34. $SO_3 + H_2O \rightarrow H_2SO_4$, sulfuric acid

35. $2\ CaO + 2\ H_2O \rightarrow H_2 + 2\ Ca(OH)_2$, calcium hydroxide

36. The nitrogen dioxide can combine with oxygen and water in the atmosphere to form nitric acid: $4\ NO_2(g) + 2\ H_2O(\ell) + O_2(g) \rightarrow 4\ HNO_3(aq)$.

Active Learning Workbook Answers

Chapter 19

Oxidation–Reduction (Redox) Reactions

1. Examples include any item that runs on batteries, such as watches, calculators, etc.

2. Yes. A galvanic cell causes current to flow through an external circuit by electrochemcial action. An electrolytic cell is a cell through which a current driven by an external source passes.

3. See the discussion in Section 19.2.

4. Oxidation: a, c, d. Reduction: b.

5. Reduction

6. Oxidation

7.
$$2\,Cr \rightarrow 2\,Cr^{3+} + 6\,e^-$$
$$3\,Cl_2 + 6\,e^- \rightarrow 6\,Cl^-$$
$$3\,Cl_2 + 2\,Cr \rightarrow 2\,Cr^{3+} + 6\,Cl^-$$

8. The second equation is the oxidation half-reaction equation.
Overall: $2\,NiOOH + 2\,H_2O + Cd \rightarrow 2\,Ni(OH)_2 + Cd(OH)_2$

9. +3, –2, +4, +6

10. +3, +5, +6, +5

11. a) Bromine reduced from 0 to –1; b) Lead oxidized from +2 to +4

12. a) Iodine reduced from +7 to –1; b) Oxygen reduced from 0 to –2

13. a) Nitrogen oxidized from +4 to +5; b) Chromium oxidized from +3 to +6

14. Chlorine is the oxidizing agent, and the bromide ion is the reducing agent.

15. Pb reduces the lead in PbO_2. PbO_2 oxidizes Pb.

16. From Table 19.2, Zn is a stronger reducer than Fe^{2+}. A strong reducer releases electrons to an oxidizer more readily than a weak reducer releases them.

17. $Na^+ < Fe^{2+} < Cu^{2+} < Br_2$

18. $Br_2 + 2\,I^- \rightleftharpoons 2\,Br^- + I_2$; forward reaction favored

19. $2\,H^+ + 2\,Br^- \rightleftharpoons H_2 + Br_2$; reverse reaction favored

20. $2\ NO + 4\ H_2O + 3\ Fe^{2+} \rightleftharpoons 2\ NO_3^- + 8\ H^+ + 3\ Fe$; reverse reaction favored

21. A strong acid releases protons readily; a strong reducer releases electrons readily. A strong base attracts protons strongly; a strong oxidizer attracts electrons strongly.

22.
$$S_2O_3^{2-} + 5\ H_2O \rightarrow 2\ SO_4^{2-} + 10\ H^+ + 8\ e^-$$
$$4\ Cl_2 + 8\ e^- \rightarrow 8\ Cl^-$$
$$S_2O_3^{2-} + 5\ H_2O + 4\ Cl_2 \rightarrow 2\ SO_4^{2-} + 10\ H^+ + 8\ Cl^-$$

23.
$$4\ NO_3^- + 8\ H^+ + 4\ e^- \rightarrow 4\ NO_2 + 4\ H_2O$$
$$Sn + 3\ H_2O \rightarrow H_2SnO_3 + 4\ H^+ + 4\ e^-$$
$$4\ NO_3^- + 8\ H^+ + Sn \rightarrow 4\ NO_2 + H_2O + H_2SnO_3$$

24.
$$2\ MnO_4^- + 16\ H^+ + 10\ e^- \rightarrow 2\ Mn^{2+} + 4\ H_2O$$
$$5\ C_2O_4^{2-} \rightarrow 10\ CO_2 + 10\ e^-$$
$$2\ MnO_4^- + 16\ H^+ + 5\ C_2O_4^{2-} \rightarrow 2\ Mn^{2+} + 4\ H_2O + 10\ CO_2$$

25.
$$Cr_2O_7^{2-} + 8\ H^+ + 6\ e^- \rightarrow Cr_2O_3 + 4\ H_2O$$
$$2\ NH_4^+ \rightarrow N_2 + 8\ H^+ + 6\ e^-$$
$$Cr_2O_7^{2-} + 2\ NH_4^+ \rightarrow Cr_2O_3 + 4\ H_2O + N_2$$

26.
$$4\ NO_3^- + 16\ H^+ + 12\ e^- \rightarrow 4\ NO + 8\ H_2O$$
$$3\ As_2O_3 + 15\ H_2O \rightarrow 6\ AsO_4^{3-} + 30\ H^+ + 12\ e^-$$
$$3\ As_2O_3 + 7\ H_2O + 4\ NO_3^- \rightarrow 6\ AsO_4^{3-} + 14\ H^+ + 4\ NO$$

27. This "property of an acid" is more correctly described as the property of an acid (hydrogen ion) acting as an oxidizing agent. The H^+ ion reacts only with those metals whose ions are weaker oxidizing agents, located below hydrogen in Table 19.2.

28. All reactants were in acidic solutions. Water is available in large amounts in any aqueous solution, as is H^+ in an acidic solution.

29. In a simple element ↔ monatomic ion redox reaction the statement is correct. The *element* oxidized or reduced can always be identified by a change in oxidation number. The oxidizing or reducing agent, however, is a *species,* which may be an element, a monatomic ion, or a polyatomic ion, such as MnO_4^-.

30. Oxidation occurs at the anode: $2\ Cl^- \rightarrow Cl_2 + 2\ e^-$. Reduction occurs at the cathode: $Na^+ + e^- \rightarrow Na$.

31. Zinc is used to prevent harmful galvanic action (corrosion) in the equipment. The conditions for galvanic action are present: two metals in contact with an electrolyte (seawater) and in metal-to-metal contact with each other. If one of the metals is going to corrode, it will be the strongest reducing agent of the group. Zinc is a stronger reducing agent than iron or copper. When it goes, it is simply and cheaply replaced.

Active Learning Workbook Answers

Chapter 20

Chemical Equilibrium

1. In a dynamic equilibrium, opposing changes continue to occur at equal rates. An equilibrium in which nothing is changing, as a book resting on a table, for example, is called a static equilibrium.

2. Both systems can reach equilibrium. At equilibrium, the salt dissolves at the same rate at which it crystallizes. Whether the container is open or closed is of no importance.

3. The system is not an equilibrium because energy must be supplied constantly to keep it in operation. Also, the water is circulating, not moving reversibly in two opposing directions.

4. There will be no reaction if the orientation of a molecule is unfavorable.

5. $\Delta E = b - c$; $E_a = a - c$

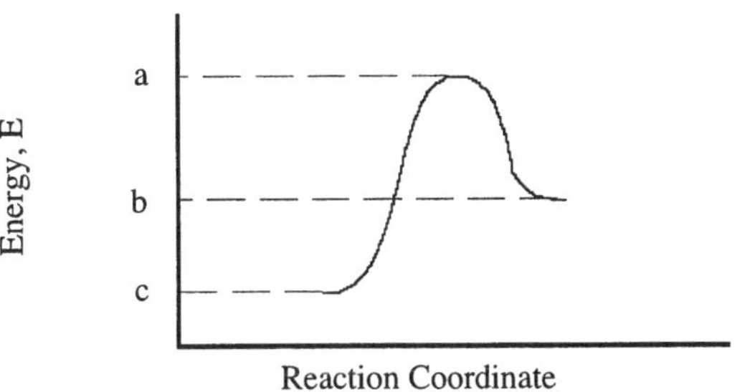

6. E_a (forward) $= a - b$; E_a (reverse) $= a - c$. Both activation energies are positive. Point a is the highest on the curve; $a > b$ and $a > c$.

7. An activated complex is an unstable intermediate species formed during a collision of two reacting particles. The properties of an activated complex cannot be described because the complex decomposes almost as soon as it forms.

8. At a higher temperature, a larger fraction of the molecules have enough kinetic energy to engage in a reaction-producing collision, so reaction rates are higher. Also, collisions are more frequent. At low temperature, a smaller fraction of the collisions produce reactions and there are fewer collisions, so the reaction rate is slower.

9. The equilibria are identical. Equilibrium will be reached more quickly in the system with the catalyst.

10. An increase in the concentration of A will increase the reaction rate, while a decrease in the concentration of B will decrease the reaction rate. The net effect depends on the size of the two changes.

11. The reverse reaction rate reaches its maximum at equilibrium.

12. See Figure 20.7.

13. If O_2 concentration is decreased, equilibrium will shift in the reverse direction, the direction in which more O_2 will be produced.

14. If O_2 concentration is increased, equilibrium will shift in the forward direction, the direction in which more O_2 will be consumed.

15. Removal of NH_3 will shift the equilibrium forward, the direction in which some additional NH_3 will be produced.

16. Increasing volume decreases the pressure. The equilibrium shifts to the left (5 mol on the left vs. 3 mol on the right) to cause a pressure increase.

17. It will not shift because there will be no change in the total number of molecules.

18. If heat is removed the equilibrium will shift in the direction that produces heat, the forward direction.

19. Cool the system. Removing heat causes the reaction to shift in the direction that produces heat, the forward direction. That increases the SO_3 yield.

20. $Ca(OH)_2(s) \rightleftharpoons Ca^{2+}(aq) + 2\ OH^-(aq)$ is the equilibrium equation. H^+ ions from the acid combine with OH^- ions to form water molecules. This reduces the OH^- ion concentration and causes a forward shift in the equilibrium. The process continues until all the $Ca(OH)_2$ is dissolved.

21. $K = \dfrac{[CO_2][H_2]}{[CO][H_2O]}$

22. $K = \dfrac{[CO][H_2]}{[H_2O]}$

23. $K = [Zn^{2+}]^3[PO_4^{3-}]^2$

24. $K = \dfrac{[H_3O^+][NO_2]}{[HNO_2]}$

25. $K = \dfrac{[Cu^{2+}][NH_3]^4}{[Cu(NH_3)_4{}^{2+}]}$

26. The equilibrium constant expression for the given equation is $K = \dfrac{[NO_2]^2}{[NO]^2[O_2]}$.

If the equation is written in reverse, the equilibrium constant expression is inverted. If different sets of coefficients are used, both the expression and its numerical value change. For example:

$NO + \frac{1}{2}\ O_2 \rightleftharpoons NO_2 \qquad K_1 = \dfrac{[NO_2]}{[NO][O_2]^{1/2}}$

$4\ NO + 2\ O_2 \rightleftharpoons 4\ NO_2 \qquad K_2 = \dfrac{[NO_2]^4}{[NO]^4[O_2]^2}$

The equilibrium constants are not equal: $K_2 = K^2 = K_1{}^4$.

27. The equilibrium will be favored in the forward direction. If K is very large, at least one factor in the denominator must be very small, indicating that at least one reactant has been almost completely consumed.

28. The equilibrium will be favored in the reverse direction. A very small equilibrium constant results when the concentration of one species on the right side of the equation is very small.

29. (a) The equilibrium is favored in the forward direction. H_2SO_3 is one of the acids that is unstable, decomposing to H_2O and SO_2, as indicated. K is large for this equilibrium. (b) The equilibrium is favored in the forward direction. $HC_2H_3O_2$ is a weak acid. Nearly all of the ions will combine to form the molecule.

30. 2.0×10^{-16}

31. 8.8×10^{-12}

32. 1.4×10^{-4} M; 4.0×10^{-3} g/100 mL

33. 2.9×10^{-5} M

34. 3.5×10^{-7} g CaC_2O_4

35. $K_a = 2.6 \times 10^{-4}$; 3.5% ionized

36. pH = 2.60

37. pH = 5.43

38. $\frac{[HC_2H_3O_2]}{[C_2H_3O_2^-]} = 3.1$

39. K = 14

40. Kinetic energies are greater at Time 1 because at Time 2 some of that energy has been converted to potential energy in the activated complex.

41. a) $Ca(OH)_2 \rightleftharpoons Ca^{2+} + 2\ OH^-$
b) (1) Adding a strong base or soluble calcium compound would increase $[OH^-]$ and $[Ca^{2+}]$, respectively, causing a shift in the reverse direction and reducing solubility of $Ca(OH)_2$.
(2) Adding an acid to reduce $[OH^-]$ by forming water; adding a cation that will reduce $[OH^-]$ by precipitation; or adding an anion whose calcium salt is less soluble than $Ca(OH)_2$ would cause a shift in the forward direction, increasing the solubility of $Ca(OH)_2$.
c) (1) Any anion with a calcium salt that is less soluble than $Ca(OH)_2$ will cause a forward shift, increasing $[OH^-]$.
(2) An acid that will form water with OH^- or a cation that will precipitate OH^- will reduce $[OH^-]$.

42. Add NO_2: R–I–I–D–I. Reduce temperature: F–D–D–I–I. Add N_2: None. Remove NH_3: R–D–I–D–D. Add a catalyst: None.

43. The "truth" of the statement depends on how you interpret it. A system *at equilibrium* is neither endothermic nor exothermic. The system is closed; heat energy neither renters nor leaves. The thermochemical *equation* for the equilibrium is endothermic in one direction and exothermic in the other direction. Both directions describe the same equilibrium.

44. $2\ SO_2(g) + O_2(g) \rightleftharpoons 2\ SO_3(g)$ and $2\ SO_3(g) \rightleftharpoons 2\ SO_2(g) + O_2(g)$ both express the equilibrium described. The K expression for one equation is the reciprocal of the other. If any fraction is greater than 1, its reciprocal is less than 1. Put another way, if the numerator is greater than the denominator, the fraction is greater than 1. In the reciprocal, the numerator is less than the denominator, so the fraction is less than 1.

45. The system can reach equilibrium only when it is closed, that is, when no water is running in the house; it cannot be reached in an open system while hard water enters the softener and soft water leaves. $[Ca^{2+}]$ is relatively high in the hard water that enters the softener. By Le Chatelier's Principle, the reaction is favored in the forward direction in which Ca^{2+} ions in the water are replaced by Na^+ ions. This also means the Na^+ ions in the resin are replaced by Ca^{2+} ions from the water. Eventually the Na^+ ions are used up and must be replenished. This is done by running water with a high sodium ion concentration through the softener. This forces the reaction in the reverse direction, again according to Le Chatelier's Principle. Na^+ ions replace the Ca^{2+} ions on the resin, and the Ca^{2+} ions are flushed down the drain.

Active Learning Workbook Answers

Chapter 21

Nuclear Chemistry

1. *Nuclide* is a general term that refers to the nucleus of any atom. An isotope is a specific kind of atom that has a specific nuclear composition.

2. Decay, when applied to a radioactive nucleus, refers to the spontaneous decomposition of the nucleus.

3. Alpha rays are helium nuclei having a mass of 4 amu and a 2+ charge. Beta rays are electrons with a mass of 0.005 amu and a 1– charge. Gamma rays are high-energy electromagnetic radiation having no mass and no charge. Penetration power increases in the order alpha < beta < gamma.

4. The collision of a radioactive emission with an atom or molecule may rearrange the electrons in the target, possibly ionizing it and causing a potentially harmful chemical change.

5. A Geiger counter is a device for detecting and measuring radiation. Figure 21.3 describes how it works.

6. A Geiger counter "clicks" when a radioactive emission is detected; a scintillation meter counts pulses of light generated by radiation. Both devices can measure radiation as well as detect it. The Geiger counter actually measures electric current; the scintillation meter measures light pulses. Both devices express their measurements as counts per unit time.

7. A gamma camera is immobile while it takes a picture, creating essentially a two-dimensional image of an object. The scanner moves as it takes many pictures. Computer enhancement and combination of many pictures allow 3-dimensional-like images to be constructed.

8. Figure 21.4 shows that smoking tobacco is the greatest source of background radiation.

9. Two half-lives have passed.

10. 1/128

11. 1.6 g

12. 1.6×10^2 g; 1.0×10^2 g

13. 25.6 hr/half-life = $t_{1/2}$

14. a) 5.7 g remain
    b) 14 minutes

15. 4.9×10^3 years

16. The original nucleus disintegrates into a $^{4}_{2}He$ nucleus and another nucleus. The final nuclide has 2 fewer protons and 2 fewer neutrons than the original. This is a transmutation, a change from one element to another, because the number of protons changes.

17. $^{228}_{89}Ac \rightarrow ^{0}_{-1}e + ^{228}_{90}Th$ $\quad$ $^{212}_{83}Bi \rightarrow ^{0}_{-1}e + ^{212}_{84}Po$

18. $^{216}_{84}Po \rightarrow ^{4}_{2}He + ^{212}_{82}Pb$ $\quad$ $^{234}_{92}U \rightarrow ^{4}_{2}He + ^{230}_{90}Th$

19. "Nuclear chemical properties of lead" is meaningless for two reasons. First, the chemical properties of all isotopes of lead are the same. Second, the nuclear properties of lead isotopes are specific for each individual isotope.

20. The count will remain at 5000/minute. The radioactivity of an element is independent of the form of the element, pure element or in a compound.

21. Radioactivity is spontaneous, while nuclear bombardment reactions are produced by projecting a nuclear particle into a target.

22. Electrons, protons, positrons, and alpha particles can be accelerated in particle accelerators. The particle must have an electrical charge to be accelerated.

23. Natural versus artificial radioactivity is not a function of atomic number. Isotopes of many elements are naturally radioactive, and artificially radioactive isotopes of many elements can be created.

24. All elements in the lanthanide series have atomic numbers less than 92, so none is a transuranium element. The elements in the actinide series with atomic numbers greater than 92 are transuranium elements.

25. $^{44}_{21}Sc$ $\quad$ $^{257}_{103}Lr$ $\quad$ $^{1}_{1}H$

26. Both fission and fusion reactions result in a release of energy. Fission reactions are those where a larger nucleus splits into smaller nuclei. In a fusion reaction, two small nuclei combine to form a larger nucleus.

27. A chain reaction is a reaction that has as a product one of its own reactants; that product becomes a reactant, thereby allowing the original reaction to continue. For a chain reaction to continue, there must be enough fissionable material to react with the neutrons given off.

28. The products of a fusion reaction are not one of the reactants, so it cannot be a chain reaction.

29. Nuclear fusion is more promising as an energy source than current fission plants because it produces more energy per given amount of fuel. Fusion fuel is more abundant, and fusion reactions generate no hazardous radioactive waster. Fusion's major drawback is the extremely high temperature needed to initiate the fusion process.

30. Natural fissionable isotopes are rare. Plutonium-234 is produced from the most abundant uranium isotope, uranium-238.

31. 1.5×10^3 tons coal

32. a) A > C > B. At the end of one half-life, half of the original A would remain, leaving the other half to be divided between B and C. Half of B disintegrates in one day, so more than half of what was produced in days 1 through 5 has passed along to C, where it is accumulated.
b) C > A > B. At the end of two half-lives, A is down to 1/4 of the starting amount. Most of the 3/4 that disintegrated has passed through B to C.

33. Emission of a beta particle would change a calcium atom into a scandium atom:
$^{47}_{20}\text{Ca} \rightarrow {}^{0}_{-1}\text{e} + {}^{47}_{21}\text{Sc}$.

34. The reaction described by Equation 21.7 is more apt to be a chain reaction. Three neutrons per uranium atoms are produced in the process in Equation 21.7; two neutrons per uranium atom are produced in the process in Question 27. The greater the number of neutrons, the more likely it is that there will be a chain reaction.

Active Learning Workbook Answers

Chapter 22

Organic Chemistry

1. The cyanide ion, CN^-, and the carbonate ion CO_3^{2-}, are not organic because they do not contain *both* carbon and hydrogen. The acetate ion, $C_2H_3O_2^-$, is organic.

2. 109.5°; tetrahedral

3. A hydrocarbon is a compound made up of carbon and hydrogen atoms. C_3H_4, C_8H_{10}, and $CH_3CH_2CH_3$ are hydrocarbons.

4. $C_{11}H_{24}$; $C_{21}H_{44}$. Alkanes have the general formula C_nH_{2n+2}.

5. Isomers are compounds having the same molecular formula but different structural formulas.

6. C_7H_{16} is the molecular formula; $CH_3CH_2CH_2CH_2CH_2CH_2CH_3$ is the line formula; $CH_3(CH_2)_5CH_3$ is the condensed formula. The structural formula is

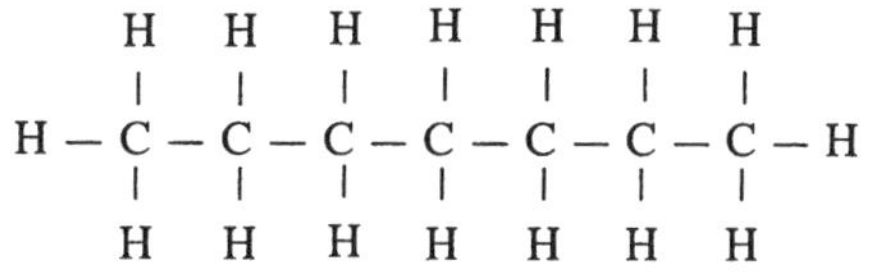

```
    H   H   H   H   H   H   H
    |   |   |   |   |   |   |
H — C — C — C — C — C — C — C — H
    |   |   |   |   |   |   |
    H   H   H   H   H   H   H
```

7. $C_2H_5—$ $C_4H_9—$

8. C_4H_{10}; decane

9. 2-methyl-4-ethylhexane

10.

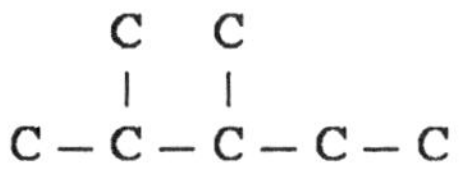

```
    C   C
    |   |
C — C — C — C — C
```

11.

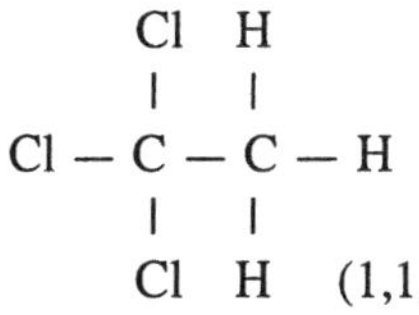

```
     Cl  H                      H   Cl
     |   |                      |   |
Cl — C — C — H             Cl — C — C — H
     |   |                      |   |
     Cl  H   (1,1,1)            Cl  H   (1,1,2)
```

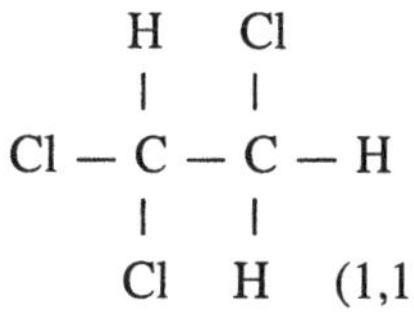

12. The general formula of a cycloalkane is C_nH_{2n}. Cyclobutane, C_4H_8, is an example:

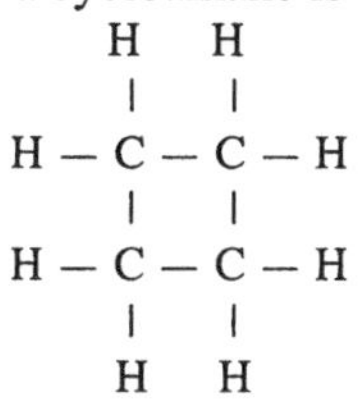

```
    H   H
    |   |
H — C — C — H
    |   |
H — C — C — H
    |   |
    H   H
```

13. 

14.

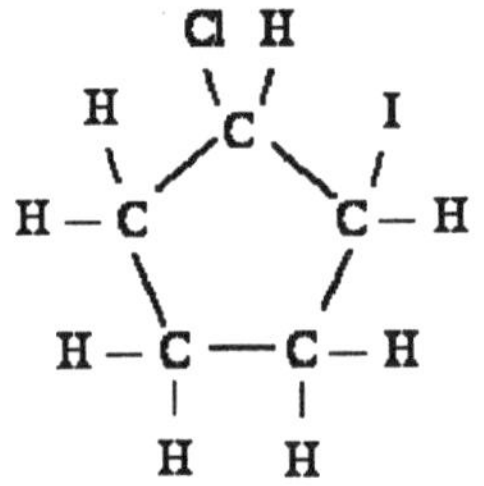

```
        Cl H
     H   \ /    I
      \  C       /
       /   \   /
H - C         C - H
      \       /
   H - C --- C - H
       |     |
       H     H
```

15. 4-ethylheptane

16. 1-bromo-3,3-dichlorobutane

17. 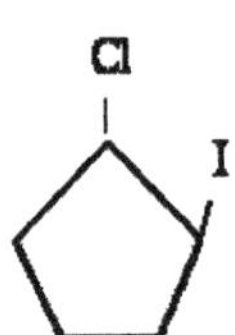

18. 1-chloro-2-ethylcyclohexane

19. An alkene has one or more double bonds; an alkane has single bonds. The general formula for an alkane is C_nH_{2n+2}, for an alkene, C_nH_{2n}.

20.

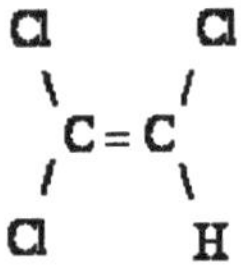

```
Cl    Cl
  \  /
  C=C
  /  \
Cl    H
```

This molecule can have only 2 additional atoms attached to each of its two carbons, so the first two chlorines go on one carbon atom. The third chlorine must go on the other carbon atom.

21.

```
   H        H                    H        CH2CH2CH3
    \      /                      \      /
     C = C                         C = C
    /      \                      /      \
CH3CH2   CH2CH2CH3            CH3CH2      H
```

Think of a line drawn in between the two lines that depict a double bond. In *cis*-3-heptene, the hydrogen atoms attached to the double bonded carbons are on the same side of that line. In *trans*-3-heptene, the hydrogen atoms attached to the double bonded carbons are on opposite sides of that line.

22.

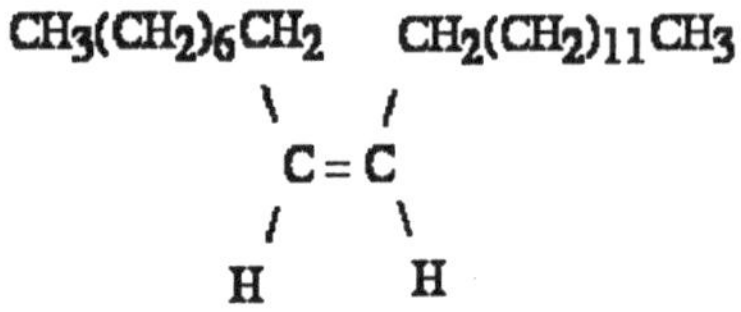

```
CH3(CH2)6CH2   CH2(CH2)11CH3
           \   /
            C=C
           /   \
          H     H
```

23. 2,2-dimethyl-3-heptyne

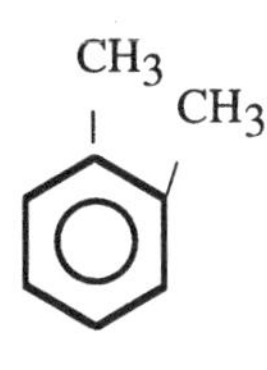

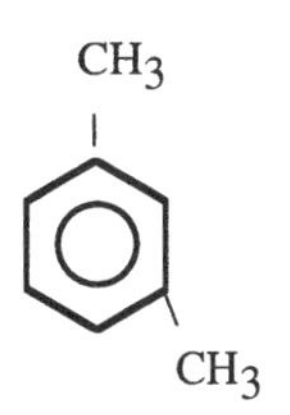

24. 1,2-dimethylbenzene 1,3-dimethylbenzene 1,4-dimethylbenzene

25. 1-bromo-4-chlorobenzene. Because both substituents are halogen atoms, the lower number is given to the first halogen in the alphabet.

26.

```
Cl                 Cl  Cl                Cl                 Cl       Cl
|                  |   |                 |                  |        |
C – C – C          C – C – C         C – C – C              C – C – C
|                                        |
Cl                                       Cl
1,1-dichloropropane  1,2-dichloropropane  2,2-dichloropropane  1,3-dichloropropane
```

27.

```
  Cl  Cl                    Cl  Cl
  |   |                     |   |
C – C – C – C           C – C – C – C
```

28.

```
CH3 – CH = CH – CH3 + H2 --catalyst--> CH3 – CH – CH – CH3
                                             |    |
                                             H    H
```

Because there is only one straight chain butane molecule, it makes no difference if the starting material is *cis* or *trans*-2-butene.

29.

```
    H   H   H   H                 H   H   H   H
    |   |   |   |                 |   |   |   |
H – C – C – C – C – OH        H – C – C – C – C – H        both alcohols
    |   |   |   |                 |   |   |   |
    H   H   H   H                 H   H   OH  H

    H  CH3  H                     H  CH3  H
    |   |   |                     |   |   |
H – C – C – C – OH            H – C – C – C – H            both alcohols
    |   |   |                     |   |   |
    H   H   H                     H   OH  H

    H   H       H   H             H   H   H       H
    |   |       |   |             |   |   |       |
H – C – C – O – C – C – H     H – C – C – C – O – C – H    both ethers
    |   |       |   |             |   |   |       |
    H   H       H   H             H   H   H       H
```

30. Although ethers have two carbon-oxygen bonds that are polar, the dipoles of these two bonds almost cancel each other by geometry. The forces of attraction in an ether are then weak dipole-dipole. In an alcohol, the hydroxyl proton on one alcohol molecule can hydrogen bond with the lone pair electrons of the oxygen atom on another alcohol molecule. The higher the forces of attraction, the higher the boiling point.

31.
```
     H   OH  H   H   H   H
     |   |   |   |   |   |
 H - C - C - C - C - C - C - H
     |   |   |   |   |   |
     H   H   H   H   H   H
```

32.
```
     H   H   H   H       H   H
     |   |   |   |       |   |
 H - C - C - C - C - O - C - C - H
     |   |   |   |       |   |
     H   H   H   H       H   H
```

33.
```
     H   H   H                H   H   H             H   H   H       H   H   H
     |   |   |                |   |   |             |   |   |       |   |   |
 H - C - C - C - OH  +  HO  - C - C - C - H  →  H - C - C - C - O - C - C - C - H + HOH
     |   |   |                |   |   |             |   |   |       |   |   |
     H   H   H                H   H   H             H   H   H       H   H   H
```

34.
```
     H   H   O                H   O   H
     |   |   ||               |   ||  |
 H - C - C - C            H - C - C - C - H
     |   |    \               |       |
     H   H     H              H       H
```

35.
```
     H   OH  H                          H   O   H
     |   |   |                          |   ||  |
 H - C - C - C - H  +  1/2 O2  →   H - C - C - C - H  +  H2O
     |   |   |                          |       |
     H   H   H                          H       H
```

36.
```
     H   H   H   H   H   O
     |   |   |   |   |   ||
 H - C - C - C - C - C - C - OH
     |   |   |   |   |
     H   H   H   H   H
```

37.
```
     H   H   O               H   H            H   H   O       H   H
     |   |   ||              |   |            |   |   ||      |   |
 H - C - C - C - OH  +  HO - C - C - H  →  H - C - C - C - O - C - C - H  +  HOH
     |   |                   |   |            |   |           |   |
     H   H                   H   H            H   H           H   H
```

The organic reaction product is ethyl propanoate, an ester.

38.
```
     H       H                       H   H
     |   ..  |                   ..  |   |
 H - C - N - C - H           H - N - C - C - H
     |   |   |                   |   |   |
     H   H   H                   H   H   H

   dimethylamine                 ethylamine
```

39. Dimethylamine is a secondary amine; ethylamine is a primary amine.

40.
```
     H   H   O                                 H   H   O
     |   |   ||             ..                 |   |   ||  ..
 H - C - C - C - OH  +  H - N - H  →  H - C - C - C - N - H  +  HOH
     |   |                  |                  |   |       |
     H   H                  H                  H   H       H
```

The organic reaction product is propanamide, an amide.

41.

```
  [H    H    H    H    H    H ]
  [|    |    |    |    |    | ]
 -[C  - C  - C  - C  - C  - C ]-
  [|    |    |    |    |    | ]
  [H    Br   H    Br   H    Br]
```

42.

```
H   H
|   |
C = C
|   |
H   CH3
```

43.

```
  [Cl   F    Cl   F    Cl   F]
  [|    |    |    |    |    |]
 -[C  - C  - C  - C  - C  - C]-
  [|    |    |    |    |    |]
  [F    F    F    F    F    F]
```

44.

```
 [O              CH3              O              CH3          ]
 [||             |                ||             |            ]
-[C - O -(C6H4)- C -(C6H4)- O - C - O -(C6H4)- C -(C6H4)- O]-
 [               |                               |            ]
 [               CH3                             CH3          ]
```

45.

```
 [O           O                              O           O                      ]
 [||          ||                             ||          ||                     ]
-[C -(C6H4)- C - NH -(C6H4)- NH - C -(C6H4)- C - NH -(C6H4)- NH]-
```

46.

```
     O          O                          H
     ||         ||                         |
HO - C -(C6H4)- C - OH       HO - CH2 - C - OH
                                           |
                                           CH3
```

47.

```
     O
     ||
HO - C - CH2 - CH2 - CH2 - CH2 - CH2 - NH2
```

48.

```
    H   H
    |   |
H - C - C - H        H       CH3     H3C      CH3      H    CH3
    |   |             \     /           \    /          \  /
H - C - C - H          C = C             C = C            C          CH2 = CH - CH2 - CH3
    |   |             /     \           /    \           / \
    H   H          H3C       H         H      H       H2C - CH2
```

49. The formula of the two-carbon alcohol is CH_3CH_2OH; the formula of the two-carbon ether is CH_3OCH_3. Both have the molecular formula C_2H_6O, and thus they are isomers. The alcohol can be transformed into the ether by moving the oxygen from a terminal carbon to a position between carbons. This will be true for any alcohol-ether isomer combination.

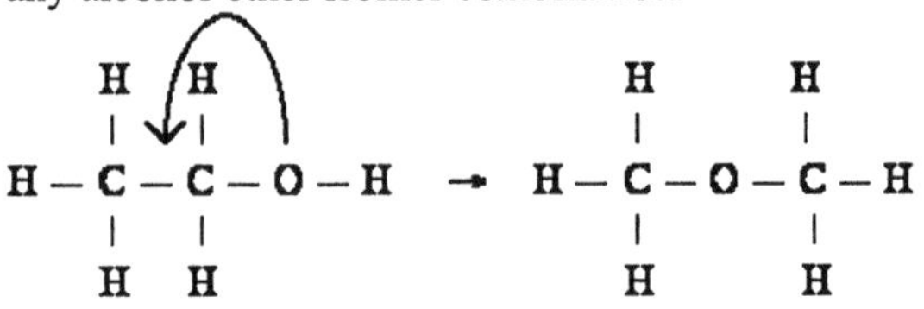

```
    H   H                    H       H
    |   |                    |       |
H - C - C - O - H   →   H -  C - O - C - H
    |   |                    |       |
    H   H                    H       H
```

50. $$\frac{1.8 \times 10^6 \text{ g}}{\text{polymer}} \times \frac{\text{molecule}}{104 \text{ g}} = 1.7 \times 10^4 \text{ molecules/polymer}$$

51. The esterification equation is

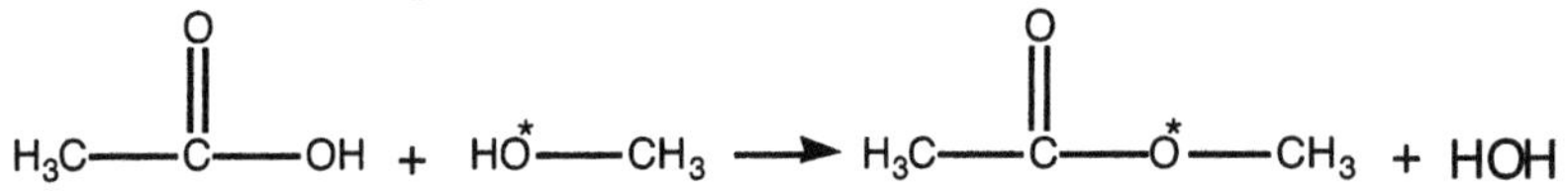

$$H_3C{-}\overset{O}{\overset{\|}{C}}{-}OH + H\overset{*}{O}{-}CH_3 \longrightarrow H_3C{-}\overset{O}{\overset{\|}{C}}{-}\overset{*}{O}{-}CH_3 + HOH$$

The asterisk on the oxygen atom in methanol identifies the oxygen-18 atom. It is the presence of the radioactive oxygen in the ester product, rather than in the water product, that shows that the water molecule is made up from a hydroxyl from the acid and only a hydrogen atom from the alcohol.

52. 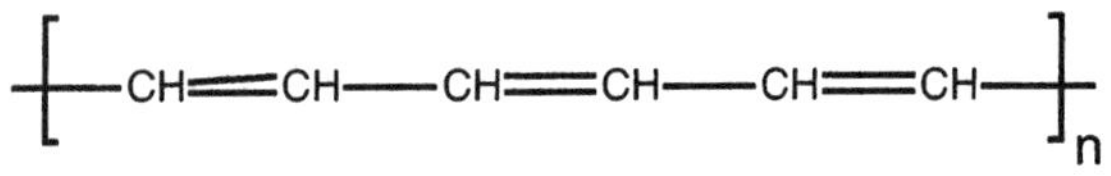 $$\left[-CH{=}CH{-}CH{=}CH{-}CH{=}CH- \right]_n$$

53. Because the acid-catalyzed condensation reaction in which amides are made is reversible, the amides can be decomposed in the presence of an acid catalyst and water. Acid rain gives precisely that combination.

Active Learning Workbook Answers

Chapter 23

Biochemistry

1.

```
        R    O
        |    ||
H2N  —  C —  C — OH
        |
        H
```

2. Phenylalanine, tyrosine, and tryptophan

3. V-T-I is valylthreonylisoleucine. I-V-T is isoleucylvalylthreonine. The C terminal acid in V-T-I is isoleucine, in I-V-T, threonine.

4.

```
                                       (C6H5, benzene ring)
       SH                                   |
       |
       CH2   O         H     O         CH2   O
       |     ||        |     ||        |     ||
NH2 — CH  —  C — NH — CH  —  C — NH — CH  —  C — OH
```

5. Tertiary protein structure tells us about the bends in a protein chain. Quaternary protein structure tells us how polypeptide chains are arranged in relation to one another.

6. The α-helix secondary structure involves hydrogen bonding between the hydrogen attached to the peptide link nitrogen and a peptide line oxygen of an amino acid further down the *same* protein chain.

7. Enzymes are usually proteins.

8. An enzyme substrate is a reactant that the enzyme helps change to product in the enzyme-catalyzed reaction.

9. Enzymes, like all catalysts, lower the activation energy of a reaction.

10. When you run a fever, enzyme-catalyzed reactions run faster than at normal body temperature. This may help the body fight off the illness more quickly.

11. Examples of aldose sugars are glucose, ribose, and deoxyribose. Aldose usually refers only to monosaccharides.

12. Find carbon-1 (the only carbon bonded to two oxygen atoms) in both structures. In α-glucose, the OH attached to carbon-1 is vertical, either pointed down or up. In β-glucose, the OH attached to carbon-1 is horizontal, or nearly so.

13. Sucrose is a disaccharide, glycogen is a polysaccharide, and fructose is a monosaccharide.

14. The simple sugars in lactose are galactose and glucose.

15. Glucose, ribose, deoxyribose, and lactose would give a positive Benedict's test, because all these sugars can have an aldchyde group in the open chain form.

16.

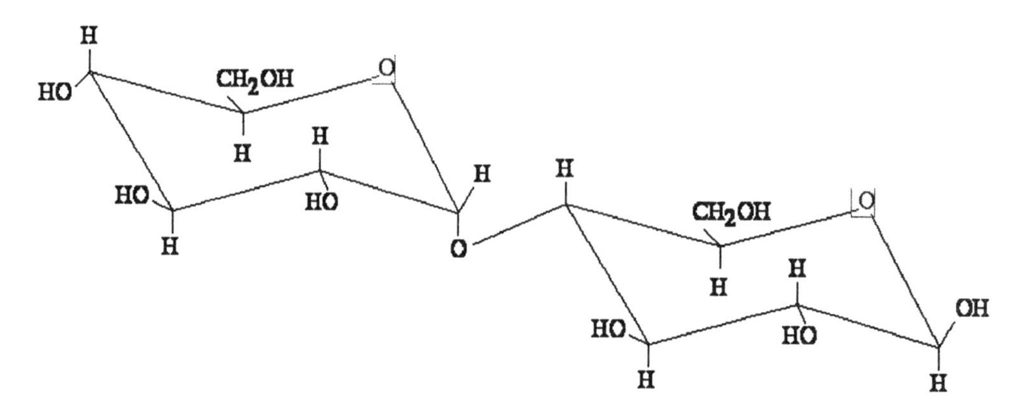

17. Starch has α(1→4) bonds; cellulose has β(1→4) bonds. Our enzymes can break the α(1→4) bonds, but not the β(1→4) bonds.

18. Immiscible in water

19. Plant sources

20.

```
A – O – CH2    A – O – CH2    B – O – CH2    B – O – CH2
        |              |              |              |
B – O – CH     C – O – CH     A – O – CH     C – O – CH
        |              |              |              |
C – O – CH2    B – O – CH2    C – O – CH2    A – O – CH2
```

21.

```
                    O                                      O
                    ||                                     ||
     CH2 – O – P – O⁻                        CH2 – O – P – O⁻
      |             |                         |             |
A – O – CH          O – X⁺           B – O – CH            O – X⁺
      |                                       |
B – O – CH2                          A – O – CH2
```

Where X^+ represents the polar head group

22. DNA is the storehouse of genetic information in all life forms. There are three types of RNA. mRNA carries instructions for protein synthesis from DNA to ribosomes; tRNA delivers specific individual amino acids to the ribosome. rRNA combines with proteins to form ribosomes.

23. See Figure 23.6

24. See Figure 23.6

25. See the Lewis diagram in Section 23.3. Ribose has a hydroxyl group at carbon 2; deoxyribose does not.

26.

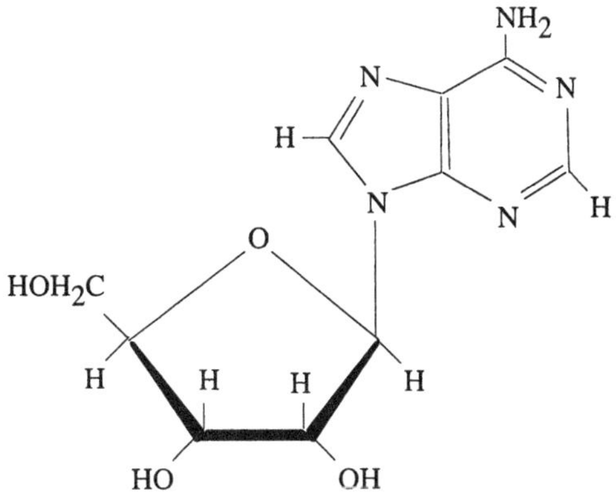

27. C-A-T:

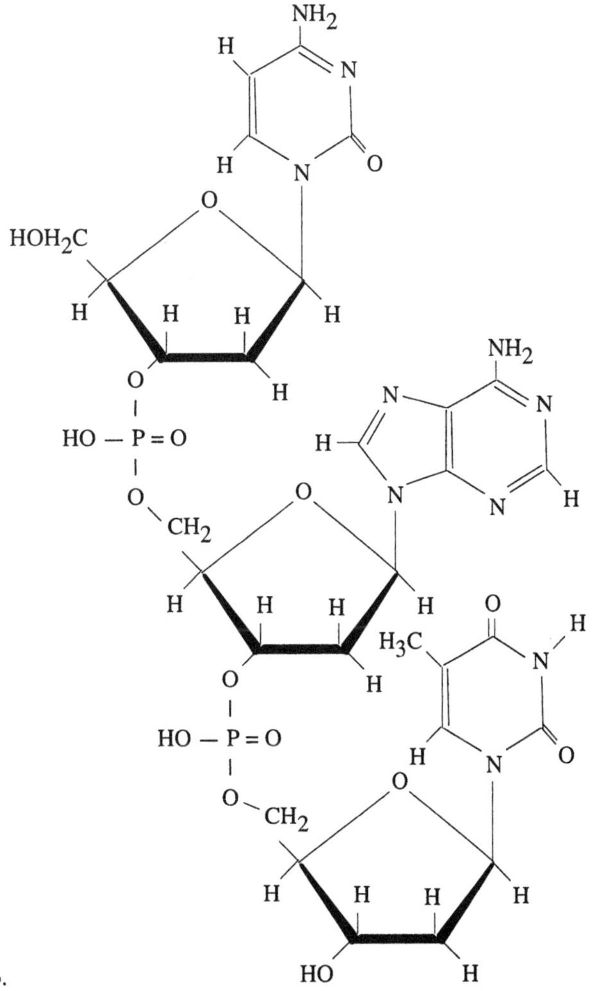

28. A-G-C

29. Transfer RNA picks up an amino acid molecule and carries it to a protein being synthesized by a ribosome.

30. $3.8 \times 10^7 \text{ g} \times \frac{1 \text{ amino acid}}{120 \text{ g}} = 3.2 \times 10^5$ amino acids

31. Glucose is the monomer in cellulose, amino acids are the monomers in proteins, nucleotides (adenine, cytosine, guanine, thymine) are the monomers in DNA, glucose in the monomer in starch, and nucleotides (adenine, cytosine, guanine, uracil) are the monomers in RNA.

32. Phosphorus

33.

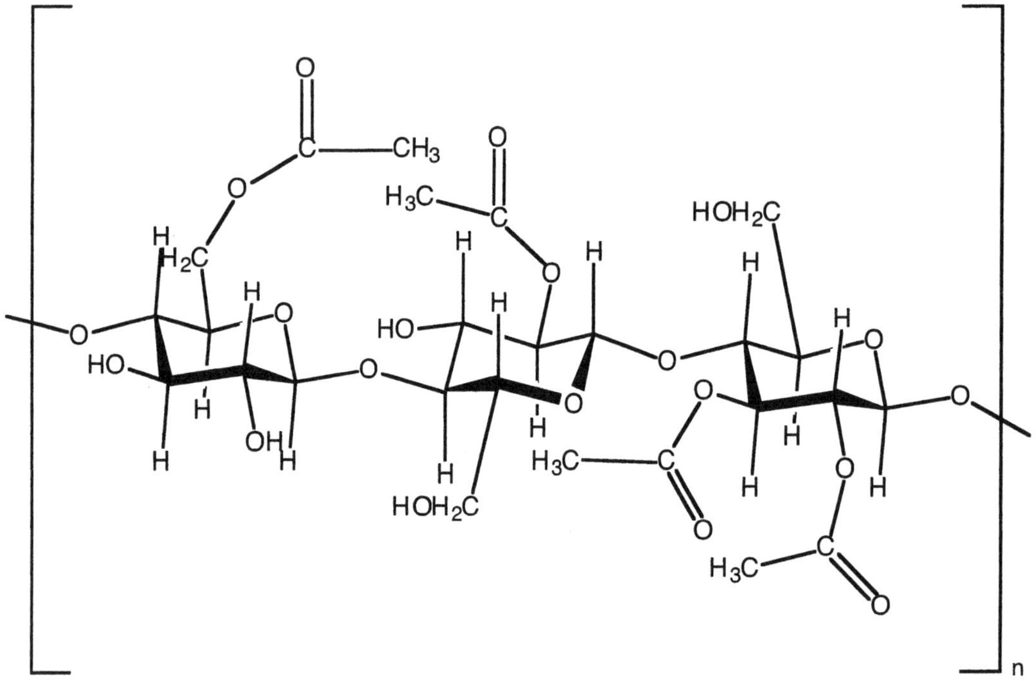

34. Because RNA is single-stranded, there is no complementary RNA strand. As a result, the percentage of guanine has no relationship to the percentage of cytosine.

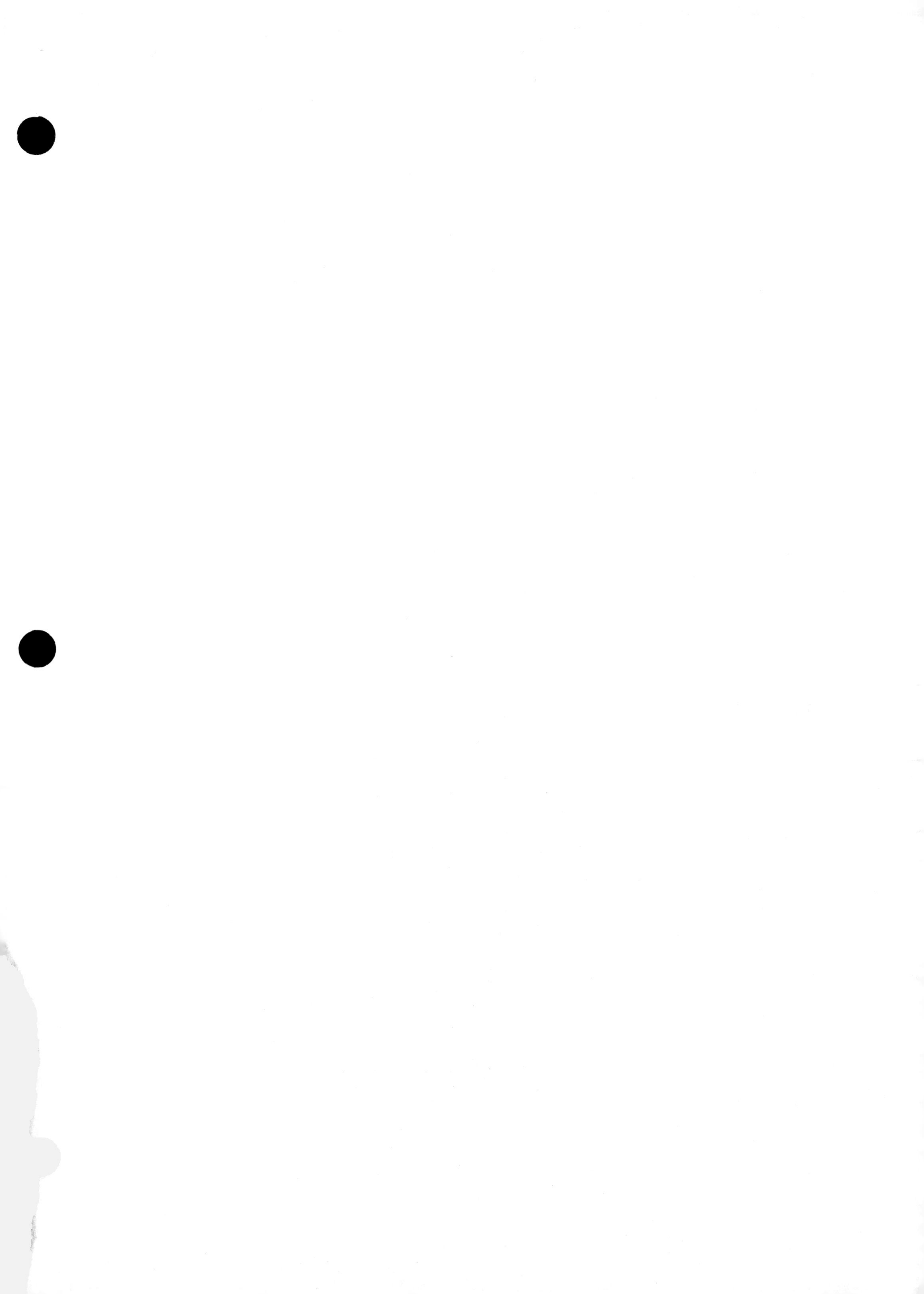